Grid Networks

Grid Networks

Enabling Grids with Advanced Communication Technology

Franco Travostino,
Nortel, USA

Joe Mambretti,
Northwestern University, USA

Gigi Karmous-Edwards,
MCNC, USA

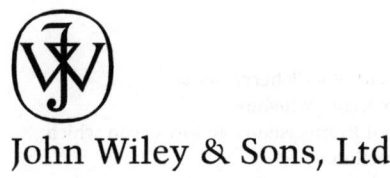

John Wiley & Sons, Ltd

Copyright © 2006   John Wiley & Sons Ltd, The Atrium, Southern Gate, Chichester,
                     West Sussex PO19 8SQ, England

                     Telephone (+44) 1243 779777

Email (for orders and customer service enquiries): cs-books@wiley.co.uk
Visit our Home Page on www.wiley.com

All Rights Reserved. No part of this publication may be reproduced, stored in a retrieval system or transmitted in any form or by any means, electronic, mechanical, photocopying, recording, scanning or otherwise, except under the terms of the Copyright, Designs and Patents Act 1988 or under the terms of a licence issued by the Copyright Licensing Agency Ltd, 90 Tottenham Court Road, London W1T 4LP, UK, without the permission in writing of the Publisher. Requests to the Publisher should be addressed to the Permissions Department, John Wiley & Sons Ltd, The Atrium, Southern Gate, Chichester, West Sussex PO19 8SQ, England, or emailed to permreq@wiley.co.uk, or faxed to (+44) 1243 770620.

This publication is designed to provide accurate and authoritative information in regard to the subject matter covered. It is sold on the understanding that the Publisher is not engaged in rendering professional services. If professional advice or other expert assistance is required, the services of a competent professional should be sought.

Other Wiley Editorial Offices

John Wiley & Sons Inc., 111 River Street, Hoboken, NJ 07030, USA

Jossey-Bass, 989 Market Street, San Francisco, CA 94103-1741, USA

Wiley-VCH Verlag GmbH, Boschstr. 12, D-69469 Weinheim, Germany

John Wiley & Sons Australia Ltd, 42 McDougall Street, Milton, Queensland 4064, Australia

John Wiley & Sons (Asia) Pte Ltd, 2 Clementi Loop #02-01, Jin Xing Distripark, Singapore 129809

John Wiley & Sons Canada Ltd, 6045 Freemont Blvd, Mississauga, Ontario, Canada, L5R 4J3

Library of Congress Cataloging in Publication Data

Grid networks : enabling grids with advanced communication technology / [edited by] Franco Travostino,
   Joe Mambretti, Gigi Karmous-Edwards.
      p. cm.
   Includes bibliographical references.
   ISBN-13: 978-0-470-01748-7
   ISBN-10: 0-470-01748-1
   1. Computational grids (Computer systems)   2. Computer networks—Design and construction.
   3. Wireless communication systems—Design and construction.   4. Optical communications—Design and construction.   I. Travostino, Franco.   II. Mambretti, Joel, 1948–   III. Karmous-Edwards, Gigi.
   QA76.9.C58G7559 2006
   004′.36—dc22   2006016095

British Library Cataloguing in Publication Data

A catalogue record for this book is available from the British Library

ISBN-13: 978-0-470-01748-7
ISBN-10: 0-470-01748-1

Typeset in 10/12pt Garamond by Integra Software Services Pvt. Ltd, Pondicherry, India
Printed and bound in Great Britain by Antony Rowe Ltd, Chippenham, Wiltshire
This book is printed on acid-free paper responsibly manufactured from sustainable forestry in which at least two trees are planted for each one used for paper production.

Dedication

For Franco, To Suely, Giulia and Matteo
For Joe, To Catherine
For Gigi, To Layah, Kamila, and Tony

Contents

Editors and Contributors xix

Contributors xxi

Foreword xxv

Acknowledgments xxvii

Introduction: The Grid and Grid Networks xxix

1 The Grid and Grid Network Services 1
Joe Mambretti
  1.1 Introduction 1
  1.2 Network Resources as First-class Grid Entities 1
    1.2.1 What is a Grid? 2
  1.3 The General Attributes of Grids 3
    1.3.1 The Grid and Design Abstraction 5
    1.3.2 The Grid as an Enabler of Pervasive, Programmable Utility
        Services 6
  1.4 Types of Grids 7
    1.4.1 Grids and Grid Networks 8
    1.4.2 Attributes of Grid Networks 9
  1.5 Grid Networks and Emerging Communication Technologies 14
  References 14

2 Grid Network Requirements and Driver Applications 17
  2.1 Introduction 17
  2.2 Grid Network Requirements for Large-scale Visualization and
    Collaboration 18
    Jason Leigh, Luc Renambot, and Maxine Brown
    2.2.1 Large-scale Visualization and Collaboration Application
        Drivers 18
    2.2.2 Current Limitations to Advanced Visualization and Collaboration 19

	2.2.3 Enabling Advanced Visualization and Collaboration with the Optiputer	20
	2.2.4 Future Challenges in Large-scale Visualization and Collaboration	22
	Acknowledgments	23
2.3	Large-scale E-science	24
	Peter Clarke	
	2.3.1 Data Management for the Large Hadron Collider Project	25
	2.3.2 Large-scale Computational Science	26
	2.3.3 Summary	30
	Acknowledgments	30
2.4	Data Mining	30
	Robert Grossman	
	2.4.1 Important Steps in the Data Mining Process	31
	2.4.2 Main Steps in Distributed Data Mining Using Commodity Networks	31
	2.4.3 Main Steps in Distributed Data Mining Using Optical Grids	32
2.5	CineGrid, a Grid for Digital Cinema	33
	Tom DeFanti, Laurin Herr, and Natalie Van Osdol	
	2.5.1 Trends	34
	2.5.2 CineGrid Cinema-centric Research	34
	2.5.3 CineGrid Consortium	36
2.6	Distributed Aircraft Maintenance Environment (DAME)	36
	Tom Jackson, Jim Austin, and Martyn Fletcher	
	2.6.1 Use Case Introduction	36
	2.6.2 Dame Customers	37
	2.6.3 Scenarios	37
	2.6.4 Resources Involved	39
	2.6.5 Functional Requirements	39
	2.6.6 Security Considerations	40
	2.6.7 Performance Considerations	40
	2.6.8 Use Case Situation Analysis	40
	Acknowledgments	41
2.7	Financial Services: Regulatory and Market Forces Motivating a Move to Grid Networks	41
	Robert B. Cohen	
	2.7.1 New Challenges for Financial Institutions and Networks	41
	2.7.2 Factors Driving Banks to Adopt Grids and High-speed Networks	41
	2.7.3 How Financial Institutions will Use Networks to Facilitate Grid Computing	43
	2.7.4 Globalization of Financial Markets	44
	2.7.5 Migration of Financial Institutions to Grid Networks	45
	2.7.6 Conclusions	46
2.8	Summary of Requirements	47
References		47

3 Grid Network Requirements and Architecture 49
Joe Mambretti and Franco Travostino
3.1 Introduction 49
3.2 Requirements 50
    3.2.1 Requirements and Coexistence of Diverse Network User Communities 50
    3.2.2 Abstraction/Virtualization 52
    3.2.3 Resource Sharing and Site Autonomy 53
    3.2.4 Flexibility Through Programmability 54
    3.2.5 Determinism 54
    3.2.6 Decentralized Management and Control 56
    3.2.7 Dynamic Integration 56
    3.2.8 Resource Sharing 57
    3.2.9 Scalability 57
    3.2.10 High Performance 58
    3.2.11 Security 59
    3.2.12 Pervasiveness 59
    3.2.13 Customization 59
3.3 Translating Requirements to Architecture 60
    3.3.1 IETF RFC 2768 60
    3.3.2 Service-oriented Architecture 61
    3.3.3 A Multitier Architecture for Grids 61
    3.3.4 Introducing Grid Network Services 63
Acknowledgment 65
References 65

4 Relevant Emerging Network Architecture from Standards Bodies 67
Franco Travostino
4.1 Introduction 67
4.2 Global Grid Forum (GGF) 68
4.3 Enterprise Grid Alliance (EGA) 69
4.4 Organization for the Advancement of Structured Information Standards (OASIS) 70
4.5 World Wide Web Consortium (W3C) 71
4.6 The IPSphere Forum 71
4.7 MPI Forum 71
4.8 Internet Engineering Task Force (IETF) 71
4.9 Distributed Management Task Force (DMTF) 73
4.10 International Telecommunication Union (ITU-T) 74
4.11 Optical Internetworking Forum (OIF) 75
4.12 Infiniband Trade Association (IBTA) 75
4.13 Institute of Electrical and Electronics Engineers (IEEE) 76
References 77

5 Grid Network Services and Implications for Network Service Design   81
Joe Mambretti, Bill St. Arnaud, Tom DeFanti, Maxine Brown, and Kees Neggers

- 5.1 Introduction   81
- 5.2 Traditional Communications Services Architecture   82
- 5.3 Grid Architecture as a Service Platform   82
  - 5.3.1 Grid Network Services Architecture   83
- 5.4 Network Services Architecture: An Overview   84
  - 5.4.1 Services Architecture Benefits   84
- 5.5 Grid Network Services Implications   86
- 5.6 Grid Network Services and Network Services   86
  - 5.6.1 Deterministic Networking and Differentiated Services   87
- 5.7 Grid Network Service Components   88
  - 5.7.1 Network Service Advertisements and OGSA   88
  - 5.7.2 Web Services   89
  - 5.7.3 Web Services Definition Language (WSDL)   89
  - 5.7.4 Universal Description, Discovery, and Integration (UDDI)   90
  - 5.7.5 Web Services-Inspection Language (WSIL)   90
  - 5.7.6 Network Service Design and Development Tools   90
- 5.8 New Techniques for Grid Network Services Provisioning   91
  - 5.8.1 Flexible Communication Services Provisioning   91
  - 5.8.2 Partitionable Network Environments   91
  - 5.8.3 Services Provisioning and Signaling   92
- 5.9 Examples of Grid Network Services Prototypes   92
  - 5.9.1 A Layer 3 Grid Network Services Prototype   93
  - 5.9.2 APIS and Signaling for Dynamic Path Provisioning   93
  - 5.9.3 A Layer 2 Grid Network Services Prototype   94
  - 5.9.4 Services-oriented Architecture for Grids Based on Dynamic Lightpath Provisioning   94
  - 5.9.5 Optical Dynamic Intelligent Network Services (ODIN)   95
  - 5.9.6 User-Controlled Lightpath Provisioning   95
- 5.10 Distributed Facilities for Services Oriented Networking   96
  - 5.10.1 Provisioning Grid Network Services   97
- References   97

6 Grid Network Services: Building on Multiservice Networks   99
Joe Mambretti

- 6.1 Introduction   99
- 6.2 Grid Network Services and Traditional Network Services   100
  - 6.2.1 The Grid and Network Quality of Service   100
- 6.3 Network Service Concepts and the End-to-end Principle   101
  - 6.3.1 Network Quality of Service and Applications Quality of Service   102
- 6.4 Grid Architecture and the Simplicity Principle   102

	6.4.1	Network Design and State Information	103
	6.4.2	Internet Best Effort Services	104
6.5	Grids and Internet Transport Layer Services		105
6.6	IETF Differentiated Services		105
	6.6.1	Diffserv Mechanisms	106
	6.6.2	Grids and Quality of Service Network Services	107
6.7	Gara and DiffServ		107
6.8	Grids and Nonrouted Networks		107
	6.8.1	Layer 2.5 Services and Quality Standards	108
	6.8.2	Grids and Layer 2.5 Services	108
6.9	Layer 2 Services and Quality Standards		108
	6.9.1	Grids and Layer 2 Quality of Service	109
6.10	Layer 1 Services and Quality Standards		109
	6.10.1	Grids and Layer 1 Quality of Service	110
6.11	The Grid and Network Services		111
References			111

7 Grid Network Middleware   113
Franco Travostino and Doan Hoang

7.1	Introduction		113
7.2	Definitions		114
	7.2.1	Network Services and Grid Network Services	114
	7.2.2	Grid Infrastructure Software	114
	7.2.3	Grid Network Infrastructure Software	114
7.3	Grid Infrastructure Software		115
	7.3.1	The Globus Toolkit	115
7.4	Grid Network Infrastructure Software		122
	7.4.1	The DWDM-RAM System	123
7.5	Components of Grid Network Infrastructure		126
	7.5.1	Network Bindings	126
	7.5.2	Virtualization Milieu	129
	7.5.3	Performance Monitoring	132
	7.5.4	Access Control and Policy	133
	7.5.5	Network Resource Scheduling	134
	7.5.6	Multidomain Considerations	135
References			139

8 Grid Networks and TCP Services, Protocols, and Technologies   145
Bartek Wydrowski, Sanjay Hegde, Martin Suchara, Ryan Witt, and Steven Low

8.1	Introduction		145
8.2	Background and Theoretical basis for Current Structure of Transport Layer Protocols		146
	8.2.1	User Datagram Protocol (UDP)	146
	8.2.2	Transmission Control Protocol (TCP)	147
	8.2.3	Window Flow Control	147

		8.2.4 Fairness	149
		8.2.5 Congestion Control Feedback System	150
		8.2.6 Congestion Control Performance	152
	8.3	Enhanced Internet Transport Protocols	157
		8.3.1 TCP Reno/NewReno	157
		8.3.2 TCP Vegas	158
		8.3.3 FAST TCP	159
		8.3.4 BIC TCP	160
		8.3.5 High-speed TCP	161
		8.3.6 Scalable TCP	162
		8.3.7 H-TCP	162
		8.3.8 TCP Westwood	162
	8.4	Transport Protocols based on Specialized Router Processing	163
		8.4.1 MaxNet	163
		8.4.2 Explicit Congestion Control Protocol (XCP)	166
	8.5	TCP and UDP	167
	Acknowledgments		168
	References		168

9 Grid Networks and UDP Services, Protocols, and Technologies 171

Jason Leigh, Eric He, and Robert Grossman

	9.1	Introduction	171
	9.2	Transport Protocols based on the User Datagram Protocol (UDP)	171
		9.2.1 UDP Transport Utility	172
		9.2.2 Reliable Blast UDP (RBUDP)	173
		9.2.3 The UDP-Based Data Transfer Protocol (UDT)	174
		9.2.4 Tsunami	178
	9.3	Lambdastream	178
	9.4	Grid Applications and Transport Protocols	179
		9.4.1 Berkley Sockets	179
		9.4.2 Future APIs	179
		9.4.3 TCP Proxies	180
	9.5	The Quanta Toolkit	180
		9.5.1 Tuning and Optimization Issues	181
		9.5.2 Communication Services Optimization	181
	9.6	Grids and Internet Transport	182
	Acknowledgments		182
	References		183

10 Grid Networks and Layer 3 Services 185

Joe Mambretti and Franco Travostino

10.1	Introduction	185
10.2	The Internet and the End-To-End Principle	185

10.3	The Internet and Layer 3 Services	186
	10.3.1 IP Concepts	186
	10.3.2 IP Components	187
	10.3.3 Differentiated Services	187
10.4	Grid Experimentation with DiffServ-Based Quality of Service	188
10.5	Internet Routing Functions	189
	10.5.1 Routing Protocols	189
	10.5.2 Communicating Routing Table Information	190
	10.5.3 Route Advertisement and Route Storage	190
	10.5.4 Routing Policies	190
	10.5.5 Routing Topologies	190
	10.5.6 Routing Metrics	191
10.6	Layer 3 Addressing and Network Address Translators (NATS)	192
10.7	IP Version 6	192
10.8	Subsecond IGP Recovery	193
10.9	Internet Security using Internet Protocol Security	193
10.10	IP Multicast	194
10.11	Internet Layer 3 Services	194
Acknowledgments		195
References		195

11 Layer 2 Technologies and Grid Networks — 197

John Strand, Angela Chiu, David Martin, and Franco Travostino

11.1	Introduction	197
11.2	Layer 2 Technologies and Grid Requirements	197
11.3	Multiprotocol Label Switching (MPLS)	198
	11.3.1 MPLS and Shared Network Infrastructure	200
	11.3.2 MPLS and Virtual Private Networks	200
	11.3.3 Grid Network Services and MPLS	201
11.4	Ethernet Architecture and Services	201
	11.4.1 Ethernet Architecture Features and Challenges	202
	11.4.2 Ethernet as a Service	204
	11.4.3 10 Gbps Ethernet and Beyond	204
11.5	Pseudo-Wire Emulation (PWE) and Virtual Private Lan Services Over MPLS (VPLS)	205
11.6	Layers 2/1 Data Plane Integration	205
	11.6.1 Sonet and TDM Extensions for Ethernet-Over-Sonet (EOS)	206
	11.6.2 Virtual Concatenation	207
	11.6.3 Link Capacity Adjustment Scheme	207
	11.6.4 Generic Framing Procedure	207
11.7	Resilient Packet Rings (RPR)	207
11.8	User–Network Interfaces	208
11.9	Optical Interworking Forum Interoperability Demonstration	210
11.10	Infiniband	211
Acknowledgments		214
References		214

12 Grid Networks and Layer 1 Services 217
Gigi Karmous-Edwards, Joe Mambretti, Dimitra Simeonidou, Admela Jukan, Tzvetelina Battestilli, Harry Perros, Yufeng Xin, and John Strand

12.1 Introduction 217
12.2 Recent Advances in Optical Networking Technology and Responses 218
    12.2.1 Layer 1 Grid Network Services 219
    12.2.2 Benefits of Grid Layer 1 Services 219
    12.2.3 The Role of Network Standards Bodies 221
12.3 Behavioral Control of Layer 1 Networks 224
    12.3.1 Management Plane 225
    12.3.2 Control Plane 225
12.4 Current Research Challenges for Layer 1 Services 229
    12.4.1 Application-Initiated Connections 229
    12.4.2 Interaction with Grid Middleware 232
    12.4.3 Integrating Novel Optical Technologies 232
    12.4.4 Resource Discovery and Coordination 233
12.5 All-Photonic Grid Network Services 235
    12.5.1 All-Photonic Grid Service 235
    12.5.2 Grid Service Scenarios for All-Photonic End-to-End Connections 236
    12.5.3 Physical Layer Quality of Service for Layer 1 Services 236
    12.5.4 Requirements for an All-photonic End-to-End Grid Service 239
    12.5.5 Open Issues and Challenges 239
12.6 Optical Burst Switching and Grid Infrastructure 240
    12.6.1 Introduction to OBS 241
    12.6.2 Grid-OBS as a Control Plane for Grid Networking 245
    12.6.3 Advances in Optical Switching Technology that make Grid-OBS a Viable Solution 246
    12.6.4 Grid-OBS use Scenario 250
References 250

13 Network Performance Monitoring, Fault Detection, Recovery, and Restoration 253
Richard Hughes-Jones, Yufeng Xin, Gigi Karmous-Edwards, John Strand

13.1 Introduction 253
13.2 Monitoring Characteristics 254
    13.2.1 The Hoplist Characteristic 255
    13.2.2 The Bandwidth Characteristic 256
    13.2.3 The Delay Characteristic 256
    13.2.4 The Loss Characteristic 257
    13.2.5 The Closeness Characteristic 257
13.3 Network Monitoring Instrumentation and Analysis 258
    13.3.1 Monitoring of Traffic Flows and Patterns 258
    13.3.2 Lightweight Monitoring 259

13.3.3 Detailed Network Investigations	261
13.3.4 Monitoring at the Application Level	262
13.4 General Considerations on Availability	262
13.5 Fault Detection	263
13.6 Recovery and Restoration	264
13.6.1 Protection for Circuit Switched Networks	266
13.6.2 Restoration for Burst/Packet-Switched Networks	268
13.7 Integrated Fault Management	272
References	273

14 Grid Network Services Infrastructure   277
Cees de Laat, Freek Dijkstra, and Joe Mambretti

14.1 Introduction	277
14.2 Creating Next-Generation Network Services and Infrastructure	278
14.2.1 End-to-End Principle	278
14.2.2 Packet-Based Data Units	279
14.2.3 Enhanced Functional Abstraction	279
14.2.4 Self-Organization	279
14.2.5 Decentralization	280
14.2.6 Distributed Service Creation	280
14.3 Large-Scale Distributed Facilities	280
14.4 Designs for an Open Services Communications Exchange	281
14.4.1 The Design of an Open Grid Services Exchange	281
14.4.2 Provisioning Implications	282
14.4.3 Exchange Facility Characteristics	282
14.5 Open Grid Optical Exchanges	283
14.5.1 Traditional Internet Exchanges	283
14.5.2 Rationale for an Open Optical Exchange	284
14.5.3 The Concept of an Optical Exchange	285
14.5.4 Interfaces and Protocols within an Optical Exchange	286
14.5.5 Optical Exchange Services	288
14.5.6 External Services	289
14.5.7 Service Matrix	289
14.5.8 Blueprint for an Optical Exchange	289
14.5.9 Monitoring in a Multilayer Exchange	289
14.6 Prototype Implementations	291
References	292

15 Emerging Grid Networking Services and Technologies   293
Joe Mambretti, Roger Helkey, Olivier Jerphagnon, John Bowers, and Franco Travostino

15.1 Introduction	293
15.2 New Enabling Technologies	294
15.3 Edge Technologies	295
15.4 Wireless Technologies	295
15.4.1 Device-Level Wireless Technologies	296

15.4.2	IEEE 802.11	296
15.4.3	Self-Organizing Ad Hoc Wireless Networks	297
15.4.4	IEEE SA 802.11b	297
15.4.5	IEEE 802.11a	298
15.4.6	IEEE 802.11g	298
15.4.7	Software-Defined Radios and Cognitive Radios	298
15.4.8	Radio Frequency Identification	299
15.4.9	Sensors	299
15.4.10	Light-Emitting Diodes (LEDS)	300
15.5 Access Technologies		300
15.5.1	Fiber to the Premises (FTTP)	300
15.5.2	Wireless Access Networks	301
15.5.3	Free Space Optics (FSO)	301
15.5.4	Light-Emitting Diodes	301
15.5.5	Broadband Over Power Lines (BPL)	301
15.6 Core Technologies		301
15.7 Photonic Integrated Circuits (PIC)		302
15.7.1	High-Performance Optical Switches	302
15.7.2	Recent Advances in High Performance Optical Switching	303
15.7.3	Optical Switch Design	304
15.7.4	Optical Switches in Core Networks	305
15.7.5	Reliability Issues	306
15.7.6	Future Advances in High-Performance Optical Switches	306
15.7.7	Implications for the Future	307
Acknowledgments		307
References		308

Appendix: Advanced Networking Research Testbeds and Prototype Implementations — 311

A.1 Introduction		311
A.2 Testbeds		312
A.2.1	OMNInet	312
A.2.2	Distributed Optical Testbed (DOT)	314
A.2.3	I-WIRE	314
A.2.4	OptIPuter	315
A.2.5	CHEETAH	316
A.2.6	DRAGON	316
A.2.7	Japan Gigabit Network II (JGN II)	317
A.2.8	Vertically Integrated Optical Testbed for Large Scale Applications (VIOLA)	318
A.2.9	StarPlane	318
A.2.10	EnLIGHTened	319
A.2.11	Lambda User Controlled Infrastructure for European Research	320
A.2.12	Global Environment for Network Innovations (GENI)	322
A.2.13	Department of Energy Ultrascience Net	323

A.3 Prototype Implementations 323
   A.3.1 StarLight 323
   A.3.2 TransLight 324
   A.3.3 NetherLight 325
   A.3.4 UKlight 326
A.4 National and International Next Generation Communications Infrastructure 326
   A.4.1 CANARIE 326
   A.4.2 SURFnet6 327
   A.4.3 National Lambda Rail 327
A.5 International Facilities 328
   A.5.1 Global Lambda Integrated Facility (GLIF) 328
   A.5.2 Global Ring Network for Advanced Application Development (GLORIAD) 329
   A.5.3 UltraLight 330
   A.5.4 GEANT2 330

Index   **333**

Editors and Contributors

Franco Travostino, Director, Advanced Technology and Research, Office of the CTO, Nortel, Inc.; Technical Advisor, IP Storage, Internet Engineering Task Force; Area Director for Infrastructure and Steering Group Member, Global Grid Forum; Chair, High Performance Networking Research Group, and Co-Chair Telecom Community Group, Global Grid Forum.

Joe Mambretti, Director, International Center for Advanced Internet Research, Northwestern University, McCormick School of Engineering and Applied Science, Northwestern University; Director, Metropolitan Research and Education Network; Co-Director, StarLight; document author, High Performance Networking Research Group, Global Grid Forum.

Gigi Karmous-Edwards, Principal Scientist, MCNC, North Carolina; Chair, Global Lambda Grid Control Plane and Grid Integration Working Group; Adjunct Professor of Computer Science, North Carolina State University; document author, High Performance Networking Research Group, Global Grid Forum.

Contributors

Jim Austin, Professor, University of York

Tzvetelina Battestilli, postdoctoral researcher, Advanced Technology Group, MCNC, North Carolina

John Bowers, Chief Technology Officer, Calient Networks

Maxine Brown, Associate Director, Electronic Visualization Laboratory, University of Illinois at Chicago

Angela Chiu, Principal Member, Research Staff, AT&T Labs – Research

Peter Clarke, Professor of e-Science, University of Edinburgh; Co-director of the National e-Science Centre; Former Member, Steering Committee, Director, Data Area, Global Grid Forum

Robert Cohen, Fellow, Economic Strategy Institute, Washington, DC; Area Director, Industrial Applications, Global Grid Forum

Tom DeFanti, Director, Electronic Visualization Laboratory; Distinguished Professor and Distinguished Professor Emeritus in the Department of Computer Science and Director of the Software Technologies Research Center, University of Illinois at Chicago; research scientist, California Institute for Telecommunications and Information Technology (Calit2), University of California, San Diego; Co-director, StarLight

Cees de Laat, Associate Professor, Faculty of Science, Informatics Institute, Universiteit van Amsterdam; Area Director for Infrastructure, Global Grid Forum

Freek Dijkstra, Jr., researcher, Advanced Internet Research Group, Informatics Institute, Universiteit van Amsterdam

Martyn Fletcher, research administrator, University of York

Robert Grossman, Professor and Director, National Center for Data Mining, University of Illinois at Chicago; Managing Partner, Open Data Group

Eric He, PhD student, Electronic Visualization Laboratory, University of Illinois at Chicago

Sanjay Hegde, research engineer, California Institute of Technology

Roger Helkey, Vice President of Research, Calient Networks

Laurin Herr, President, Pacific Interface, Inc.

Doan Hoang, Professor and Director, Advanced Research in Networking Laboratory (ARN), UTS Advanced Research Institute for Information and Communication Technology, University of Technology, Sydney

Richard Hughes-Jones, Lead, e-Science Grid Network Research and Development, Particle Physics Group, Manchester University; member, Trigger/DAQ Group, ATLAS Experiment, Large Hadron Collider Program; Secretary, Particle Physics Network Coordinating Group, United Kingdom; Co-chair, Network Measurements Working Group, Global Grid Forum; Co-chair, PFLDnet

Tom Jackson, Professor, University of York

Olivier Jerphagnon, Network Consulting Manager, Calient Networks

Admela Jukan, Associate Professor, Énergie, Matériaux et Télécommunications – INRS, Universite du Quebec, Montréal

Jason Leigh, Director, Electronic Visualization Laboratory, and Associate Professor, Computer Science, University of Illinois at Chicago; document author, High Performance Networking Working Research Group, Global Grid Forum

Steven Low, Associate Professor, California Institute of Technology

David Martin, Program Manager, Internet Standards and Technology, IBM; Visiting Research Associate, International Center for Advanced Internet Research, Northwestern University; document author, Grid Forum Steering Group, Global Grid Forum

Kees Neggers, Managing Director, SURFnet bv; Director, GigaPort Network Project; Chairman, RIPE NCC Executive Board; former President, Reseaux Associes pour la Recherche Europeenne; Emeritus Trustee, Internet Society; Chair, Global Lambda Integrated Facility

Harry Perros, Alumni Distinguished Graduate Professor, Computer Science Department, North Carolina State University

Luc Renambot, PhD student and research associate, Electronic Visualization Laboratory, University of Illinois at Chicago

Bill St. Arnaud, Senior Director, Advanced Networks, CANARIE, Canada

Martin Suchara, student, California Institute of Technology

John Strand, Consultant, AT&T Labs – Research

Dimitra Simeonidou, Professor, University of Essex; Chair, High Performance Networking Research Working Group, Global Grid Forum

Natalie Van Osdol, Vice-President, Pacific Interface, Inc.

Ryan Witt, student, California Institute of Technology

Bartek Wydrowski, research engineer, California Institute of Technology

Yufeng Xin, senior scientist, MCNC, North Carolina

Foreword

New telecommunications and information technologies are making it possible *for the first time* to create large-scale, distributed infrastructure that tightly and dynamically integrates multiple resources – *in real time*. Such distributed fabrics will be used to create multiple new types of communication services with enhanced capabilities that far exceed those commonly used in the world today and, perhaps more importantly, enable the creation of powerful new applications.

Already, early prototypes of these capabilities are transforming science and engineering, enabling studies across space and timescales far beyond what has been possible and what is still common today: These prototypes are enabling integration not only among observational, experimental, and computational methodologies in a given discipline, but also among different environments such as the seafloor, land mass, and the atmosphere. They are making possible sharply vivid collaborative environments that transcend traditional visual communication modes, allowing for realistic remote presence in which virtual collaborators appear almost as true to life, as if they were physically present in the room.

These powerful applications are destined to migrate quickly from the realms of early adopters to wider communities. Perhaps the ultimate "application" these technologies will support is "borderless collaboration" among multiple countries, ethnic cultures, languages, and disciplines to address increasingly complex problems facing the world, a scenario which, in turn, has the potential of leading to greater global understanding and harmony.

Larry Smarr, *Director of the California Institute for Telecommunications and Information Technology and Harry E. Gruber Professor at the Jacobs School's Department of Computer Science and Engineering at University of California at San Diego*

Acknowledgments

Tomonori Aoyama, Eric Bernier, Jim Chen, Andrew Chien, Peter Dinda, Tiziana Ferrari, Ian Foster, Samrat Ganguly, Wolfgang Gentzsch, Rachel Gold, Aaron Johnson, Peter Kaufmann, John Lange, David Lillethun, Greg Nawrocki, Reza Nejabati, Mort Rahimi, Larry Smarr, Valerie Taylor, Malathi Veeraraghavan, Jeremy Weinberger, Linda Winkler, Fei Yei, and Oliver Yu.

Introduction

The Grid and Grid Networks

The term "Grid" designates a fundamentally new architecture that provides powerful capabilities for creating advanced information technology services. The Grid was designed to provide services that cannot be supported with traditional architecture. Grid architecture allows resources to be gathered and customized to support an almost unlimited number of services, which can be individualized to meet precise requirements. Grid architecture has been designed to incorporate virtually any type of resource into a common, shared environment.

The Grid provides scalable, reliable, secure mechanisms for discovering, assembling, integrating, utilizing, reconfiguring, and releasing multiple heterogeneous resources. These resources can include compute clusters, specialized computers, software, mass storage, data repositories, instruments, and sensors. However, Grid environments are not restricted to incorporating common information technology resources. The architecture is highly expandable and can extend to virtually any type of device with a communications capability.

One objective of Grid architectural design is to enable distributed collaborative groups, which can be termed "virtual organizations," to accomplish their common goals more effectively and efficiently by using specialized shared distributed infrastructure. Consequently, communities world-wide are designing and implementing Grid environments to support many new types of interactive collaboration. In addition, general and specialized Grids are being used for an increasingly broader range of applications and services.

OBJECTIVES OF THIS BOOK

From the time of its earliest implementations, the Grid has incorporated communications services as key resources, especially services based on the Internet and web technology. A basic premise motivating Grid architectural development is that these low-cost, highly scaleable, ubiquitous communications services are critical to the Grid vision of allowing multiple distributed communities to share resources within a common environment. These services have been especially important in addressing the communications requirements of data-intensive Grid applications.

Until recently, almost all Grid environments have been based on a fairly narrow range of communication services, and most of those services were not fully integrated into the Grid environments that they supported. In the last few years, multiple initiatives have been established to investigate the potential for new and emerging communication services architecture and technologies to significantly expand Grid capabilities. These research efforts are exploring methods that more closely integrate network resources into Grid environments, that allow for enhanced control of those resources and the services that they support, that expand the number of communication services that are directly accessible, and which provide for additional communications capabilities.

The book was developed to provide an overview of the key concepts related to these trends. It provides an overview of the emerging networking concepts, architectures, and technologies that are allowing Grids to fulfill their promise of becoming a truly ubiquitous distributed service environment, with fully integrated controllable communication services.

For example, several initiatives are designing and developing a services-oriented architecture for network resources, based on the same standard Web Services models that are being explored for other types of Grid services, using the Open Grid Services Architecture (OGSA). Other research initiatives are incorporating into Grid environments new innovative networking services and technologies. Today, many emerging new technologies related to traditional communications services are being tested in advanced Grid environments, including those related to high-performance transport and packet routing, switching, and dynamic optical provisioning.

These initiatives are complemented by developments in networking architecture, services, and technologies, which are being driven by forces of change external to Grid communities. As Grid design continues to evolve, networking architecture, services, and technologies are also continuing to advance rapidly at all levels. These innovations are highly complementary to directions in Grid evolution. These new advanced architecture and techniques allow for additional opportunities enabling customized communications services to be directly integrated within Grid environments. When implemented within Grid environments, these innovations provide versatile and powerful tools for services development, implementation, and delivery.

BOOK STRUCTURE

This book has four major sections and an appendix.

Part I. The first part of the book, Chapters 1 through 4, presents basic concepts in Grid networks, case studies that drive requirements, key attributes of Grid networks, and the standards organizations and initiatives that are developing architecture and protocols important to Grid network design and development.

Part II. The second part of the book, Chapter 5 through 7, describes key concepts related to the "services" that are provided by Grid networks. This part of the book describes the specialized meaning of service in a Grid network context, as well as the software components that are used to create and support those services, such as Grid middleware. These chapters also describe the rationale of the architecture of that software.

Part III. The third part of the book, Chapters 8 through 12, extends this general discussion of network services to a presentation of the specific services that are associated with the first four layers of the most common network service model.

Part IV. The fourth part of the book, Chapters 13 through 15, addresses several special subjects, including issues related to performance and operations, a discussion of architecture optimized for Grid network services, and an overview of emerging network technologies of particular interest to Grid communities.

Appendix: Advanced Networking Research Testbeds and Prototype Implementations. The appendix provides an overview of advanced networking research testbeds that are exploring concepts related to Grid networks. The appendix also describes a number of facilities that are beginning to provide large-scale Grid network services, nationally and globally.

Chapter 1

The Grid and Grid Network Services

Joe Mambretti

1.1 INTRODUCTION

This chapter presents a brief overview of Grid concepts, architecture, and technology, especially as they relate to Grid networks. Because many excellent explications of Grids have been published previously, including seminal descriptions and definitions [1–3], only a few basic concepts are discussed here.

This chapter also introduces the theme of extending Grid architectural principles to incorporate a wider consideration of communication services and network technologies. This chapter provides a context for subsequent discussions of architecture and methods that allow network resources to be implemented as components that are completely integrated with other resources within Grid environments, rather than as separate, external resources. This integration requires network services to be characterized by a set of attributes that characterize other Grid resources. These attributes are introduced in this chapter, and they are further described in Chapter 3.

1.2 NETWORK RESOURCES AS FIRST-CLASS GRID ENTITIES

Networks have always been essential to Grid environments and, consequently, network services and technologies have been a particularly important topic of discussion within the Grid community. However, network resources have not always been considered what has been termed "full participants" within Grid environments, or "first-class entities": They have not been generally provisioned in accordance with

---

Grid Networks: Enabling Grids with Advanced Communication Technology Franco Travostino, Joe Mambretti, Gigi Karmous-Edwards © 2006 John Wiley & Sons, Ltd

many key Grid architectural principles; they have not been closely integrated with other components Grid environments; they have rarely been implemented as directly addressable, reconfigurable resources.

In part, this circumstance results from the challenge of implementing core network nodes that can be fully administered by Grid processes. This situation is a direct consequence of traditional network architecture and practice, which places an emphasis on stacking services and functions within hierarchical layers, governed by centralized control and management processes. Grid environments are created not as rigid functional hierarchies, but more as collections of modular capabilities that can be assembled and reassembled based on dynamically changing requirements.

Recently, new network architecture and complementary emerging techniques and technologies have provided means by which these challenges can be addressed. These new approaches enable implementations of Grid network resources to be shaped by the same principles as other Grid resources. They also allow dynamic interactions among Grid network services and other components within Grid environments.

1.2.1 WHAT IS A GRID?

Grid networking services are best presented within the context of the Grid and its architectural principles. The Grid is a flexible, distributed, information technology environment that enables multiple services to be created with a significant degree of independence from the specific attributes of underlying support infrastructure. Advanced architectural infrastructure design increasingly revolves around the creation and delivery of multiple ubiquitous digital services. A major goal of information technology designers is to provide an environment within which it is possible to present any form of information on any device at any location. The Grid is an infrastructure that highly complements the era of ubiquitous digital information and services.

Traditionally, infrastructure has been developed and implemented to provide a carefully designed, and relatively limited, set of services, usually centrally managed. Grid architecture has been designed specifically to enable it to be used to create many different types of services. A design objective is the creation of infrastructure that can provide sufficient levels of abstraction to support an almost unlimited number of specialized services without the restrictions of dependencies inherent in delivery mechanisms, local sites, or particular devices. These environments are designed to support services not as discrete infrastructure components, but as modular resources that can be integrated into specialized blends of capabilities to create multiple additional, highly customizable services. The Grid also allows such services to be designed and implemented by diverse, distributed communities, independently of centralized processes. Grid architecture represents an innovation that is advancing efforts to achieve these goals.

As with multiple other technology designs, including those that led to the Internet, the Grid evolved from a need to address the information technology requirements of science and engineering. The Grid is a major new type of infrastructure that builds upon, and extends the power of, innovations that originally arose from addressing the requirements of large-scale, resource-intensive science and engineering applications.

The original architectural concepts developed to address these types of requirements led to the initial formulations of prototype Grid architecture. Early Grid infrastructure was developed to support data and compute intensive science projects. For example, the high-energy physics community was an early adopter of Grid technology. This community must acquire extremely high volumes of data from specialized instruments at key locations in different countries. They must gather, distribute, and analyze those large volumes of data as a collaborative initiative with thousands of colleagues around the world [4].

These requirements, as well as multiyear development implementation cycles, lead to architectural designs that eliminate dependencies on particular hardware and software designs and configurations. Scientists designing and developing major applications cannot become dependent either on static infrastructure, given the rate of technology change, or on infrastructure that is subject to continual changes in basic architecture. They require a degree of separation between their applications and specific, highly defined hardware and configurations.

Many computationally intensive science applications are supported by parallel processing. This degree of separation between applications and infrastructure enables the computational scientists to develop methods for distributed parallel processing on different types of computational systems across multiple distributed domains. Such distribution techniques allows applications to take advantage of large numbers of diverse distributed processors.

Although the Grid was initially developed to support large-scale science projects, its usefulness became quickly apparent to many other application communities. The potential of its architecture for abstracting capabilities from underlying infrastructure provide a means to resolve many issues related to information technology services.

1.3 THE GENERAL ATTRIBUTES OF GRIDS

Many characteristics of Grids are common to other information technology environments, and other characteristics differ by only small degrees. However, some attributes are generally implemented only within Grid environments. Grid environments are usually not designed as rigid hierarchies of layered functions. Instead, they resemble collections of modular capabilities that can be assembled and reassembled based on dynamically changing requirements. This approach does not preclude creating hierarchical stacks of functions; however, its basic design is oriented horizontally across resources rather than vertically. Usually, basic design considerations are the prerogative of the infrastructure designer. Using Grids, these choices can be determined by the application and service designers.

Decisions about the placement of capabilities within specific functional areas are particularly important when creating an architectural model. Recently, architectural designs have tended to allow an increasing number of capabilities within any functional area. Also, determining the layers at which to place functions has been a challenge, and such placement has been a subject of considerable debate. For example, it has been strongly argued that the optimal approach is placing functionality at the edge rather than at the core of systems. Grid environments do not necessarily predetermine or presume a "right answer" with regard to placement of capabilities within functional areas or functional areas within predefined layers. They provide options and allow the communities using the environment to make these determinations.

General Grid characteristics include the following attributes. Each of these attributes can be formally expressed within an architectural framework. Within Grid environments, to a significant degree, these determinations can be considered more art than craft. Ultimately, it is the application or service designer who can determine the relationship among these functions.

(a) *Abstraction/virtualization*. Grids have exceptional potential for abstracting limitless customizable functions from underlying information technology infrastructure and related resources. The level of abstraction within a Grid environment enables support for many categories of innovative applications that cannot be created with traditional infrastructure, because it provides unique methods for reducing specific local dependencies and for resource sharing and integration.

(b) *Resource sharing*. One consequence of this support for high levels of abstraction is that Grid environments are highly complementary to services based on resource sharing.

(c) *Flexibility/programmability*. Another particularly important characteristic of the Grid is that it is a "programmable" environment, in the sense of macro-programming and resource steering. This programmability is a major advantage of Grid architecture – providing flexibility not inherent in other infrastructure, especially capabilities made possible by workflow management and resource reconfigurability. Grids can enable scheduled processes and/or continual, dynamic changing of resource allocations and configurations, in real time. Grids can be used to support environments that require sophisticated orchestration of workflow processes. Much of this flexibility is made possible by specialized software "toolkits," middleware that manages requests and resources within workflow frameworks.

(d) *Determinism*. Grid processes enable applications to directly ensure, through autonomous processes, that they are matched with appropriate service levels and required resources, for example through explicit signaling for specialized services and data treatments.

(e) *Decentralized management and control*. Another key feature underlying Grid flexibility is that its architecture supports the decentralization of management and control over resources, enabling multiple capabilities to be evoked independently of processes that require intercession by centralized processes.

(f) *Dynamic integration*. Grids also allow for the dynamic creation of integrated collections of resources that can be used to support special higher level environments, including such constructs as virtual organizations.

(g) *Resource sharing*. Grid abstraction capabilities allow for large-scale resource sharing among multiple, highly distributed sites.

(h) *Scalability*. Grid environments are particularly scalable – they can be implemented locally or distributed across large geographic regions, enabling the reach of specialized capabilities to extend to remote sites across the world.

1.3 The General Attributes of Grids

(i) *High performance.* Grids can provide for extremely high-performance services by aggregating multiple resources, e.g., multiple distributed parallel processors and parallel communication channels.

(j) *Security.* Grids can be highly secure, especially when segmentation techniques are used to isolate partitioned areas of the environment.

(k) *Pervasiveness.* Grids can be extremely pervasive and can extend to many types of edge environments and devices.

(l) *Customization.* Grids can be customized to address highly specialized requirements, conditions, and resources.

Grid environments can provide these capabilities if the design of their infrastructure is developed within the context of a Grid architectural framework (described in Chapter 3). Increasingly, new methods are being developed that allow for the integration of additional resources into Grid environments while preserving, or extending, these capabilities. For such resources to "fully participate" within a Grid environment, they must be able to support these attributes.

Grids are defined by various sets of basic characteristics, including those that are common to all information technology systems, those that are common to distributed systems, and those that define Grid environments. The general characteristics of a Grid environment described here are those that define basic Grid environments. These characteristics are made possible by the way that resource components are implemented and used within a Grid environment. These individual resource components contribute to the aggregate set of capabilities provided by the Grid. A Grid environment comprised multiple types of resources that can be gathered, integrated, and directly managed as services that can perform defined tasks.

1.3.1 THE GRID AND DESIGN ABSTRACTION

Two key attributes of Grids described in the previous section are those related to abstraction and pervasive programmability. The principle of abstraction has always been fundamental to information technology design. Many important new phases of technology development have been initiated by an innovation based on providing enhanced levels of abstraction. As another phase in this evolution, the Grid builds on that tradition. For example, this abstraction capability makes the Grid particularly useful for creating common environments for distributed collaborative communities.

Grids are used to support virtual organizations. An important benefit of the Grid is its capability for supporting not only individual applications and services but also complete large-scale distributed environments for collaborative communities, thereby "enabling scalable virtual organizations" [2]. Grid developers have always stressed the need to create an environment that can support "coordinated resource sharing and problem solving in a dynamic, multi-institutional virtual organization" [2].

This defining premise has been one of the motivations behind the migration of the Grid from science and engineering to more industrial implementations as well as to other more general domains. Grids can be used to create specialized environments for individuals, large groups, organizations, and global communities. Grids are even being used to support groups of individuals world-wide who are collaborating as if they were all within the same local space – sharing customized global virtual environments.

Grid services abstractions are expressed through standard services definition, middleware, protocols, application programming interfaces, software tools, and reconfigurable infrastructure. These abstraction capabilities are made possible primarily by a set of sophisticated Grid middleware, toolkit suites, which reside between services and infrastructure – separating upper level end-delivered service functionality from lower level resources such as system software, data, and hardware within specific configurations.

Grid application requirements preclude traditional workflow and resource usage, such as those that utilize components as discrete production units. Traditional information technology components have been used as separate components, e.g., computer processors, storage, instruments, and networks. Although these components are connected, they are not integrated.

Grid developers have designed methods, based on services abstractions, for creating environments within which it is possible to discover, gather, and integrate multiple information technology components and other resources from almost any location. Grid architecture provides for an extremely open and extensible framework that makes it possible to create distributed environments using these methods of collecting and closely integrating distributed heterogeneous resources.

1.3.2 THE GRID AS AN ENABLER OF PERVASIVE, PROGRAMMABLE UTILITY SERVICES

The term "Grid" was selected to describe this environment as an analogy to the electric power grid, that is, a large-scale, pervasive, readily accessible resource that empowers multiple different devices, systems, and environments at distributed sites. However, this metaphoric description of the Grid as a set of ubiquitous utility services may overshadow its versatility – its potential for flexibility and reconfigurability. General utility infrastructure is usually designed to deliver a single service, or a narrow range of services. Those services are to be used in the form in which they are delivered. The power Grid is based on a relatively fixed infrastructure foundation that provides a fairly limited set of services, and its underlying topology certainly cannot be dynamically reconfigured by external communities.

In contrast, the information technology Grid can be used to create an almost unlimited number of differentiated services, even within the same infrastructure. The Grid is an infrastructure that provides a range of capabilities or functions, from which it is possible for multiple distributed communities, or individuals, to create their own services.

The Grid is "programmable," in the sense of high-level macro-programming or "resource steering" – providing capabilities for dynamically changing underlying infrastructure. This potential for dynamic change is a primary benefit of Grid environments, because it provides an almost endless potential for creating new communication services as well as for expanding and enhancing existing services. Grid services are self-referential in that they include all information required to find, gather, use, and discard resources to accomplish goals across distributed infrastructure. Grid services are also highly modularized so that they can be advertised to other Grid services and related processes and combined in ad hoc ways to accomplish various tasks. This flexibility is being extended to all Grid resources, including Grid networks.

1.4 TYPES OF GRIDS

Grid architecture continues to evolve as the overall design concepts continue to improve and as it is employed for additional tasks. Grids are often associated with high-performance applications because of the community in which they were originally developed. However, because Grid architecture is highly flexible, Grids have also been adopted for use by many other, less computationally intensive, application areas. Today, many types of Grids exist, and new Grids are continually being designed to address new information technology challenges.

Grids can be classified in various ways, for example by qualities of physical configuration, topology, and locality. Grids within an enterprise are called intra-grids, interlinked Grids within multiple organizations are called inter-grids and Grids external to an organization are called extra-grids. Grids can have a small or large special distribution, i.e., distributed locally, nationally or world-wide. Grids can also been classified by their primary resources and function, for example computational Grids provide for high-performance or specialized distributed computing. Grids can provide modest-scale computational power by integrating computing resources across an enterprise campus or large-scale computation by integrating computers across a nation such as the TeraGrid in the USA [5].

Data Grids, which support the use of large-scale distributed collections of information, were originally developed for the distributed management of large scientific datasets. Many data Grids support the secure discovery, utilization, replication, and transport of large collections of data across multiple domains. For most data Grids, the primary design consideration is not access to processing power but optimized management of intensive data flows. Data Grids must manage and utilize data collections as a common resource even though those collections exist within multiple domains, including those at remote locations [4,6].

Grids continue to integrate new components and innovative methods, to meet the needs of existing and new applications. Application Grids are devoted to supporting various types of applications. Examples include those which support visualization, digital media, imaging, and collaborative communication (such as the Access Grid, a specialized communications environment), storage grids (which support massive data repositories), services grids (which are devoted to general or specialized services), sensor grids, Radio Frequency Identification Systems (RFID) Grids, and security grids. Grids can even exist on a very small scale, for example, across collections of tiny devices, such as electronic motes.

At the same time, new types of world-wide network facilities and infrastructure are being created and implemented to support global high-performance services. For example, "Global Lambda Grids," which are based on high-performance optical networks, are supporting major science projects around the world [7]. One research project is exploring new tools for scientific research based on large-scale distributed infrastructure that uses advanced, high-performance optical technologies as a central resource [8].

1.4.1 GRIDS AND GRID NETWORKS

Extending general Grid attributes to communication services and network resources has been an evolutionary process. A key goal has been to ensure that these services and resources can be closely integrated with multiple other co-existent Grid services and resources. This close integration is one of the capabilities that enable networks to become "full participants" within Grid environments, as opposed to being used as generic, accessible external resources.

Almost all Grids are implemented as distributed infrastructure. Therefore, from the earliest days of their design and development, Grids have always utilized communications services, especially those based on TCP/IP (transmission control protocol/Internet protocol). Grids could not have been developed without the Internet, a widely deployed, inexpensive data communications network, based on packet routing. As discussed elsewhere in this book, the Internet and Grids share a number of basic architectural concepts.

Many fundamental Grid concepts incorporated new approaches to networking created for specialized projects, such as the innovative I-WAY project (Information Wide Area Year), which was based on an experimental broadband network implemented for Supercomputing 95 [9]. The I-WAY project demonstrated for the first time that a national network fabric could be integrated to support large-scale distributed computing. The software created for that project became the basis for the most widely implemented Grid software used today [10].

However, until recently, the mechanisms that allow networks to be fully integrated into Grid environments did not exist, in part because Grid architectural concepts differ from those that have governed the design of traditional networks. Before the Internet, traditional networks were designed specifically to support a narrow range of precisely defined communication services. These services were implemented on fairly rigid infrastructure, with minimal capabilities for ad hoc reconfiguration. Such traditional networks were designed with the assumptions that target service requirements are known, and that the supporting infrastructure would remain relatively unchanged for many years. Traditional networks were provisioned so that they could be used as resources external to other processes, with minimal capabilities for dynamic configurations or ad hoc resource requests. They have been centrally managed and controlled resources.

The Internet design has been a major benefit to Grid deployments. Unlike legacy telecommunications infrastructure, which has had a complex core and minimal functionality at the edge, the Internet places a premium on functionality at the edge supported by a fairly simple core. This end-to-end design principle, described in Chapter 10, enables innovation services to be created and implemented at the edge of the network, provides for high-performance network backbones, and allows for significant service scalability.

Because the Internet generally has been provisioned as an overlay on legacy communications infrastructure, its potential to support Grid communications services has not yet been completely realized. To enable networks to be utilized with the same flexibility as other Grid resources, Grid networks should incorporate the design goals that shape the larger Grid environment within which they are integrated. Currently, various initiatives are creating frameworks that allow for Grid network resources to accomplish this goal. These initiatives are also beginning to create capabilities that provide for interactivity among multiple high-level Grid services, processes, and network resources. These methods can be used to integrate network resources much more closely with other resource components of Grid environments.

1.4.2 ATTRIBUTES OF GRID NETWORKS

The architecture and methods that are being created for enabling network resources to be more closely integrated into Grid environments are directed at enabling those resources to have the same characteristics as the general Grid environment. The key attributes of Grid network features comprise basic themes for this book, such as capabilities for abstraction, programmability, services oriented architecture, and related topics.

1.4.2.1 Abstraction

One of the most important features of a Grid is its potential for abstracting capabilities from underlying resources and enabling those capabilities to be integrated to support customized services. The Grid architectural model presupposes an environment in which available modular resources can be detected, gathered, and utilized without restrictions imposed by specific low-level infrastructure implementations. This architecture does not specify the complete details of all possible resources, but instead describes the requirements of classes of Grid components. For example, one class of components comprises a few basic abstractions and key protocols that are closest to applications. Another set consists of capabilities for discovering, scheduling, gathering, interlinking, coordinating, and monitoring resources, which can be physical or logical. Another set comprises the actual resources, sometimes termed the Grid "fabric."

The virtualization of resources is as powerful a tool for creating advanced data network services. A major advantage to the virtualization of Grid network functionality through abstraction techniques is increased flexibility in service creation, provisioning, and differentiation. It allows specific application requirements to be more directly matched with network resources. Virtualization also enables networks with very different characteristics to be implemented within a common infrastructure and enables network processes and resources to be integrated directly with other types of Grid resources. For example, low-level functionality within the core of a network can be extended directed into individual applications, allowing applications to signal directly for required network resources.

Using high-level abstractions for network services and integrating network capabilities through Grid middleware provides a flexibility that it is not possible to achieve with traditional data networks. Traditional data networks support only a limited range of services, because they are based on rigid infrastructure and topologies, with restricted abstraction capabilities. General network design and provisioning is primarily oriented toward provisioning highly defined services on specific physical infrastructure, making enhancements and changes difficult, complex, and costly.

1.4.2.2 Resource sharing and site autonomy

The Global Grid Forum (GGF), described in Chapter 4, is engaged in specifying the open Grid services architecture and leveraging the Web Services framework, one component of which is the Web Service Resource Framework (WSRF), also described in Chapter 4. The Grid development communities are engaged in implementing Grid infrastructure software with Web Services components. These components provide access to sets of building blocks that can be combined easily into different service combinations within classes, based on multiple parameters. They can be used to customize services and also to enable shared resources within autonomous environments.

Within a Grid network services context, these capabilities provide new mechanisms for network services design and provisioning, especially new methods for directly manipulating network resources. This approach allows for the creation of customized services by integrating different services at different network layers, including through inter-layer signaling, to provide precise capabilities required by categories of applications that cannot be deployed, or optimized, within other environments. Using these techniques, novel network services can be based on multiple characteristics, e.g., those based on policy-based access control and other forms of security, priority of traffic flows, quality of service guarantees, resource allocation schemes, traffic shaping, monitoring, pre-fault detection adjustments, and restoration techniques.

1.4.2.3 Flexibility through programmability

An important characteristic of the Grid is that it is a programmable environment. However, until recently, Grid networks have not been programmable. This programmability provides a flexibility that is not characteristic of common infrastructure. As noted, network infrastructure has traditionally been designed to support fairly static services with fixed parameters. As a result, network services are costly to deploy and reconfigure, because major changes are primarily accomplished through time-consuming physical provisioning and engineering.

To date, almost all Grids have been based on communication services provided by statically provisioned, routed networks, and the common accessible data service has been a single, undifferentiated, "best effort" service, with minimal potential for service determinism, flexibility, and customization.

In the last few years, several initiatives have been established to create a Grid network services architecture that enables communication services to be substantially more flexible. Using these new methods, Grid network services can be provisioned as "programmable," allowing continually dynamic changing of service and resource allocations, including dynamic reconfigurations. Similarly, these methods make it is possible to initiate processes that can implement instantiations of Grid network services, for short or long terms, with static attributes or with continually changing attributes.

1.4.2.4 Determinism in network services

Because the Grid is flexible and programmable, it allows applications to be matched with the precise resources required. This ability to request and receive required resources and to define precisely matching service levels is called "determinism." Determinism is especially meaningful to Grid networking. Grids have usually been based on common "best effort" data communication services, not deterministic services. Often, the networks on which Grids are based do not provide consistent levels of service, and there have not been any means by which specific levels of service could be requested or provided.

A primary goal of Grid network research is to create more diverse communication services for Grid environments, including services that are significantly more deterministic and adjustable than those commonly used. New methods are being created that allow individual applications to directly signal for the exact levels of network service required for optimal performance. Network service responsiveness, such as its delivered performance, is determined by the degree to which network elements can be adjusted – managed and controlled – by specialized explicit signaling.

Deterministic networking is important to achieving optimal applications performance. It is also a key enabling technology for many classes of applications that cannot be supported through traditional network quality of service mechanisms. This capability includes mechanisms both for requesting individual network services that have specific sets of attributes and also, when required, for reconfiguring network resources so that those specific services can be obtained. This capability is critical for many classes of applications. For example, Grid technology is used to support many large-scale data-intensive applications requiring high-volume, high-performance data communications. Currently, this type of service is not well supported within common Internet environments: large data flows disrupt other traffic, while often failing to meet their own requirements.

1.4.2.5 Decentralized management and control

An important capability for Grid environments is decentralized control and management of resources, allowing resource provisioning, utilization, and reconfiguration without intercession by centralized management or other authorities. During the last few years, various technologies and techniques have been developed to allow decentralized control over network resources. These methods allow Grid networks to be "programmed," significantly expanding Grid network services capabilities. Today, methods are available that can provide multiple levels of deterministic, differentiated services capabilities not only for layer 3 routing, but also for services at all other communication layers.

Some of these methods are based on specialized signaling, which can be implemented in accordance with several basic models. For example, two basic models can be considered two ends of a spectrum. At one end is a model based on predetermining network services, conditions, and attributes, and providing service qualities in advance, integrated within the core infrastructure. At the other end is a model based on mechanisms that continually monitor network conditions, and adjust network services and resources based on those changes. Between these end points, there are techniques that combined pre-provisioning methods with those based on dynamic monitoring and adjustment. Emerging Grid networking techniques define methods that provide for determinism by allowing applications to have precision control over network resource elements when required.

1.4.2.6 Dynamic integration

Grid architecture was designed to allow an expansive set of resources to be integrated into a single, cohesive environment. This resource integration can be accomplished in advance of use or it can be implemented dynamically. Traditionally, the integration of network resources into environments requiring real-time ad hoc changes has been a challenge because networks have not been designed for dynamic reconfiguration. However, new architecture and techniques are enabling communication services and network resources to be integrated with other Grid resources and continually changed dynamically.

1.4.2.7 Resource sharing

A primary motivation for the design and development of Grid architecture has been to enhance capabilities for resource sharing, for example, utilizing spare computation cycles for multiple projects [11]. Similarly, a major advantage to Grid networks is that they provide options for resource sharing that are difficult if not impossible in traditional data networks. Virtualization of network resources allows for the creation of new types of data networks, based on resource sharing techniques that have not been possible to implement until recently.

1.4.2.8 Scalability

Scalability for information technology has many dimensions. It can refer to expansion among geographic locations, enhanced performance, an increase in the number of services offered and in the communities served, etc. Grid environments are by definition highly distributed and are, therefore, highly scalable geographically. Consequently, Grid networks can extend not only across metro areas, regions, and nations but also world-wide. The scalability of advanced Grid networks across the globe has been demonstrated for the last several years by many international communities, particularly those using international networks.

Currently, the majority of advanced Grid networks are being used to support global science applications on high-performance international research and education networks. This global extension of services related to these projects has been demonstrated not only at the level of infrastructure but also with regard to specialized services and dynamic allocation and reconfiguration capabilities.

1.4.2.9 High performance

Because many Grid applications are extremely resource intensive, one of the primary drivers for Grid design and development has been the need to support applications requiring ultra-high-performance data computation, flow, and storage. Similarly, Grid networks require extremely high-performance capabilities, especially to support data-intensive flows that cannot be sustained by traditional data networks. Many of the current Grid networking research and development initiatives are directed at enhancing high-performance data flows, such as those required by high-energy physics, computational astrophysics, visualization, and bioinformatics.

For Grid networks, high performance is measured by more than support for high-volume data flows. Performance is also measured by capabilities for fine-grained application control over individual data flows. In addition, within Grid networks, performance is defined by many other measures, including end-to-end application behavior, differentiated services capabilities, programmability, precision control responsiveness, reconfigurability, fault tolerance, stability, reliability, and speed of restoration under fault conditions.

1.4.2.10 Security

Security has always been a high-priority requirement that has been continually addressed by Grid developers [12]. New techniques and technologies are currently being developed to ensure that Grid networks are highly secure. For example, different types of segmentation techniques used for Grid network resources, especially at the physical level, provide capabilities allowing high-security data traffic to be completely isolated from other types of traffic. Also, recently, new techniques using high-performance encryption for Grid networks have been designed to provide enhanced security to levels difficult to obtain on traditional data networks.

1.4.2.11 Pervasiveness

Grid environments are extensible to wide geographic areas, including through distributed edge devices. Similarly, Grid network services are being designed for ubiquitous deployment, including as overlay services on flexible network infrastructure. Multiple research and development projects are focused on extending Grids using new types of edge technologies, such as wireless broadband and edge devices, including consumer products, mobile communication devices, sensors, instruments, and specialized monitors.

1.4.2.12 Customization

Just as Grids can be customized to address specialized requirements, new Grid network architecture and methods provide opportunities for the creation and implementation of multiple customized Grid communication services that can be implemented within a common infrastructure. Grid networks based on capabilities for adaptive services, resource abstraction, flexibility, and programmability can be used to create many more types of communication services than traditional networks. New types of communication services can be created through the integration and combination of other communication services.

For example, such integration can be accomplished by integrating multiple types of services at the same network layer, and others by integration services across layers. New services can be also created by closely integrating Grid network services with other Grid resources.

1.5 GRID NETWORKS AND EMERGING COMMUNICATION TECHNOLOGIES

This chapter describes basic Grid environment attributes that could be used as general design goals for infrastructure development, implementation, or enhancement. These attributes are being formalized through architecture being created by standards committees (described in Chapter 4). Various initiatives, including those established by standards committees, are extending these attributes to Grid network resources. Currently, Grid research communities are creating a new architectural framework that will enable network resources to be used within Grid environments as easily as any other common resource. These initiatives are extending current Grid architectural principles, inventing new concepts, and creating new protocols and methods.

These research and development efforts are taking advantage of recent network innovations to accomplish these goals, especially innovations related to data network services and those that are allowing the integration of services across all of the traditional network layers. Given the importance of the Internet, many advanced techniques have been developed for more sophisticated routed packet-based services and for Internet transport. These topics are discussed in Chapters 8, 9, and 10. Advanced architecture is also being designed to take advantage of innovations related to other types of transport networking, including techniques for high-performance switching and dynamic path provisioning. These topics are discussed in Chapter 11.

Other important recent research and development activities have focused on the potential for Grid communications based on lightpath services, supported by agile optical networks, especially those based on dynamically provisioned lightpaths. A variety of optical technologies are being tested in advanced Grid environments to demonstrate how their capabilities can be used to complement traditional network services. These topics are discussed in Chapter 12.

In addition, many emerging communications technologies are being investigated for their potential for supporting Grid networking environments. Various research projects are developing powerful, advanced communication technologies based on innovative technologies, including those related to wireless, free space optics, LED technology, and optical switches. These topics are discussed in Chapter 15.

REFERENCES

[1] I. Foster and C. Kesselman (eds) (2004) *The Grid: Blueprint for a Future Computing Infrastructure*, 2nd edn, Morgan Kaufmann Publishers.
[2] I. Foster, C. Kesselman, and S. Tuecke (2001) "The Anatomy of the Grid: Enabling Scalable Virtual Organizations," *International Journal of Supercomputer Applications*, 15(3), 200–222.
[3] F. Berman, A. Hey, and G. Fox (2003) *Grid Computing: Making The Global Infrastructure a Reality*, John Wiley & Sons, Ltd.
[4] I. Foster and R. Grossman (2003) "Data Integration in a Bandwidth-Rich World," special issue on "Blueprint for the Future of High Performance Networking," *Communications of the ACM*, 46(1), 50–57.

References

[5] www.teragrid.org.

[6] H. Newman, M. Ellisman, and J. Orcutt (2003) "Data-Intensive E-Science Frontier Research," special issue on "Blueprint for the Future of High Performance Networking," *Communications of the ACM*, 46(11), 68–75.

[7] T. DeFanti, C. De Laat, J. Mambretti, and B. St. Arnaud (2003) "TransLight: A Global Scale Lambda grid for E-Science," special issue on "Blueprint for the Future of High Performance Networking," *Communications of the ACM*, 46(11), 34–41.

[8] L. Smarr, A, Chien, T. DeFanti, J. Leigh, and P. Papadopoulos (2003) "The OptIPuter," special issue on "Blueprint for the Future of High Performance Networking," *Communications of the ACM*, 46(11), 58–67.

[9] T. DeFanti, I. Foster, M. Papka, R. Stevens, and T. Kuhfuss (1996) "Overview of the I-WAY: Wide Area Visual Supercomputing," *International Journal of Supercomputer Applications and High Performance Computing*, 10(2/3), 123–130.

[10] I. Foster, J. Geisler, W. Nickless, W. Smith, and S. Tuecke (1997) "Software Infrastructure for the I-WAY High Performance Distributed Computing Project," *Proceedings of 5th Annual IEEE Symposium on High Performance Distributed Computing*, pp. 562–571.

[11] K. Czajkowski, S. Fitzgerald, I. Foster, and C. Kesselman (2001) "Grid Information Services for Resource Sharing." *Proceedings of the 10th IEEE International Symposium on High Performance Distributed Computing (HPDC-10)*, IEEE Press.

[12] I. Foster, C. Kesselman, G. Tsudik, and S. Tuecke (1998) "A Security Architecture for Computational Grids," *Proceedings of the 5th ACM Conference on Grid and Communications Security Conference*, pp. 83–92.

Chapter 2

Grid Network Requirements and Driver Applications

2.1 INTRODUCTION

This chapter presents several specialized Grids, which are described here to assist in explaining Grid network requirements. Although since the initial inception of Grid architecture many individual types have been defined, they tend to fall within several representative classes, as noted in the previous chapter. Reflecting their origin within the large-scale science community, one class of Grids, computational Grids, is optimized for computational performance. They are designed primarily as tools for computationally intensive problems.

Another class of Grids, data Grids, consists of environments that address the particularly problem of managing and analyzing large-scale distributed data. Service Grids are those that have been designed to meet the specific requirements of specialized services within distributed environments. Many other types of Grids exist. The following section, which presents examples of specialized Grid environments, is followed by an overview of the network requirements inherent in general Grid environments.

Grid Networks: Enabling Grids with Advanced Communication Technology   Franco Travostino, Joe Mambretti, Gigi Karmous-Edwards   © 2006 John Wiley & Sons, Ltd

2.2 GRID NETWORK REQUIREMENTS FOR LARGE-SCALE VISUALIZATION AND COLLABORATION

Jason Leigh, Luc Renambot, and Maxine Brown

2.2.1 LARGE-SCALE VISUALIZATION AND COLLABORATION APPLICATION DRIVERS

Doctors want to better study the flow dynamics of the human body's circulatory system. Ecologists want to gather, analyze, and distribute information about entire ecosystems in estuaries and lakes and along coastlines. Biologists want to image and progressively magnify a specimen on a remote electron microscope, zooming from an entire system, such as a rat cerebellum, to an individual spiny dendrite. And crisis management strategists want an integrated joint decision support system across local, state, and federal agencies, combining massive amounts of high-resolution imagery, highly visual collaboration facilities, and real-time input from field sensors.

Remote colleagues want to interact visually with massive amounts of data as well as high-definition video streams from live cameras, instruments, archived data arrays, and real-time simulations. In essence, these applications want situation rooms and research laboratories in which the walls are seamless ultra-high-resolution tiled displays fed by datastreamed over ultra-high-speed networks, from distantly located visualization and storage servers, enabling local and distributed groups of researchers to work with one another while viewing and analyzing visualizations of large distributed heterogeneous datasets, as shown in Figure 2.1.

Computational scientists, or e-scientists, want to study and better understand complex systems – physical, geological, biological, environmental, and atmospheric – from the micro to the macro scale, in both time and space. They want new levels

Figure 2.1. Conceptualization of a high-definition research environment showing a several hundred megapixel data fusion screen on the right screen, super-high-definition videoteleconferencing on the left screen, and an auto-stereo 3D image of earthquake data, seemingly suspended in "mid-air."

2.2 Grid Network Requirements

of persistent collaboration over continental and transoceanic distances, coupled with the ability to process, disseminate, and share information on unprecedented scales, immediately benefiting the scientific community and, ultimately, everyone else as well. These application drivers are motivating the development of large-scale collaboration and visualization environments, built on top of an emerging global "LambdaGrid" cyberinfrastructure that is based on optical networks [1].

The OptIPuter, a specific cyberinfrastructure research effort that couples computational resources over parallel optical networks in support of data-intensive scientific research and collaboration, is discussed.

2.2.2 CURRENT LIMITATIONS TO ADVANCED VISUALIZATION AND COLLABORATION

What has been missing, to date, is the ability to augment data-intensive Grid computing – primarily distributed number crunching – with high-throughput or LambdaGrid computing. A Grid is a set of networked, middleware-enabled computing resources; a LambdaGrid is a Grid in which the lambda networks themselves are resources that can be scheduled like any other computing, storage, and visualization resource. Recent major technological and cost breakthroughs in networking technology have made it possible to send multiple lambdas down a single length of user-owned optical fiber. (A lambda, in networking parlance, is a fully dedicated wavelength of light in an optical network, capable of bandwidth speeds of 1–10 Gbps.) Metro and long-haul 10-Gbps lambdas are 100 times faster than 100T-base fast Ethernet local area networks used by PCs in research laboratories. The exponential growth rate in bandwidth capacity over the past 12 years has surpassed even Moore's law, as a result, in part, of the use of parallelism in network architectures. Now the parallelism is in multiple lambdas on single-strand optical fibers, creating supernetworks, or networks faster (and some day cheaper) than the computers attached to them [2].

A supernetwork backbone is not the only limitation. One needs new software and middleware to provide data delivery capabilities for the future lambda-rich world. The OptIPuter project is providing middleware and system software to harness the raw capabilities of the network hardware in a form that is readily available to and usable by applications [2]. More specifically, the OptIPuter seeks to overcome the following bottlenecks:

2.2.2.1 Portability

Today, applications written for one graphics environment have to be redesigned before they can run under other environments. For example, visualization tools developed for desktop computers can rarely take advantage of the processing power of a cluster of graphics computers; conversely, visualization tools developed for clusters rarely function on desktop computers. Also, users cannot easily add their own video sources, compression modules, and/or network protocols to the systems, which are needed for high-bandwidth wide-area networks [3].

2.2.2.2 Scalability

The ability of visualization software and systems to scale in terms of the amount of data they can visualize, and the resolution of the desired visualization, is still an area of intensive visualization research.

2.2.2.3 Networking

Demanding visualization applications need large amounts of bandwidth to interactively access and visually display remote datasets. Even so, long-distance photonic networks are limited by the speed of light, causing latency issues that impede interaction.

2.2.2.4 Collaboration

Collaborators want to interact with visualizations of massive amounts of data and high-definition video streams from live cameras, instruments, archived data arrays, and real-time simulations, without having to make modifications to source and/or destination machines. There are no commercial solutions for easily accessing and displaying vast datastreams from multiple sources, both locally and remotely. More specifically, it is difficult to synchronize individual visualization streams to form a single larger stream, scale and route streams generated by an array of $M \times N$ nodes to fit an $X \times Y$ display, and exploit a variety of transport protocols, such reliable blast User Datagram Protocol (UDP) and IP multicast.

2.2.3 ENABLING ADVANCED VISUALIZATION AND COLLABORATION WITH THE OPTIPUTER

The OptIPuter, so named for its use of optical networking, IP, computer storage, and processing and visualization technologies, is an infrastructure research effort that tightly couples computational resources over parallel optical networks using the IP communication mechanism. The OptIPuter is being designed as a "virtual" parallel computer in which the individual "processors" are widely distributed clusters; the "memory" is in the form of large distributed data repositories; "peripherals" are very large scientific instruments, visualization displays, and/or sensor arrays; and the "motherboard" uses standard IP delivered over multiple dedicated lambdas that serve as the "system bus." The goal of this new architecture is to enable scientists who are generating terabytes and petabytes of data to interactively visualize, analyze, and correlate their data from multiple storage sites on high-resolution displays connected over optical networks.

To display high-resolution images, OptIPuter partner Electronic Visualization Laboratory at the University of Illinois at Chicago developed "LambdaVision," a 100-megapixel ultra-high-resolution display wall built from stitching together dozens of LCD panels, shown in Figure 2.2.

This wall is managed by a software system called SAGE (Scalable Adaptive Graphics Environment), which organizes the large-screen's "real estate" as if it were one continuous canvas [4]. SAGE allows the seamless display of various networked applications

2.2 Grid Network Requirements

Figure 2.2. "LambdaVision" 100-megapixel (55-LCD panel) tiled.

over the whole display. Each visualization application – whether archived data images, pre-rendered movies, real-time simulations, or video from live cameras or instruments – streams its rendered pixels (or primitives) to a virtual high-resolution frame buffer, allowing for any given layout onto the display, similar to how users open and position windows on their desktops.

2.2.3.1 Portability

The SAGE architecture allows multiple rendering nodes or clusters to access a virtual frame-buffer across the network. The framework intelligently partitions the graphics pipeline to distribute the load. Factors such as the computing and rendering capabilities of the participating machines are needed to decide the load distribution. The framework will ultimately supports the notion of multiple collaborators simultaneously accessing a display space through a shared "window" manager. SAGE is designed to support data fusion for very large datasets [4].

2.2.3.2 Scalability

To achieve real-time visualization and collaboration, the OptIPuter uses large pools of computing resources (such as remote clusters equipped with high-performance graphics processors) and streams the results to participating endpoints. The pooling of computing resources increases utilization, especially when they are cast as Grid services that are combined with other services to form a pipeline that links large-scale data sources with visualization resources. And, since the costs of networking costs are lower than those of computing and storage, it is more cost-effective for users to stream images from costly rendering farms or storage repositories to low-cost thin clients for display. The next-generation LambdaVision display may be configured with small-form-factor computers equipped with gigabit network interfaces that are directly strapped to the backs of the displays, with a simple software system that manages the routing of the visualization streams to the appropriate displays to form seamless images.

2.2.3.3 Networking

The OptIPuter project's two main visualization applications, JuxtaView and Vol-a-Tile, run on clusters of computers and stream the results to LambdaVision [3,5,6]. JuxtaView is a tool for viewing and interacting with time-varying large-scale 2D montages, such as images from confocal or electron microscopes or satellite and aerial photographs. Vol-a-Tile is a tool for viewing large time-varying 3D data, such as seismic volumes. JuxtaView and Vol-a-Tile are unique in that they attempt to anticipate how the user will interact with the data, and use the available network capacity to aggressively *pre-fetch* the needed data, thus reducing the overall latency when data is retrieved from distantly located storage systems.

2.2.3.4 Collaboration

Using SAGE, collaborators at multiple remote sites with heterogeneous displays will eventually be able to share and simultaneously visualize multiple datasets. For instance, users will be able to simultaneously view high-resolution aerial or satellite imagery, volumetric information on earthquakes and ground water, as well as high-definition video teleconferencing. For video streaming, SAGE can simultaneously decode multiple compressed high-definition streams and display them on the tiled displays.

2.2.4 FUTURE CHALLENGES IN LARGE-SCALE VISUALIZATION AND COLLABORATION

Owing to impedance mismatches among storage, visualization, and display systems, the management of multiple parallel flows at each point in the path is ultimately where the challenges lie. Parallel disk system throughputs need to be balanced against the rendering capabilities of the visualization systems. The output of each visualization system needs to be coordinated to control the amount of bandwidth received at each computer that drives a tile of the display wall so as not to overload it.

Visualization streams need to be transmitted over reliable high-bandwidth, low-latency channels to permit remote steering of the application. The bandwidth needed to support these streams is dependent on the resolution of the source. For example, a single panel of an image (of resolution 1600 × 1200) on a tiled display at 24-bit color depth, updating at 15 frames per second, requires approximately 660 Mbps (assuming the stream is not compressed.) The bandwidth required to fill the 55-panel LambdaVision display is approximate 35 Gbps.

Interactions between applications such as JuxtaView and Vol-a-Tile should occur over low-latency and low-jitter *reliable* network channels because low-bandwidth

2.2 Grid Network Requirements

control messages, unlike the actual data being sent, need to be transmitted quickly to ensure that applications are responsive to users' actions.

For high-definition video and its accompanying audio, unlike visualization streams, the use of an *unreliable* transmission channel is sufficient for most cases. Visualization streams require reliability because the viewer should not perceive any artifacts in the images that may cause misinterpretations of the data. Whereas commodity audioconferencing tools today typically use unreliable transport protocols for audio delivery, extensive prior research in computer-supported cooperative work has shown that the quality of a collaboration is affected first and foremost by the quality of the voices that are transceived. Audio streaming in high-performance Grid environments should therefore take advantage of the larger amounts of available bandwidth to dramatically raise the quality of audioconferencing.

Security is also an issue. In situations where viewers are interacting with sensitive information, the visualization and audio/video streams need to be encrypted. The goal is to encrypt and decrypt at multi-10 Gbps rates while incurring minimal delay and expenditure of computing resources.

Collaboration among remote sites means that people need to see and share the same visualizations at the same time, creating the need for multicast. Although a variety of techniques exist for supporting reliable multicast [7], *high-bandwidth (in the order of multi-10 Gbps) and low-latency, reliable* multicast is an unsolved problem and an active area of research within the Grid community [8]. It is a particularly challenging problem at the endpoints that must receive the multicast traffic, and must be solved for any applications using networked tiled displays. In the SAGE model, the bandwidth utilized is dependent on the image resolution of the rendering source. As windows on the tiled display are resized or repositioned on the walls, SAGE must reroute the multiple streams from the rendering source to the PC nodes that drive the displays.

When the windowed area is smaller than the original source of image, the source must be rescaled so that the combined network traffic does not overwhelm the display node (limited to 1 Gbps). This problem becomes much more complex when multicasting is needed to support independent window operations (such as position or scaling) at each display site with different configurations. Possible approaches include utilizing a layered multicast approach to interleave each source stream over multiple multicast streams and coordinating the receivers so that each display node knows which layered stream to subscribe in order to assemble the images for the desired screen resolution, or using high-performance PCs equipped with 10-Gbps interfaces to intercept each multicast stream and perform resizing and rerouting operations on behalf of the thin-client displays.

ACKNOWLEDGMENTS

The Electronic Visualization Laboratory (EVL) at the University of Illinois at Chicago specializes in the design and development of high-resolution visualization and virtual reality display systems, collaboration software for use on multi-gigabit networks, and advanced networking infrastructure. These projects are made possible by major funding from the National Science Foundation (NSF); specifically, LambdaVision is funded by NSF award CNS-0420477 to EVL (see (www.evl.uic.edu/cavern/lambdavision), and the OptIPuter is funded by NSF cooperative agreement OCI-0225642, to Calit2 at UCSD (see <www.optiputer.net), with EVL being one of the lead partners. EVL also receives funding from the State of Illinois, General Motors Research, the Office of Naval Research on behalf of the Technology Research, Education, and Commercialization Center (TRECC), and Pacific Interface Inc. on behalf of NTT Optical Network Systems Laboratory in Japan.

2.3 LARGE-SCALE E-SCIENCE
Peter Clarke

E-science can be defined as the invention and application of advanced computing infrastructures and techniques to achieve greater capability, quality, and volume in scientific research. In the UK, e-science was the basis of a nationwide initiative supported by government and research funding councils to develop infrastructure and promote interdisciplinary research. The program was widely recognized as being remarkably successful, leading to a large number of projects, which enhanced the acceptance of advanced computing techniques in research. The program also motivated the forging of collaborations, which were, until then, difficult to constitute under traditional funding methods [9]. The UK e-science initiative also provided one of the two conditions necessary for a sea change in the provision of advanced networks for research in the UK. The other factor was the emergence of a significant body of network-aware "e-scientists," who provided the crucially necessary scientific lobby. As a result, the UK now has a pervasive advanced R&D network facility known as UKLight, which is described in more detail in the appendix.

Apart from a technical infrastructure, UKLight signaled a change in basic thinking in respect of networks – from the perspective of the users through to the policy makers. Previously, a national R&D network within the UK had never existed. As a result, a "chicken and egg" situation ensued – because no R&D network existed, no projects were proposed which could use it. Because no pressure developed for an R&D network, no resources were ever allocated to implement it. After UKLight was established, all parties became aware of the importance of advanced networking to the UK scientific research position in the context of the global community, especially with regard to enabling specific disciplinary domains. Thereafter, this type of resource became a fundamental part of conceptualizing about scientific research, and of the planning for future generations of the National Research and Education Network (SuperJANET). For example, lightpath provisioning was specifically built into the SuperJANET5 rollout in 2006.

A set of projects was initiated to pilot the benefit of the UKLight facility to several different large scale e-science research domains: (i) particle physics; (ii) radio astronomy; (iii) condensed matter simulation at high performance computing facilities; and (iv) integrative biology. The following three case studies describe the projects which have emerged within the UK.

2.3 Large-scale E-science

2.3.1 DATA MANAGEMENT FOR THE LARGE HADRON COLLIDER PROJECT

Particle physics is one of the founder disciplines in the use of leading-edge networks. This circumstance is related to a historical need to run "in-house networks" before the Internet became truly established, and recently has been motivated by the need to handle the many petabytes of data that will be generated by the Large Hadron Collider (LHC) project at CERN in Geneva [10]. The LHC program will collide protons together at 7 Teraelectronvolts (TeV). The detectors surrounding the interaction point have upward of 10^8 electronic data channels, which leads to an annual data volume of some ~20 petabytes aggregated across experiments and including simulated datasets. The need to manage this data has led to the emergence of a massive Grid computing infrastructure, distributed throughout the world – the LHC Computing Grid (LCG) [11]. The LCG will link national data centers (so called "Tier-1" centers) in participant countries via a virtual private optical network to CERN.

To prepare for the start of data taking in 2007, the LCG project undertook a set of "service challenges," designed to prove the capability of the distributed infrastructure in a realistic way (i.e., not just a set of artificial one-off demonstrations). CERN and the Tier-1 sites were required to sustain data transfer rates both to and from disk and tape at successively higher rates and for successively longer sustained periods. For example, in April 2005, a series of tests were conducted to determine whether reliable file transfer from disk to disk could be achieved, with an aggregate throughput of 500 MB/s from Tier-1 centers to CERN being sustained for one week. In September 2005, a one-week period of sustained 400 MB/s transfer from disk at CERN to tape at the Tier-1 sites was followed by a three-month period of full service operation during which experimenters were able to prove their software chains and computing models.

As an example of the achievements at that time, Figure 2.3 shows the throughput obtained from the Tier-1 centre in Oxfordshire (the CCLRC Rutherford Appleton

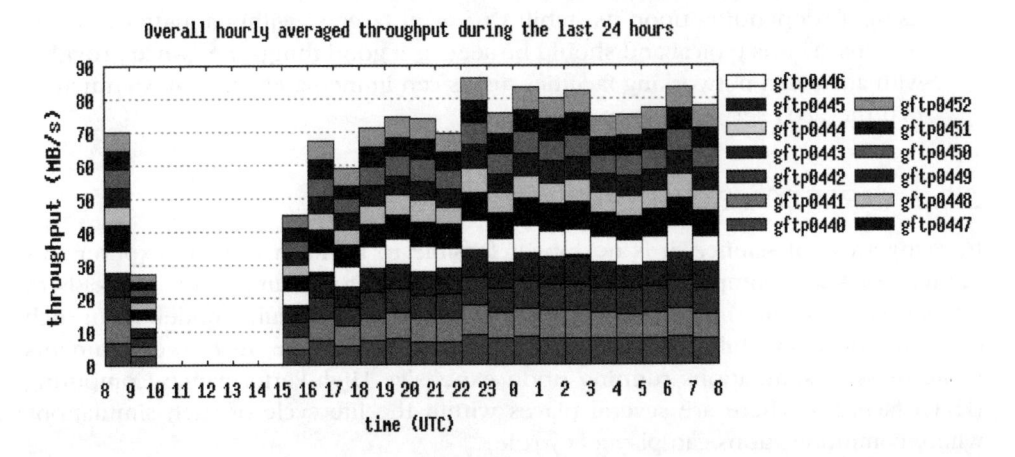

Figure 2.3. The graph shows data rates obtained as part of an LCG service challenge by the Tier-1 center in the CCLRC Rutherford Appleton Laboratory near Oxford. Two 1-Gbps Ethernet links had been provisioned from CERN to the CCLRC Rutherford Laboratory in Oxford. An aggregate data rate of over 80 MB/s was achieved between production machines through the use of parallel file transfers using GridFTP.

Laboratory [12]), which is part of the UK particle physics Grid (GridPP [13]), to CERN in Geneva using Grid File Transfer Protocol (FTP) [14] over a 2-Gbps dedicated link.

The users in this case (i.e., the particle physicist researchers) did achieve something they could not otherwise have done using the R&D network, for such sustained rates could not in practice have been achieved over the production network given various potential bottlenecks and the potential effect on other users. In fact, the pressure was immediately applied to raise the capacity to 10 Gbps. However, several key messages emerged from the service challenges from the UK point of view.

(1) Most users of advanced networking are, rightly, not interested in network "magic." Figure 2.3 shows that practical constraints did not allow the use of anything other than "out of the box" data transport applications, hence 13 separate FTP streams. Although demonstrations of transport applications are often (TCP and non-TCP) able to achieve much greater single-stream rates than this, these are often performed by network experts (magicians!) and there is a long way to go before "average" users will benefit.

(2) Even if the users had wished to use advanced transport applications, in practice most do not have the resources to devote to integrating them into preexisting production software suites. Such suites are also often not in the control of any one group, making modification a complex and drawn-out process. In addition, it can prove to be very difficult to modify any code (e.g., TCP stack) within the operating system of a heavily used production machine. The message is that high-speed transport must become part of standard operating systems and transport applications before it will be of widespread use.

(3) The computing resource providers needed almost production-quality services (in terms of persistence and reliability) from the word go. This led to disappointment in some cases as the R&D network was not built for this purpose. In this case the particle physics community went from "no use of UKLight" to using it "depending upon its stability to meet targets" within a matter of a few months. This is typical and should be seen as a good thing – i.e., once provided with advanced networking facilities, users can immediately benefit, so much so that the R&D service is soon inadequate.

2.3.2 LARGE-SCALE COMPUTATIONAL SCIENCE

In many areas of science, it is no longer feasible to perform true-life experiments because of scale, complexity, and time. This is true, for example, in the fields of astrophysics, climate science, condensed matter and biomolecular modeling. In such cases, the only possibility of achieving insight is to undertake *in silico* experiments using massive simulations running upon expensive High-Performance Computing (HPC) facilities. There are several places within the life cycle of such simulations where communications can play a key role.

The first is when parameters must be optimized at the beginning of a simulation. This is achieved through computational steering, which may involve visualizing the state of a simulation while a researcher changes parameters either conventionally through a mouse or keyboard or perhaps through a haptic device. The visualization

2.3 Large-scale E-science

process may involve sending a large simulation output stream to a remote rendering engine, requiring the communications dependability which lightpaths can provide. The human feedback mechanism also requires deterministic low-jitter paths in order to operate smoothly. If the network is inadequate in any way this can lead to loss of expensive time through simulation stalling and human frustration resulting from to jerky response and parameter overshoot.

There are also several reasons why it becomes necessary to checkpoint and migrate or clone a simulation to another facility. For example, once an interesting regime has been reached in a simulation running on a smaller machine, it is moved to a much larger machine for a full production run. To do this a checkpoint file must be transferred. Given the nature and expense of large HPC facilities, dependability on scheduling is important and hence schedulable network links can help to maximize efficiency.

2.3.2.1 TeraGyroid

The TeraGyroid [15] project was a transatlantic collaborative experiment that federated the UK e-Science Grid [9] and the US TeraGrid [17] using dedicated connections between the UK and Chicago. The project was first undertaken during the SC '03 supercomputing meeting in Phoenix and won the HPC Challenge Award for most innovative data-intensive application. It was later adapted for the ISC2004 supercomputing conference in Heidelberg in 2004, winning an award in the category "integrated data and information management." The scientific objective was to study the self-assembly, defect pathways, and dynamics of liquid crystalline cubic gyroid mesophases using the largest set of Lattice-Boltzmann (LB) simulations ever performed. TeraGyroid addressed a large-scale problem of genuine scientific interest at the same time as showing how intercontinental Grids could be interconnected to reduce the time to insight. The application was based upon RealityGrid [14] and used computational steering techniques developed to accelerate the exploration of parameter spaces.

In terms of Grid technology the project demonstrated the migration of simulations (using Globus middleware [12]) across the Atlantic. Calculations exploring the parameter space were steered from University College London and Boston. Steering required reliable near-real-time data transport across the Grid to visualization engines. The output datasets were visualized at a number of sites, including Manchester University, University College London, and Argonne National Laboratory using both commodity clusters and SGI Onyx systems. Migrations required the transfer of large (0.5-TB) checkpoint files across the Atlantic. The computational resources used are listed in Table 2.1.

The network used was essentially a pre-UKLight configuration and is shown in Figure 2.4. Within the UK traffic was carried on the SuperJANET development network (MB-NG), which provided a dedicated 2.5-Gbits/s IP path. British Telecommunications (BT) donated two 1-Gbps links from London to Amsterdam, which, in conjunction with the high-bandwidth SurfNet [19] provision and the TeraGrid backbone, completed the circuit to the SC '03 exhibition floor at Phoenix.

The project was a great success overall. From the scientific point of view, the enhanced communications paths enabled a process that led to the anticipated

Table 2.1 Experimental computational resources

Site	Architecture	Processors	Peak (TF)	Memory (TB)
HPCx (Daresbury)	IBM Power 4 Regatta	1024	6.6	1.0
CSAR (Manchester)	SGI Origin 3800	512	0.8	0.5 (shared)
CSAR (Manchester)	SGI Altix	256	1.3	0.38
PSC (Pittsburgh)	HP-Compaq	3000	6	3.0
NCSA (Urbana-Champaign)	Itanium 2	512	2.0	4.0
SDSC (San Diego)	Itanium 2	256	1.3	1.0

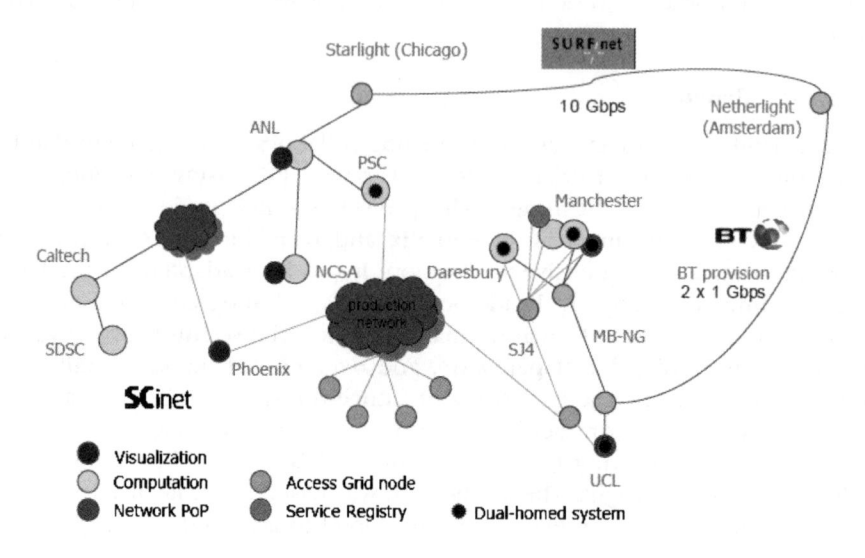

Figure 2.4. Experimental topology.

reduced time to insight. However, several interesting lessons in respect of communications were learned.

(1) The event demonstrated the complexity in establishing an effective collaboration between all of the relevant network bodies in the chain. This is a larger number than one might initially guess [University of Manchester Computing Service, Daresbury Laboratory Networking Group, UKERNA, UCL Computing Service, BT, SurfNet (NL), Starlight (US), Internet-2 (US)] for all of whom coordination was required to establish configuration, complex routing and commissioning. Dual homed machines (i.e. with two network interfaces) required careful configuration. The interaction of all authorities concerned took far more time than naively anticipated even though all parties were working with good will toward a common goal. This left less time than would have been ideal when the network was stable and the scientists could commission the application.

2.3 Large-scale E-science

(2) The data transport rates which were achieved were less than maximal. Over a 1-Gbps Ethernet path between Manchester and Chicago a UDP rate of 989 Mbps was achieved for memory-to-memory transfers. However. during SC '03, the maximum achieved throughput for actual checkpoint files (using TCP) was only 656 Mbps, and often much less than this. Several factors accounted for this.

  (a) TCP stacks on end systems. It is well known that standard TCP experiences performance problems in a high-bandwidth, high-latency environment. The machines involved in the tests were made by Silicon Graphics Inc. (SGI) and used the IRIX operating system, and no kernel patches were available for trying anything other than the standard TCP stack.

  (b) Maximum Transmission Unit (MTU) mismatch. There were issues with configuring the hosts with large MTUs.

  (c) Disks on end systems. There is a performance hit on the end-to-end transfer when standard end-system disks are used. High-performance disk systems are required for optimum performance.

  (d) Network interface card. The configuration of the interface card driver is significant in achieving the best end-to-end performance and it was not possible to investigate all the different options for the SGI end hosts.

  This outcome is typical, and demonstrates practically that which is well known – that practical transport rates are limited by the end systems. All of these issues need to be addressed well in advance in order to take advantage of multi-Gbps network links. The important message is that, even though these facts are well known, many conditions must be satisfied before such knowledge can be practically deployed on expensive production facilities.

(3) It was often difficult to identify whether a problem is caused by middleware or networks. The team was used to the symptoms caused by the presence of firewalls, and were generally able to resolve these without undue difficulty once the relevant administrators were identified. However, even more baffling problems arose. One problem, which manifested as an authentication failure, was ultimately traced to a router unable to process jumbo frames. Another was eventually traced to inconsistent results of reverse Domain Name System (DNS) lookups on one of the TeraGrid systems. A third problem manifesting as grossly asymmetric performance between FTP get and put operations was never completely resolved, but MTU settings appear to have been implicated.

2.3.2.2 SPICE

A subsequent trial endeavor was undertaken by the same group. The SPICE [20] project was the winner of the SC '05 Analytics Challenge. The project computed the free energy of a DNA strand as it penetrated through a cell membrane. The computational technique (not described here) was novel, and was required to reduce the computational time drastically compared with a traditional full-time simulation approach. The technique involved "pulling" the DNA with an artificial force, and spawning of multiple parallel simulations. A priori the optimum pulling parameters are not known, and must be chosen carefully to avoid large systematic errors. It is for the parameter optimization that high-quality network connections are required.

Scientists set a simulation running and render and visualize the simulation at a remote visualization engine in a normal way. The scientists would physically pull the DNA through the membrane pore using either a conventional mouse or a haptic force feedback device. The applied force would be computed in the visualization engine, which would then send an update back to the simulation for inclusion in the next time-step. A two-way communication path resulted, i.e., the pulling force update being sent from visualizer to simulation and the simulation being sent to the visualizer at each time step. Early investigations showed that the quality of the network connection had a significant effect. In one situation simulation was as the CSAR (Computer Services for Academic Research) facility in Manchester and visualization was at University College London. When a dedicated 500-bps link was used the simulation ran 25% faster than when the production network was used.

2.3.3 SUMMARY

As a result of a set of convergent circumstances a pervasive R&D network known as UKLight was built in the UK in 2005. A set of early adopter projects was ready and waiting, including particle physics, radio astronomy, and computational science applications. These were able to link remote sites both within and outside the UK. The experiences of each show that they were all able to achieve to achieve "new ways of working" that were previously impossible, but practical problems had to be overcome in each case. Through these and other applications which subsequently arose, the case for lightpaths became accepted within the UK and was then incorporated into the next generation of the UK National Research and Education Network (NREN) as a widely available service.

ACKNOWLEDGMENTS

The author wishes to acknowledge the assistance of Professor P. Coveney, Dr. S. Pickles, Dr. R. Blake, Dr. A. Sansum, Dr. J. Gordon, and Dr. S. Jha for information supplied and for permission to use images.

2.4 DATA MINING
Robert Grossman

This section describes some of the ways in which Grids powered by optical networks and applications powered by user-controlled lightpaths will change data mining.

Data mining is the semiautomatic discovery of patterns, changes, associations, anomalies, and other statistically significant structures in large datasets. Distributed data mining is concerned with mining remote and distributed data.

2.4 Data Mining

It is sometimes helpful to think of data mining as applying a data mining algorithm, specified by a data mining query, to a learning set to produce a data mining or statistical model, which can be thought of very concretely as a file, say an XML (Extensible Markup Language) file, containing a description of the model [21]. For example, a data mining query might specify that a tree-based classifier or tree-based regression be computed from a processed dataset. From this perspective, in a distributed data mining computation, one has the freedom (i) to move the data, (ii) to move the data mining query to the data, (iii) to move the models, or (iv) to move some combination of these (see, for example, ref. 22).

Distributed data mining often makes use of what is called an ensemble of models, which is simply a collection of data mining models together with a rule for combining them or selecting one or more of them. For example, classification models may be combined using a majority vote, while regression models may be combined by averaging the predictions of each individual model.

It is common to distinguish between data mining algorithms per se and the more general process of data mining. If data mining is thought of broadly as the process of finding interesting structure in data, then there are three important steps in this process, all of which are changed in fundamental ways with the growing availability of wide-area, high-performance optical networks and user-controlled lightpaths.

2.4.1 IMPORTANT STEPS IN THE DATA MINING PROCESS

The data mining process comprises three important steps:

(1) The first step is to explore and prepare the data. This results in a dataset of features. This is also called a learning set.
(2) The second step is use the learning set as the input to a statistical or data mining algorithm to produce a statistical or data mining model, as described above.
(3) The third step is take the resulting model and to use it for scoring or otherwise processing data, either as a stand-alone application or as an embedded application.

First, it is useful to examine at the impact of Grids on step 2 of the data mining process. For more details, see ref. 23. A traditional rule of thumb when mining large distributed datasets to produce a data mining model was to leave the data in place and proceed as follows:

2.4.2 MAIN STEPS IN DISTRIBUTED DATA MINING USING COMMODITY NETWORKS

The main steps in distributed data mining using commodity networks can be summarized as follows:

(1) Scatter the data mining query to the distributed data.
(2) Run the data mining query at each local site to produce a data mining model.

(3) Gather the resulting data mining models at a central site.
(4) Combine the data mining models to produce a single model, usually an ensemble model.

One might, in addition, then move the ensemble model back to each of the different locations, run the ensemble locally to produce results, move the results to a common location, and then combine the results.

Producing a data mining model with Grids enabled by optical networks and user-controlled lightpaths employs a different procedure (see below).

2.4.3 MAIN STEPS IN DISTRIBUTED DATA MINING USING OPTICAL GRIDS

In this cases, the main steps are as follows:

(1) Distribute portions of the data between the various sites.
(2) Scatter the data mining query to the distributed data.
(3) Run the data mining query at each local site to produce a data mining model.
(4) Gather the resulting data mining models at a central site.
(5) Combine the models to produce a single or ensemble model.

Repeat steps 1–5 as required by iteratively redistributing data and models.

Distributed data mining sometimes distinguishes between vertically partitioned data (different sites contain different attributes) and horizontally partitioned data (different sites contain different records).

With traditional networks having long-bandwidth-delay products, the effective bandwidth might be 1–5 Mbps. In practice, this limited the amount of data that was moved to about 10–100 Mbps, and so effectively limited distributed data mining to small distributed datasets or horizontally partitioned data. With Grids powered by optical networks and applications that can access user-controlled lightpaths, it is now practical to mine gigabyte- and terabyte-sized vertically partitioned distributed datasets.

From the discussion so far can be summarized as follows and three evident truths can be concluded.

(1) *With optical networks it also now makes sense to move the data.*

With optical networks and some of the new and emerging protocols described in this book, one can easily move data over a network with high-bandwidth-delay product at 900 Mbps over a 1-Gbps network and 9 more Gbps over a 10-Gbps network. This effectively means that a rule of thumb is that it now makes sense to move 10–100 Gbit/s of data.

Next, the first step in the data mining process was considered, which is to explore, transform, and prepare data prior to the application of a data mining query. Traditionally, this step has been accomplished by moving all the data to a central location, building a data mart, and then exploring and preparing the data. With high-performance wide-area networks, alternative approaches are possible.

First, it is possible to explore directly remote and distributed data. For example, a system to do this is described in ref. 24. The ability to directly explore remote data is important since reducing the overhead to explore data makes it more likely that third-party data will be used. As an analogy, the web reduced the complexity of viewing remote documents and substantially increased browsing of documents by casual users; in the same way, systems for exploring remote data utilizing high-performance networks have the potential for substantially increasing the use of third-party data in data mining applications.

Grids enabled with optical networks also enable virtual warehouses to be developed. The terms "virtual data warehouse" and "data mart" mean a system in which remote and distributed data is left in place but integrated data is available on demand as required. For example, one table with key K may be in location A, while another table with key K may be available in location B. When needed at a third location, C, the tables are joined on demand using the key K and provided to a user or application at location C (see, for example, ref. 25).

This concept can be summarized as follows:

(2) *With optical Grids, remote and distributed data can be explored, transformed and integrated on demand, while leaving the data in place.*

Finally considered here is the third main step in the data mining process – applying a data mining model to data. Sometimes this process is called scoring. It is a fundamental challenge simply to scale up scoring to the higher speeds required when working with optical Grids. With some of the new protocols described in this book, it is now possible for the first time to score data at 1 Gbps or higher over wide area networks. Until recently, scoring datastreams over wide-area networks with high-bandwidth-delay products was effectively limited in most cases to a few megabits per second. With the type of optical networks described in this book, datastreams can often be filtered, selected, and scored at line speed.

These principles can be summarized as follows:

(3) *With optical networks, scoring, selecting, filtering, and performing continuous queries on datastreams can be done at line speed, even over wide area networks with high bandwidth delay products.*

2.5 CINEGRID, A GRID FOR DIGITAL CINEMA

Tom DeFanti, Laurin Herr, and Natalie Van Osdol

CineGrid is an international research initiative to create scaled-up optical testbeds and middleware for advanced digital media, scientific research, and higher education applications. With the support and participation of institutions in San Diego, Los Angeles, San Francisco, Seattle, Chicago, Tokyo, and Amsterdam, CineGrid is an emerging global facility for the community of researchers, artists, scientists, and educators working with ultra-high-performance digital media.

CineGrid's goal is to conduct research into the feasibility and trade-offs of different networking approaches for secure distribution of commercial-grade compressed and encrypted Digital Cinema Packages (DCPs) like those suitable for releasing completed digital movies to theaters around the world. In July 2005, a new global specification for theatrical release of digital cinema up to 4K resolution using JPEG 2000 compression was recommended by the Digital Cinema Initiatives (DCI), a partnership of the seven major Hollywood studios.

The digital components (picture, sound, and subtitles) typically stay uncompressed through multiple stages of post-production workflow, making storage and network transfer a state-of-the-art affair. Further complications arise from the very high intellectual property value of these "raw" assets, indicating maximum security through encryption and other techniques. The DCI recommendations are currently in process within SMPTE DC28 to officially become open standards for global adoption.

2.5.1 TRENDS

CineGrid addresses three eruptive trends:

- Hollywood cinema post-production and distribution will become overwhelmingly digital, adapting and extending technologies developed over the past 25 years for high-quality image sensing, image processing, and displays in science and education.
- A new generation of 1- and 10-gigabit optical networks, capable of handling massive file transfers and even super-high-definition (super-HD) motion pictures in real time is fast becoming available at reasonable cost.
- Grid computing and OptIPuter efforts are yielding new tools to securely manage distributed visual resources over networks.

CineGrid is the newest application of the OptIPuter, a new digital architecture that exploits the emergence of user-controlled, parallel optical gigabit and 10-gigabit networks. The OptIPuter is funded by the National Science Foundation, and led by Dr. Larry Smarr, director of the California Institute of Telecommunications and Information Technology (Calit2) at the University of California San Diego (UCSD) and University of California Irvine, and the Electronic Visualization Laboratory at the University of Illinois at Chicago; UIC). UCSD, UIC, and Pacific Interface, Inc. (PII) are working with Keio University's Research Institute for Digital Media and Content (DMC) and NTT Network Innovation Labs in Japan to test the feasibility of secure digital transmission via IP-based networks of various forms of DCI-compliant digital cinema content and other forms of 4K content over great distances.

2.5.2 CINEGRID CINEMA-CENTRIC RESEARCH

CineGrid cinema-centric key research efforts include:

- secure content distribution at highest quality;
- pre-production collaborative design and planning;
- digital dailies, interactive viewing and mark-up;
- distributed post-production;
- digital film scanning and restoration using networks;
- digital media archiving using networks;
- quality assessment and quality control;

- education of next-generation cinema professionals;
- remote High-Definition Television (HDTV) and super-HD lectures and interactive teaching;
- refinement of tools and requirements;
- facile exchange of student works;
- joint development of training methods.

CineGrid is a testbed devoted to the following key technical issues in digital cinema:

- secure distribution – file transfer and streaming, scheduling, session management, protocols, security, anti-piracy encoders/decoders;
- interactive videoconferencing over IP using HDTV or super-HD;
- local, regional, and wide-area computational architectures based on switched 1-Gbit Ethernet and 10 Gbit Ethernet;
- workgroup architectures for fast, huge, distributed storage;
- service-oriented architectures supporting standards;
- networked digital archives for education and media research;
- image evaluation techniques and research into perception of very high-quality media, including calibration and color management;
- advanced displays: direct view and projection, mono and stereo displays on single and tiled screens.

The first international realization of CineGrid was at iGrid 2005, September 26–30, 2005, in San Diego, as part of a project entitled "International Real-time Streaming of 4K Digital Cinema." This was an experiment using both switched and routed connectivity between the Digital Cinema Lab within the DMC at Keio University in Tokyo and Calit2 in San Diego. This demonstration sent streaming "4K" motion pictures (3840 × 2160 pixels/frame) compressed using NTT's JPEG 2000 codecs. The transmitted images were shown on the 32 × 16′ screen in the Calit2 200 seat auditorium using a SONY 4K SXRD projector. This experiment was a joint effort of Keio University, UCSD, UIC/EVL, NTT Labs, and Pacific Interface. Valuable cooperation, equipment loans, and staffing came from Sony Electronics, NTT Labs, SGI, Olympus, Toppan, and Astrodesign.

The CineGrid experiments at iGrid 2005 involved a variety of real-time streaming transfers originating at Keio University in Tokyo using JGN2/NICT and GEMnet2/NTT links across the Pacific, connecting in Chicago and Seattle, respectively, via the CAVEwave, PacificWave, and CENIC networks across the US to the Calit2 auditorium in San Diego. Over the three days of iGrid 2005, more than six hours of 4K full motion content, including interactive scientific visualization, prerendered computer animation, digital cinema scanned from film, digital motion picture, and still cameras, "digital dailies," and live 4K videoconferencing, was shown. As a result of iGrid 2005, CineGrid has built a persisting, always available 1-Gbit Ethernet testbed from San Diego through Los Angeles, the Bay Area, and Seattle, to Tokyo, with scheduled connectivity available at higher speeds and to other places. Also reachable now through scheduled CineGrid circuits are key groups in Chicago and Amsterdam.

2.5.3 CINEGRID CONSORTIUM

CineGrid founding members providing the core infrastructure, facilities, and leadership include:

- Calit2 at UCSD;
- EVL at UIC;
- Pacific Interface, Inc.;
- The School of Cinema–Television at the University of Southern California;
- The National Center for Supercomputing Applications, University of Illinois at Urbana–Champaign;
- The Research Channel at the University of Washington;
- The Research Institute for Digital Media and Content (DMC) at Keio University;
- NTT Network Innovations Labs;
- The Digital Cinema Technology Forum (DCTF);
- The Digital Cinema Consortium of Japan (DCCJ).

CineGrid researchers are enabling new kinds of distributed media production, remote mentoring, remote-controlled scientific research, networked collaboration, and international cultural exchange. Areas of investigation range from visual special effects for theatrical-quality digital cinema production to scientific use of remote high-definition instrumentation.

2.6 DISTRIBUTED AIRCRAFT MAINTENANCE ENVIRONMENT (DAME)*

Tom Jackson, Jim Austin, and Martyn Fletcher

2.6.1 USE CASE INTRODUCTION

The Distributed Aircraft Maintenance Environment (DAME) [19] provides a Grid-based, collaborative, and interactive workbench of remote services and tools for use by human experts. This environment currently supports remote analysis of vibration and performance data by various geographically dispersed users – local engineers and remote experts. The diagnosis environment is built around a workflow system and an extensive set of data analysis tools and services, which can provide *automated* diagnosis for known conditions. Where automated diagnosis is not possible, DAME provides remote experts with a *collaborative and interactive* diagnosis and analysis environment.

---

* Copyright © Global Grid Forum (2003). All Rights Reserved. This document and translations of it may be copied and furnished to others, and derivative works that comment on or otherwise explain it or assist in it implementation may be prepared, copied, published and distributed, in whole or in part, without restriction of any kind, provided that the above copyright notice and this paragraph are included on all such copies and derivative works.

2.6.2 DAME CUSTOMERS

The customers for the operational system are the primary actors: maintenance engineers, maintenance analysts, and domain experts who use the system to assist in the analysis and diagnosis of engine conditions. The system comprises applications that perform the analysis of very large datasets (requiring possible transport), in the order of terabytes, from remote locations.

The DAME tools and services, which are used to provide diagnosis and prognosis of engine conditions, are the "network customers" and these include:

- signal processing tools and services – to detect known conditions;
- search tools and services – to allow the searching of historical fleet archives and current data archives for similar data amounting to terabytes of data;
- case-based reasoning tools and services – to provide advice based on symptoms and cases.

2.6.3 SCENARIOS

Figure 2.5 shows the actors and interactions with the system, which is shown as the cloud in the center.

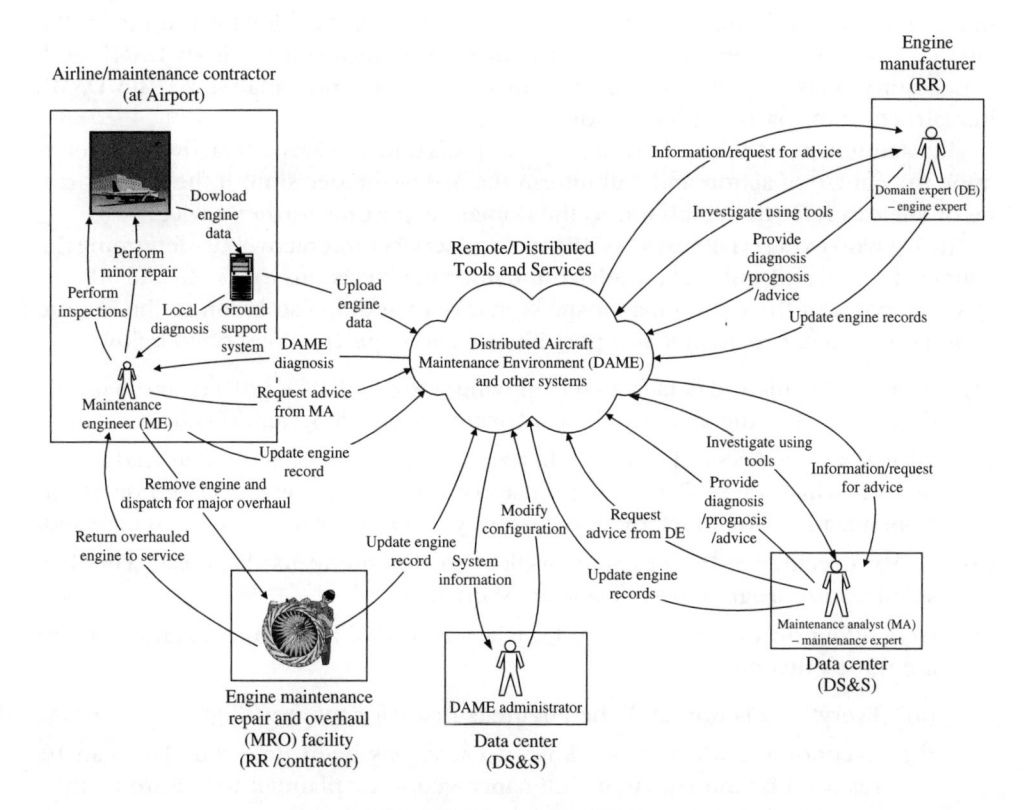

Figure 2.5. DAME, actors and interactions.

The diagnostic process includes the following actors:

- maintenance engineers (located at airports): carry out engine diagnosis and maintenance activities;
- remote maintenance analysts (located at the data centre): provide technical advice, coordinate analysis and brokerage, and use the DAME tools and services interactively;
- remote domain experts (located at the engine manufacturer's site): act as a repository of knowledge, provide expert diagnostic advice on unidentified anomalies, and use the DAME tools and services interactively.

The role of the DAME administrator shown in Figure 2.5 is to manage the systems facilities; this person plays no active part in the diagnostic process. The Maintenance, Repair, and Overhaul (MRO) facility is an internal company facility providing maintenance, repair, and overhaul of engines while they are off the aircraft. This facility is used only when an engine has to be removed from the aircraft, for example when a planned major overhaul is required.

The interactions between the various users can be complex, particularly in situations where additional tests are requested by the remote experts in order to confirm a diagnosis.

The Maintenance Engineer (ME) uses the automated features of DAME in carrying out diagnosis and minor engine maintenance in preparation for turn-round of the aircraft. The ME will receive diagnosis information automatically from DAME and, if necessary, may request assistance from the maintenance analyst via the DAME infrastructure for particular conditions.

The Maintenance Analyst (MA) uses DAME tools and services interactively to determine the course of action and will inform the ME of the decision. If the MA desires, he or she can refer the condition to the domain expert for further advice.

The Domain Expert (DE) uses DAME tools and services interactively to determine the course of action and will provide advice and informed decisions to the MA and ME.

An overview of the typical diagnostic scenario, including escalation to the remote experts (maintenance analyst and possibly domain expert), is described below.

(1) An aircraft lands and data from the on-wing system (QUICK [20]) is automatically downloaded to the associated local Ground Support System (GSS).

(2) QUICK and its GSS indicate whether any abnormality (this is a detected condition for which there is a known cause) or novelty (this is a detected deviation from normality for which there is currently no known cause) has been detected.

(3) DAME executes an automatic workflow to determine its diagnosis. This is a standard preprogrammed diagnostic sequence.

(4) Depending on the result of the QUICK and DAME automatic diagnoses there are three outcomes:

   (a) Everything is normal – the engine is ready for the next flight.
   (b) A condition which has a known cause has been detected. This can be resolved by immediate maintenance action or planned for future maintenance action, as appropriate.

(c) A condition which currently does not have a clear cause has been detected or there is some ambiguity about the cause. This case is referred to the remote experts (a maintenance analyst and possible a domain expert) for consideration.

2.6.4 RESOURCES INVOLVED

In the deployment of DAME, GridFTP may be used in the download of raw data.

The most intensive computing application is the use of the engine simulations – the number of simulations will depend eventually on the demands of the domain experts using the system.

All activities will all take place under control of the workflow system, which orchestrates the DAME services and tools and transfers data between them.

2.6.5 FUNCTIONAL REQUIREMENTS

Remote experts may want to consider and review the current data, search and review historical data in detail, and run various tools including simulations and signal processing tools in order to evaluate a particular situation. Each aircraft flight can produce up to 1 GB of data per engine, which, when scaled to the fleet level, represents a collection rate of the order of terabytes of data per year. The storage of this data also requires vast data repositories that may be distributed across many geographic and operational boundaries. DAME provides a diagnostic infrastructure to allow enhanced condition monitoring and diagnosis using geographically dispersed data, services, and experts.

The significant requirements on the infrastructure are as follows:

- The diagnostic processes require collaboration between maintenance engineers, maintenance analysts, and domain experts from different organizations: airline, support contractors, and engine manufacturer. These individuals are geographically dispersed and need to deploy a range of different engineering and computational tools to analyze the problem.
- The infrastructure must allow users appropriate access to the data from the engine under consideration.
- The infrastructure must allow users appropriate access to the terabytes of historical data from the engine under consideration and other similar engines.
- The ability to search the vast stores of historical vibration and performance data is required. Advanced pattern matching and data mining methods are used to search for matches to novel features detected in the vibration data. These services are able to operate on the large volumes of data and must be able to return the results in a time that meets operational demands.
- There is a need to provide signal processing and engine simulation tools using data from the engine or historical data.
- The diagnostic process must be completed in a timely and dependable manner commensurate with the turn-round times of the aircraft.

The DAME system is advisory so the data and results transfer must be reliable and must have integrity to be effective for the business case.

2.6.6 SECURITY CONSIDERATIONS

Elements of the raw data and data transferred between tools and services have confidentiality requirements that must be protected during transfer.

2.6.7 PERFORMANCE CONSIDERATIONS

Significant network scenarios are:

(1) Download of raw engine data from the aircraft to the ground – this starts the automatic workflow for the DAME diagnosis for the maintenance engineer.

(2) The movement of raw data to its store on the ground – for various reasons there are two options in early deployments:

   (a) Deployments in which all data is moved to one centralized store. Potentially, tens of terabytes of data per day from a world-wide operation would eventually need to be transported over networks to the central store and data management center. Here the raw data (high bandwidth) is effectively taken to the tools and services, with results (low bandwidth) being passed to the workflow system and then on to the geographically dispersed users.

   (b) Deployments in which a distributed store (at the airports) would be used. Larger airport data stores would, potentially, need to store around 1 TB per day, if all aircraft landing produced data. However, the transfer would take place only over the local networks within the airport. Tools and services would be locally resident at the airport. In this case, the tools and services are taken to the raw data, with results being passed back to a centralized workflow system and then to the geographically dispersed users.

All diagnoses and escalation through all experts must take place within the turn-round time of the aircraft (including all executed workflow and all service execution), and data transit times also have to take place within the turn-round time of the aircraft.

2.6.8 USE CASE SITUATION ANALYSIS

The system has been developed with industrial customers (Rolls-Royce plc and Data Systems and Solutions LLC) and a proof of concept demonstrator produced. Industrial deployment is now in progress as part of the BROADEN (Business Resource Optimization for Aftermarket and Design on Engineering Networks) project. This is a UK Department of Trade and Industry (DTI) Inter Enterprise Computing Technology Call Project.

ACKNOWLEDGMENTS

This work was undertaken as part of the DAME project, with welcome assistance from Rolls-Royce plc, Data Systems & Solutions LLC, and Cybula Ltd and the teams at the Universities of York, Leeds, Sheffield, and Oxford. This research was supported by the UK Engineering and Physical Sciences Research Council (grant GR/R67668/01) and through contributions from Rolls-Royce plc and Data Systems and Solutions LLC.

2.7 FINANCIAL SERVICES: REGULATORY AND MARKET FORCES MOTIVATING A MOVE TO GRID NETWORKS

Robert B. Cohen

2.7.1 NEW CHALLENGES FOR FINANCIAL INSTITUTIONS AND NETWORKS

Many new forces are reshaping the way in which banks use networks. Some of these are regulatory, such as the Sarbanes–Oxley law and the Bank for International Settlements' Basle II requirements. Others are changes in the nature of financial services business, owing to the fact that riskier and more profitable operations need more robust analytical capabilities to support them. In some cases, financial business is becoming more dependent on leveraged activities. This is the case when a bank's investment in a financial instrument such as a hedge depends upon a number of conditions, for instance prices of commodities and currencies.

As a result, an entirely new level of accountability is required that is not possible with traditional methods. In these instances, accountants may require banks to change the way in which they evaluate risk and press for the use of more powerful and more rapid number crunching. In addition, financial institutions face declining margins on traditional operations, such as transactions for trading or investments, and one way they try to improve the profitability of these operations is to increase their volume. Many banks and brokerage houses have used more high-powered computing to scale up the number of transactions they support. But this has also required an increase in networking to support these transactions, send information about the transactions to required backup sites, and analyze the transactions for illegal activity, such as money laundering.

2.7.2 FACTORS DRIVING BANKS TO ADOPT GRIDS AND HIGH-SPEED NETWORKS

These changes certainly affect how banks use networks. Why? Regulatory requirements require banks and other financial institutions to go far beyond merely analyzing local investments in local markets. Banks, brokerage houses, and many insurance firms do business and make investments in global markets. As a consequence, they need to analyze the value they have at risk not only in the US market, but also in overseas markets. If a major bank invests in markets in the New York, London, and Tokyo, it must be able to analyze the state of its investments in all of these markets. It must be able to create a full view of its entire investment portfolio, particularly those parts of its portfolio that are exposed to the greatest risk.

Consequently, banks need to be able to analyze investments, especially leveraged financial instruments that they have purchased in the major financial centers [28]. This requires networks that can transmit information about new positions that were taken in the past few minutes and integrate them with existing information about positions in such high-risk instruments as collateralized debt obligations. The resulting information flows provide banks and other financial institutions with the ability to review the risk level of investments in their portfolios several times during a trading day. Such updates must encompass positions in a number of financial markets. They are also reviewed before new investments are made in other high-risk instruments.

This requirement to do far more detailed, and compute-intensive, analysis of investments, particularly of what are called "exotics," the more risky financial instruments, such as hedges and collateralized debt obligations, is one of the major transformations that is driving change in the financial sector. While most large banks and brokerage houses first adopted clusters in the mid-1990s to speed up their compute-intensive operations, they focused their analytical expertise on fixed investments or other markets that did not involve such esoteric investments. As the nature of banking has changed, banks have made a large share of their profits from taking positions in the market that involve more risk [29]. They are not just investing in a certain number of stocks. They are making investments in financial instruments whose value can change depending upon how the value of a stock or a group of stocks or bonds may change over time, or values of commodities or currencies can change over time.

These investments can gain or lose value depending upon the rate of change in debt or equity values, rather than just changes in equity values themselves. As a consequence, banks can achieve big gains if they estimate the direction of the change in value correctly. The risk of unexpected, large drops in the value of these investments, due to default or financial problems, although small, needs to be evaluated carefully to know the exact risk embodied in a portfolio of funds that a bank has invested in.

And because banks are global, these risks cannot just be measured in a single market, they need to be measured globally. Consequently, positions that are established by various branches of a single bank in all major and minor financial centers need to be evaluated as though there were a single trader in a single location. To meet the requirements of the Sarbanes–Oxley regulations, financial institutions are required to ensure access to comprehensive data across the enterprise for financial reporting, seamlessly feed data to workflow applications across the enterprise, and feed real-time executive dashboards for detection and alerting of material changes in a company's financial status.

The USA Patriot Act requires financial institutions to extend the ability to access and integrate data across an entire enterprise for analysis and detection of money laundering activities and to enhance an organization's ability to respond to government requests for account and account holder information. The Act requires that banks be able to obtain customer and account data transparently in real time across a distributed enterprise [30]. The operative phrase here is "across the enterprise" because financial institutions can have trading or investing operations around the US and across the globe.

2.7.2.1 Scaling traditional operations

Another factor driving change is the need to scale traditional operations such as transactions, including sales of stocks and bonds and financial instruments. While such transactions represent a somewhat small percentage of profits for financial institutions, they are important. For the past decade, as entry into financial markets by new firms has become easier and involved less regulation, the profitability of transactions has dropped considerably. In response, banks and brokerage houses have sought to increase the number of transactions that their systems can handle daily. This has resulted in an increase in the use of Grids to scale to the new, higher levels of transactions and to provide better reporting.

With transactions, as is noted in the Patriot Act, financial institutions must be able to examine customer and account transactions in real time across an enterprise that stretches from Frankfurt and London to New York and Tokyo. Networks will provide the way to gain more insight into transactional data on a global level. Today, banks and brokerage houses are preparing to meet this challenge. They are among the first to deploy Services-Oriented Architectures (SOAs) that facilitate sharing data and applications that track and analyze transactions. This is a first step to creating a broad global view of what is happening in individual branches of banks and at individual traders' desks.

What are the consequences of not having the connectivity between parts of banks or brokerage houses that let them "see" where investments have been made and the type of transactions of stocks, bonds, and other financial instruments? Traders or investors at financial institutions need to be able to understand the value and price of financial instruments in real time. If they do not, they may not be able to identify a weak spot in a portfolio that can result in hundreds of millions of dollars worth of losses or they may miss gaining "wallet share" from customers who have made profitable investments with a bank or brokerage house.

Networks are necessary to gain this view of investments and transactions because they are the infrastructure that compiles a global view from widely distributed information. But there are issues that many banks face in building up such a global view. Many systems that banks are using to describe their derivatives, for instance, do not let traders analyze data when there are heterogeneous derivative instruments involved and do not provide a way to estimate risk accurately from these instruments. As a result, traders do not have clear visibility about where there are risks in their portfolios. One way to solve this problem is to implement XML repositories to store a wide number of financial trade structures and capture the complexity of derivative transactions [31]. The resulting analysis of these repositories could be compiled if there are high-speed networks between bank branches that handle derivative transactions.

2.7.3 HOW FINANCIAL INSTITUTIONS WILL USE NETWORKS TO FACILITATE GRID COMPUTING

Financial institutions face a number of pressures that will force them to change their use of networks. In most cases, banks and brokerage houses need to track critical information better to manage risk and exposure to risk. These pressures may begin with very simple financial operations that must be coordinated over a global financial network, for instance the need to track credit card transactions. Today, a card that is stolen in London may be used to charge items through any merchant around the globe. Most credit card issuers have only begun to chisel together the functional parts of such a tracking operation. In addition, once the data can be gathered, it needs to be analyzed and evaluated to see if there is fraudulent activity. In essence, credit cards are likely to create pressure for banks to build larger information networks and to expand their ability to evaluate the charges on cards that they have issued.

One move that banks are likely to make to meet this credit card challenge is the creation of more extensive and robust networks. The large number of transactions can create sizable data flows from major markets to a bank's main offices. The information will be analyzed by large clusters or Grids, but it is possible that, during peak business periods, banks will need to share compute and data resources and may begin to construct early financial Grids to support the need for surges in compute needs. Some of these Grids will provide utility computing from new service providers. Some initial managed Grid services announced by communication service providers are very likely serving large banks. The need to integrate information across a financial institution suggests some of the directions in which banks and brokerage houses may move in the next few years.

A second stage in the move to Grid networks will probably be reached when banks and brokerage houses deploy systems to manage global risk. Initially, these Grid networks may only connect large international offices within a bank, such as the London and New York trading centers, where much of the business with financial instruments such as derivatives is done. The networks would let banks transfer vast amounts of computing power between such centers to meet surges in demand, such as when accounts are settled and analyzed at the end of the trading day. At some banks and brokerage houses, the compute power required to settle accounts can be several times the average daily demand, perhaps as high as 10 times the normal demand. Since senior executives want to see the results and the risk analysis as soon as possible, this creates pressure to shorten processing times and complete reports by the end of the business day.

While these Grid networks would be "rudimentary" because they would not connect a large number of a bank's important international offices, they would be the initial step in further expansion. Once compute resources and data were shared between New York and London, for instance, a bank could begin to link Frankfurt, Paris, Tokyo, and offshore banking centers into such a network. Additional pressure from regulators to implement Basle II or provide more statistical reporting for Sarbanes–Oxley might speed up this move to Grid networks. High costs on international networks or an inability of global telecommunications networks to support Grid traffic could slow the deployment of these networks.

2.7.4 GLOBALIZATION OF FINANCIAL MARKETS

A third stage in the move to Grid networks could occur as banks become more global. Most banks remain primarily those from a certain region – US banks have a limited presence in Europe, European banks have a limited presence in the US. As banks and brokerage firms build more global enterprises, the ability to meld these enterprises and take advantage of open systems will offer considerable benefits. If two large banks have Grid computing in place to manage risk and control credit card operations, as they move to more Open Source standards, it will be easier to integrate large banking operations. So a significant stage would be reached for financial institutions once they build upon the collaborative and integrating possibilities inherent in Grid computing and Grid networks.

For example, merging a large European bank with a large US bank might be much easier, with the Open Source-based environment able to move resources from one division of the acquired bank to another division of the acquiring bank in just a few hours. This type of transfer has already been accomplished by a major US firm, with the entire operation being shut down and brought to life under new ownership in less than 24 hours. The ability to interconnect existing Grid networks to support the integration of the two banks' Grid computing facilities would be part of the acquisition. It would also create a scale of banking that would force international rivals to match the size and scale of the new bank or lose certain sophisticated business that only the new banking entity would be prepared to offer at attractive rates to customers. As a consequence, this would spark an international race to consolidation, largely supported by Grid networks and the ability to share Grid computing resources and data resources. Part of the move to consolidation might result in a broader adoption of utility computing by financial institutions.

Creating such large financial institutions would be a boon for banks because it could help them rationalize costs. At the same time, it would concentrate risk within an even larger institution, raising the threat that, if it failed, there might be catastrophic consequences for the world's financial markets. This could result in regulators forcing even stricter requirements on financial investments that involve considerable risk. Regulators (and other investors) might ask such large financial entities to provide them with credit risk analyses (and investment performance reports) several times a day. (It is common for smaller investment firms to give some of their funds to larger investment banks to manage. In recent years, some firms have demanded much more frequent accounts of the profits made on their investments and the risk taken by the firm managing their funds.)

This might create a next phase of Grid network build-out to meet regulators' and investors' requirements. If this happens around 2010, it may be assumed that prices for broadband networks continue to fall and that more equipment in networks supports collaborative computing. If this is true, increased demand for accountability might motivate financial institutions to build larger Grid networks, connecting not only primary and secondary financial centers around the globe, but also partners and important business customers that want to have the banks provide them with better ways of managing their money. So as business becomes more global, another round of Grid network construction would begin.

2.7.5 MIGRATION OF FINANCIAL INSTITUTIONS TO GRID NETWORKS

Today, banks are moving to a new generation of services based upon their experience with Grids, exploiting virtualization to create (SOAs. These SOAs not only respond to scalability and resiliency requirements, but establish Grid networks that will support banks' responses to Sarbanes–Oxley and Basle II. Here is a description of a few of the ways in which two banks, Wachovia and JP Morgan Chase (JPMC), are moving closer to implementing Grid networks to support risk management.

In Wachovia's case, Grid computing is serving as the basis for creating a "general-purpose transactional environment" [32]. When Wachovia successfully handled "value-at-risk" analyses [33], it moved to create the first parts of this environment to focus its Grid on pricing financial instruments that require Monte Carlo simulations [34]. Wachovia has used an SOA "platform" to make its Grid a "virtualized application server" [35] that will be the foundation for utility computing. The Grid that Wachovia has will track all transactions and compile the detailed transaction information needed to comply with Sarbanes–Oxley. Since it includes most of the important risk analytics that the derivatives and credit operations at Wachovia have used, the bank will use the Grid as the basis for an internal system to comply with Basle II. This will very likely require linking operations in North Carolina with those in New York.

At JPMC, the bank's Grid permits the management and control needed to support the bank's compliance with Sarbanes–Oxley and Basle II. The bank's Grid "runs on a high-speed private network that is separate from the bank's corporate network" [35]. Over the last two years, JPMC has added a virtualization project to its compute backbone.

In this project, the bank has, in its main office, application request support from any number of available resources. In addition, JPMC has moved to virtualize applications and databases in addition to compute resources in credit derivatives, where the IT group created a CDIR (Credit Derivatives Infrastructure Refresh) solution [36]. This scalability solution provided traders with on-demand computing resources [37] and supported automating repairs and "fixes" for an "event-driven infrastructure" [38] that provided bank executives with an integrated view of the IT infrastructure for the credit group. Now, for end-of-day processing, traders can request far more resources than they could previously. The bank can now begin to use this system globally. The next phase is likely to see the systems at various international locations linked to each other to share virtualized resources.

2.7.6 CONCLUSIONS

Banks have faced significant business and regulatory challenges that have spurred them to adopt Grid computing and resulted in them taking the first steps to deploy Grid networks. As this section notes, these challenges are creating even greater pressures to employ Grid networks as the main way in which banks can comply with business demands and meet the greater need to evaluate and assess risks and the need to grow even larger on a global scale. There are likely to be several stages in the build-out of Grid networks over the rest of this decade, largely tied to regulatory and business scale issues. Two cases, Wachovia and JP Morgan Chase, illustrate how rapidly banks are moving to adopt Grid computing and virtualize resources, steps that are preliminary to moving to Grid networks that will span a bank's global reach. Thus, banks are likely to be among the first to deploy extensive Grid networks for business and risk assessment purposes.

2.8 SUMMARY OF REQUIREMENTS

The use cases described in this chapter exemplify the potential for innovative applications and services when they can benefit from capabilities that are abstracted from individual characteristics of specific hardware and software environments. These Grids require a flexible environment that can be directly manipulated as opposed to one that compromises the potential of the application. They require access to repositories of resources that can be gathered, integrated, and used on demand, and which can be readjusted dynamically, in real time. An essential requirement is that the control and management of these resources be decentralized, in part, because constant interactions with centralized management processes generate unacceptable performance penalties and cannot scale sufficiently. These requirements are further described in the next chapter.

REFERENCES

[1] M. Brown (2003) "Blueprint for the Future of High-Performance Networking (Introduction)," *Communications of the ACM*, 46(11), 30–33.

[2] L. Smarr, A. Chien, T. DeFanti, J. Leigh, and P.M. Papadopoulos (2003) "The OptIPuter," special issue "Blueprint for the Future of High-Performance Networking," *Communications of the ACM*, 46(11), 58–67.

[3] L. Renambot, A. Rao, R. Singh, B. Jeong, N. Krishnaprasad, V. Vishwanath, V. Chandrasekhar, N. Schwarz, A. Spale, C. Zhang, G. Goldman, J. Leigh, and A. Johnson (2004) *SAGE: the Scalable Adaptive Graphics Environment*, WACE.

[4] B. Jeong, R. Jagodic, L. Renambot, R. Singh, A. Johnson, and J. Leigh (2005) "Scalable Graphics Architecture for High-Resolution Displays," *Proceedings, Using Large, High-Resolution Displays for Information Visualization Workshop*, IEEE Visualization 2005, Minneapolis, MN, October 2005.

[5] N. Krishnaprasad, V. Vishwanath, S. Venkataraman, A. Rao, L. Renambot, J. Leigh, A. Johnson, and B. Davis (2004) "JuxtaView – a Tool for Interactive Visualization of Large Imagery on Scalable Tiled Displays," *Proceedings of IEEE Cluster 2004*, San Diego, September 20–23, 2004.

[6] N. Schwarz, S. Venkataraman, L. Renambot, N. Krishnaprasad, V. Vishwanath, J. Leigh, A. Johnson, G. Kent, and A. Nayak (2004) "Vol-a-Tile – A Tool for Interactive Exploration of Large Volumetric Data on Scalable Tiled Displays" (poster), *IEEE Visualization 2004*, Austin, TX, October 2004.

[7] M. Barcellos, M. Nekovec, M. Koyabe, M. Dawe, and J. Brooke (2004) "High-Throughput Reliable Multicasting for grid Applications," *Fifth IEEE/ACM International Workshop on Grid Computing* (Grid '04), pp. 342–349.

[8] M. den Burger, T. Kielmann, and H. Bal (2005) *Balanced Multicasting: High-throughput Communication for grid Applications*, SC '05, Seattle, WA, November 12–18, 2005.

[9] http://www.rcuk.ac.uk/escience/.
[10] http://www.cern.ch/.
[11] http://lcg.web.cern.ch/LCG/.
[12] http://www.cclrc.ac.uk/.
[13] http://www.gridpp.ac.uk/.
[14] W. Allcock (2003) *GridFTP: Protocol Extensions to FTP for the Grid*, Grid Forum Document, No. 20, April 2003.

[15] http://www.realitygrid.org/.
[16] http://www.ngs.ac.uk/.
[17] http://www.teragrid.org/.
[18] http://www.globus.org/.
[19] http://www.surfnet.nl/info/en/home.jsp.
[20] http://www.realitygrid.org/Spice.
[21] R.L. Grossman, S. Bailey, A. Ramu, B. Malhi, P. Hallstrom, I. Pulleyn and X. Qin (1999) "The Management and Mining of Multiple Predictive Models Using the Predictive Model Markup Language (PMML)," *Information and Software Technology*, 41, 589–595.
[22] A.L. Turinsky and R.L. Grossman (2006) *Intermediate Strategies: A Framework for Balancing Cost and Accuracy in Distributed Data Mining, Knowledge and Information Systems* (in press). Springer.
[23] R.L. Grossman, Y. Gu, D. Hanley, X. Hong, and G. Rao (2003) "Open DMIX – Data Integration and Exploration Services for Data Grids, Data Web and Knowledge Grid Applications," *Proceedings of the First International Workshop on Knowledge Grid and Grid Intelligence (KGGI 2003)* (edited by W.K. Cheung and Y.Ye), IEEE/WIC, pp. 16–28.
[24] P. Krishnaswamy, S.G. Eick, and R.L Grossman (2004) *Visual Browsing of Remote and Distributed Data, IEEE Symposium on Information Visualization (INFOVIS'04)*, IEEE Press.
[25] A. Ananthanarayan, R. Balachandran, R.L. Grossman, Y. Gu, X. Hong, J. Levera, and M. Mazzucco (2003) "Data webs for Earth SCIENCE data," *Parallel Computing*, 29, 1363–137.
[26] J. Austin, T. Jackson, *et al.* (2003) "Predictive Maintenance: Distributed Aircraft Engine Diagnostics," *The Grid*, 2nd edn (edited by Ian Foster and Carl Kesselman), MKP/Elsevier.
[27] A. Nairac, N. Townsend, R. Carr, S. King, P. Cowley, and L. Tarassenko (1999) "A System for the Analysis of Jet Engine Vibration Data," *Integrated Computer-Aided Engineering*, 6(1), 53–65.
[28] J. Sabatini (2003) "Leveraging Scenario Analysis in Operational Risk Management," Federal Reserve Bank of New York, May 2–30, 2003, Conference on Leading Edge Issues in Operational Risk Measurement.
[29] M. Hardy (2004) "Calibrating Equity Return Models," *GARP 2004*, February 25, 2004.
[30] L. Lipinsky de Orlov (2005) "Grid Technologies: State of the Marketplace," Presentation to Israel Grid Technology Association, March 2005.
[31] D. Poulos (2005) "As It Happens: How To Harness Technology To Manage Derivatives Investment Risk, Real-time," *Hedge Funds Review*, October, 33.
[32] "Buzz Over grid Computing Grows," *Network World*, October 6, 2005.
[33] Line 56, DataSynapse brochure on Wachovia, July 2002.
[34] R. Ortega, "The Convergence of Grid and SOA: 8 Reasons To Make grid Part of Your SOA Strategy," DataSynapse webcast highlighting Wachovia.
[35] C. Davidson (2002) "JP Morgan unveils Project Compute Backbone,"watersonline.com, 9(18), October.
[36] S. Findlan (2005) Panel on "Leveraging Sun's grid Architecture for High Performance in the Financial Market," 2005 Conference on High Performance on Wall Street, September 26, 2005.
[37] E. Grygo (2005) "JPMorgan's Virtual Reality," Sun Microsystems Services and Solutions, November 2005 http://www.sun.com/solutions/documents/articles/fn_jpmorgan_virtual_aa.xml.
[38] S. Findlan (2005) Panel on "Leveraging Sun's grid Architecture for High Performance in the Financial Market," 2005 Conference on High Performance on Wall Street, September 26, 2005.

Chapter 3

Grid Network Requirements and Architecture

Joe Mambretti and Franco Travostino

3.1 INTRODUCTION

Chapter 1 describes attributes that are common to Grid environments and that are currently being extended to Grid network resources. Chapter 2 presents selected Grid use cases, which, along with many other Grid activities, are driving the development of the requirements that motivate Grid design. This chapter provides more detailed descriptions of basic Grid network requirements and attributes, for example by providing examples related to network technologies.

This chapter also presents an overview of basic components of Grid network architecture. As noted in Chapter 1, decisions about placing capabilities within specific functional areas have particular significance when designing an architectural model. These decisions essentially define the model. Chapter 1 also notes that recent design trends allow for increasing degrees of freedom with regard to such placements. However, in the following discussions, the descriptions of capabilities with functional areas will present basic concepts and will not describe an exhaustive list of potential capabilities. This chapter also introduces the theme of services-oriented architecture and relates that topic to Grid network services.

Grid design models are formalized into architectural standards primarily by the Global Grid Forum (GGF), in cooperation with the efforts of other standards organizations described in Chapter 4. These standards organizations translate

Grid Networks: Enabling Grids with Advanced Communication Technology Franco Travostino, Joe Mambretti, Gigi Karmous-Edwards © 2006 John Wiley & Sons, Ltd

requirements, attributes, and capabilities into an architectural framework that is used by Grid designers and developers.

3.2 REQUIREMENTS

3.2.1 REQUIREMENTS AND COEXISTENCE OF DIVERSE NETWORK USER COMMUNITIES

The Internet has been an extremely successful technology innovation. Currently, there are approximately one billion users of the common Internet. This widely accessible Internet is increasingly considered "the network." However, network designers must look beyond this common assumption and examine deeper issues related to network requirements and appropriate technology responses. For example, it may be useful to recognize that the one billion users of the current Internet can be described as constituting a single community with a set of basic requirements. However, it may also be useful to consider this group as an aggregate of multiple communities with varying requirements. Such distinctive requirements dictate different technology solutions.

As a conceptual exercise, it is instructive to segment network users into three general categories. For example, network users can be classified into three communities, as illustrated in Figure 3.1.

The first group, class A, includes typical home users with services provided by digital subscriber line (DSL) or cable modems, who may have access at rates around 1 Mbps, who use commodity consumer services: good web access, e-mail with megabyte attachments, downloads of streaming media, messaging, and peer-to-peer (music, gaming) applications. Class A users typically need full Internet routing. Their

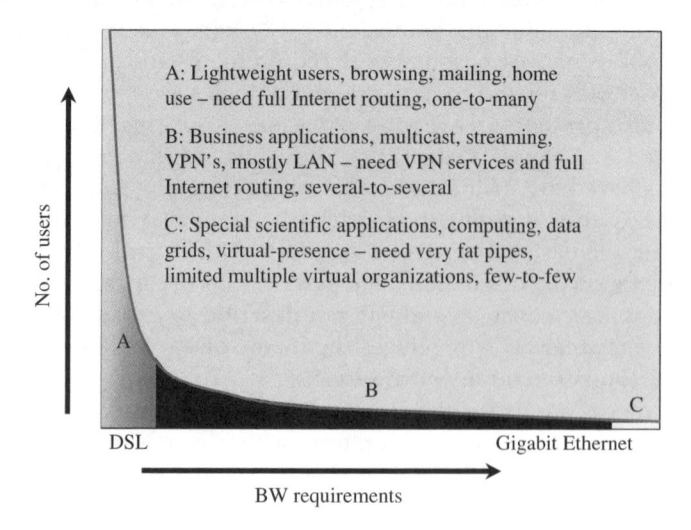

Figure 3.1. Numbers of class A, B, and C users compared with their bandwidth appetite. A taxonomy developed by De Laat [1].

individual traffic flows are generally small and short-lived, and they can be routed from anywhere to anywhere (and back).

The second community, class B, consists of corporations, enterprises, universities, Grid-based virtual organizations, and laboratories that operate at gigabit per second local-area network (LAN) speeds. Class B connectivity uses many switched services, virtual private networks (VPNs), and full Internet routing uplinks, often through firewalls. This community typically needs protected environments, many-to-many connectivity, and collaboration support. The majority of the traffic typically stays within the virtual organization. However, class B users are also connected to perhaps several thousand other sites via routed high-performance networks, some of which are dedicated to specific communities.

The third community, class C, represents a few hundred truly high-end applications currently being developed, which need transport capacities of multiple gigabits per second for a duration of minutes to hours, originating from a few places, destined for a few other places. Class C traffic often does not require routing, as it usually takes the same route from source to destination. However, it requires dynamic path provisioning because most of these applications require the gathering and utilization and releasing of resources at multiple sites.

Assuming that the total backbone traffic of the total sum of class A users is the same order of magnitude as class B traffic in a region, approximately 1 Gbps, then the needs of a 5-Gbps class C user constitute a distinctly disruptive requirement.

The network traffic generated by many Grid applications spans these communities – it generally can be supported within the range of medium- to high-bandwidth in zone B with some peaks in zone C. Other Grid applications may exist only within zone C.

This issue of multiple communities and diverse requirements is particularly important at this point in the development of the Internet. Currently, the communities that are concerned about advancing Internet technology have noted that the large current installed base of Internet services, equipment, and providers has slowed research and innovation, in part because of a need to be "backwardly compatible" with this installed base. The mere development of a technology innovation does not advance the state of networking. For an innovation to provide a measure of progress, it must be adapted and widely deployed. However, because of the existing large installed base, it is difficult today to introduce highly advanced, disruptive technology into the network. Therefore, many Internet research and development projects are focused only on incremental improvements within existing architecture, technology, and infrastructure. The abstraction and virtualization capabilities of Grid environments in general, and Grid networks in particular, may assist in addressing this issue.

3.2.1.1 Requirements and the Open Systems Interconnect (OSI) reference model

For over 20 years, since it was first introduced by the International Organization for Standardization (ISO), the OSI reference model [2] has been a de facto lingua franca concept among networking researchers and practitioners. The OSI model (Figure 3.2) is a practical tool for describing areas of network functionality and their relationships.

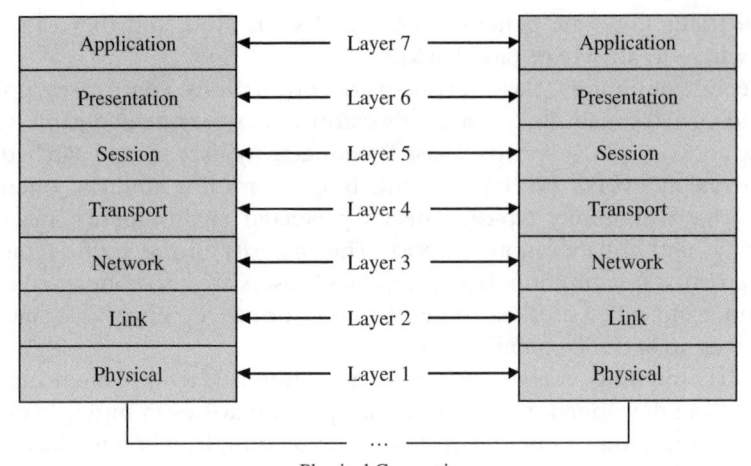

Figure 3.2. The OSI reference model.

This book uses references to the OSI model layers to describe basic concepts. Because this chapter relates requirements to architectural concepts, it refers to several OSI layers. Also, as Chapter 14 indicates, next-generation network designs are have recently begun to move away from this classic view of functionally separate network functional layers to one that is more integrated. These new concepts are particularly useful for Grid network design.

3.2.2 ABSTRACTION/VIRTUALIZATION

Grid environments provide powerful methods for reducing specific local dependencies and for resource sharing and integration. These Grid mechanisms enable developers to abstract limitless customizable functions from supporting information technology infrastructure. This level of abstraction can extend to network resources. Network virtualization is a means to represent selected abstracted network capabilities to upper-layer software, including applications. Network virtualization is a discipline in its own right, separate from Grid network virtualization. In general, as with other types of information technology virtualization, these methods allow for the use of capabilities within functional areas without having to address the specific dependencies of certain types of low-level protocols and hardware.

Furthermore, network virtualization can be used to manipulate the functionality of the underlying network resources. For example, using these methods, a generic virtual network can be considered as having a number of "knobs," "buttons," and "dials" for the provisioning and/or control of the network's behavior. This virtual instrument panel is actually a software module that resides within the network infrastructure. Responding to a user's or application's request, this module can reach out to one or more network elements through any number of standard network signaling protocols or service interfaces, e.g., the Simple Network Management

3.2 Requirements

Protocol (SNMP), Transaction Language 1 (TL1), Web Based Enterprise Management (WBEM), User–Network Interface (UNI) and many others. These signaling protocols and service interfaces are generalized and they can be employed in a number of scenarios, e.g., an operator's console, automated network fault surveillance software, root-cause failure analysis software, network node-to-node signaling, control plane software, specialized application programming interface (API), etc.

Because of the heterogeneity of network protocols, systems, equipment, and technologies, many standard and proprietary representations of command languages and protocols exist. Also, the syntax and semantics of the communications used vary greatly. To attempt to expose network capabilities without virtualization would lead to unmanageable complexity. The abstraction of these capabilities, implemented through some type of a virtual instrument panel, can address this diversity by significantly reducing this complexity.

For example, only selected knobs and buttons related to meaningful capabilities need to be exported to upper-layer software. This virtual instrument panel would provide for greater uniformity. The exposed knobs could appear to be isomorphic, presenting an image that would makes it appear as if only one set of commands and a minimal number of protocols could provide the required functionality. Therefore, the network virtualization layer is a key component for to dynamic, automated, adaptive provisioning.

This virtualization layer is also an important mechanism for migrating from legacy architecture. By placing a higher layer abstraction over legacy architecture, it is possible to transition to new technology more easily, by implementing co-existent sets of functionality under that abstraction layer. Using network virtualization, legacy and new technology can co-exist until the older components are removed from the environment.

Network virtualization is being advanced independently of Grid development. However, the practices and principles of network virtualization are being brought into areas of Grid architectural development. Similarly, initiatives related to Grid network abstraction are assisting in the conceptualization of virtual networks.

3.2.3 RESOURCE SHARING AND SITE AUTONOMY

Within a Grid environment, a participating site will rarely relinquish the ownership of its resident assets to any other site. Instead, a site may publish a set of service stipulations, while retaining full autonomy to match its assets to the implementation of such service stipulations. This approach is seen as a key enabler to scaling a Grid to a very large footprint. It contrasts with the prior common practice of sharing distributed systems and clusters, which usually involved processes for explicitly donating or "exporting" assets.

Using these service stipulations, multiple services providers, organizations, and individuals can devise common agreements for services exchanges and guarantees. During a Grid's life cycle, many different such stipulations, or service level agreements (SLAs), can be routinely established and terminated between resource providers and consumers. Furthermore, these SLAs are often composed in complex relationships, with some SLAs depending upon others. SLAs can be transferred among virtual organization participants, or even across virtual organizations.

The autonomy principle, i.e., autonomy by way of service versus implementation decoupling, applies to individual resource components within a site as well. At this finer level of granularity, this approach can be used to obtain services that can be recombined to form other services, resulting in benefits such as such code reuse, scalable resource management, easier failure troubleshooting, etc.

All of these concepts apply to network services as well, especially for Grids that require explicitly defined communication performance. As obtainable resources within Grid environments, they can be integrated with other Grid resources to create new types of ad hoc services.

3.2.4 FLEXIBILITY THROUGH PROGRAMMABILITY

Grid architecture provides for a "programmable" environment instead of a fixed infrastructure. Grid flexibility is enabled by software suites consisting of toolkits and other request management middleware, by continual resource allocation and reconfigurability, and by options for organizing operations through workflow management. However, processes can also be scheduled through automated provisioning techniques. These techniques allow for processing through predetermined steps. For many complex, routine, repetitive tasks, it is best to use predetermined methods, which are less labor-intensive and less error-prone, quicker to execute, and often easier to integrate into other processes.

Many Grid processes are based on the sophisticated orchestration of a workflow within defined workflow frameworks. A workflow can be considered as an equivalent to a flight plan in aviation. The workflow defines a set of tasks that must occur or which can occur when designated conditions are met and depending on the order of occurrence, as well as the decision points that may result in different forms of execution and termination demarks. This workflow approach to process management generalizes and extends the notion of a standard periodic duty cycle.

This approach can be useful for large-scale network service provisioning, for example to manage unusually large flows. If a large-scale Grid data burst over a network can be anticipated and coordinated with precision, appropriate provisioning can be accomplished in advance. This type of activity can be incorporated into an ongoing "Grid duty cycle," which may be periodic or aperiodic, governed by the Grid's workflow parameters, which anticipate when large-scale data bursts are expected and prepares to support them.

3.2.5 DETERMINISM

One of the most powerful capabilities of a Grid consists of the ability to match applications requirements with appropriate resources to support precise service levels, i.e., determinism. Recently, a number of initiatives have begun to address the issue of determinism in Grid networks. The greater the level of network resource determinism, the greater the level of the confidence with which Grid applications can rely on underlying network elements in the infrastructures for predictable performance. Mechanisms are currently being created that allow applications to meet their end-to-end requirements for data communications, while effectively and efficiently utilizing network resources.

3.2 Requirements

The ideal data flow for an application often is not the same as an ideal data flow for a network. Therefore, the term sometimes used for the application ideal is "goodput" (the data flow characteristics that are meaningful to the application) rather than the more general network term "thoughput" (multiples of bits per second measured at any point in the network and thus inclusive of header, payload, retransmissions, etc.). Both types of flows use the same common measures, for example bit error rate, flow control, congestion control, protocol verbosity, total data copies, reordering, copying of data, and others. However, acceptable throughput for standard network operations may not be sufficient to appropriately support an applications.

This issue is easiest to observe in delay-intolerant applications, such as digital media. However, determinism is important to almost all Grid applications. Whether a Grid application is delay tolerant (e.g., intensive data migration) or nondelay tolerant (e.g., real-time visualization), determinism is equally important for many application essential functions, e.g., placing an upper bound to the time to successful completion for a particular Grid task. A delay-tolerant Grid depends on the stability of throughput metrics and bit error statistics. A nondelay tolerant Grid depends on low variance of round trip time figures (jitter). In almost all cases, the lack of determinism can undermine the success of an application. Within Grid environments, specific deterministic performance instantiations can be established through SLAs, or it can be inherent in its middleware, which can dynamically detect requirements in real time and adjust accordingly.

The capability for providing resource determinism through dynamic provisioning is another important advantage of Grid environments. This capability is also one that has generally not been accessible to the networks that support Grid environments. Without dynamic provisioning, network designers are forced to statically provision network paths for a Grid and make arbitrary decisions, often well ahead of any practical experimentation, as to how to right size such network paths. To some applications, these decisions may seem draconian, because they do not relate to application requirements. The bursty nature of Grids makes the latter task particularly challenging, with high risks of underprovisioning (thus, greatly limiting a Grid's potential) or else requiring overallocations, forcing network resources to be underutilized.

However, within Grid networks, dynamic provisioning can provide capabilities for allocating and releasing network resources in either space (e.g., from point X to point Y) or time (e.g., from time t_0 to time t_1) or both. Dynamic provisioning can be used within a context that tracks a Grid's demand over time and satisfies it with step increments/decrements directly related to a Grid's own budget of network resources. With regard to the time dimension, dynamic provisioning can be further described as being just in time or well ahead of time (e.g., time of day reservations).

Adaptive provisioning further refines dynamic provisioning of a network. This capability is a critical tool for achieving determinism. Often, to achieve the best result for an application, resources must be allocated and adjusted in real time. Adaptive provisioning provides Grid infrastructure with the capability to autonomously alter the implementation of network resources used by a Grid in response to various circumstances (e.g., new capacity becoming available, bargain pricing for resources at specific hours, etc.). Adaptive provisioning is predicated upon establishing real-time feedback loops between network, Grid infrastructure, and Grid application(s). A Grid infrastructure can continuously monitor how often applications meet or fail some published targets. Should failures reach some configured thresholds, the Grid infrastructure can provide for adjustments. This process can be undertaken by negotiating a network SLA that is more favorable to the application, or it can be accomplished by preset parameters inherent in the infrastructure. Also, to fulfill the service level of an application request, the Grid infrastructure activates particular subsystems – for instance, a replica manager can provision circuits to clone and publish copies of working datasets at some strategic locations.

3.2.6 DECENTRALIZED MANAGEMENT AND CONTROL

Another key feature of Grid architecture which enhances its overall flexibility is that it provides for decentralization of management and control over resources. This decentralization feature enables multiple capabilities to be used independently, without requiring intercession by centralized processes. Decentralized management and control is especially important for dynamic provisioning. Although general Grid environments have usually been highly decentralized, many Grid networks have been dependent on central management and control functions. This type of implementation places restriction on Grid use. Grid network attributes such as determinism and dynamic and adaptive provisioning cannot be accomplished if every resource allocation or change has to be governed by a central authority.

In the last few years, the Grid community has been developing new methods for Grid network architecture to provide for highly decentralized resource management and control mechanisms for all major aspects of network resource allocation and reconfiguration. These mechanisms allow Grids to create, deploy, and adjust large-scale virtual networks dynamically. They provide for capabilities at all networks layers 1 through 7 of the OSI reference model, and for integration among layers.

3.2.7 DYNAMIC INTEGRATION

Another primary feature of Grid architecture is that it allows for the dynamic creation of integrated collections of resources that can be used to support special higher level environments, including such constructs as virtual organizations. Another term used to describe this capability is "resource bundling." Recently, various initiatives have developed capabilities that enable the bundling of network resources with other Grid resources, such as computation, storage, visualization, sensors, etc. Using this resource bundling or integration, the orchestration of network resources becomes an integral part of a larger ecosystem wherein resources with different characteristics can be best utilized. Using a high-level software entity such as a community scheduler (also referred to as meta-scheduler), it is possible to describe the resource blueprint of a Grid by way of Boolean relationships among all the Grid resources, as in the following example:

$R_{CPU,i}$ AND $R_{NET,j}$

($R_{CPU,i}$ AND $R_{NET,j}$) OR ($R_{CPU,i}$ AND $R_{NET,z}$ AND $R_{STORAGE,m}$)

with $R_{TYPE,instance}$ being an instance of resource of a given type.

Given such blueprints, a community scheduler can make scheduling decisions that reflect co-dependencies among heterogeneous resources.

3.2.8 RESOURCE SHARING

A major motivation for the development of the Grid architecture was to develop new capabilities for resource sharing. Grid abstraction capabilities allow for large-scale resource sharing among multiple distributed locations at sites world-wide. Grid resource sharing enables collaborative groups and virtual organizations to be established at any point that has access to a Grid environment.

Current initiatives in Grid networking are providing extensions to methods for large-scale network resource sharing that allow for better end-delivered communication services as well as for better utilization of resources, while maintaining overall flexibility for applications and infrastructure. These methods are also addressing a number of inherent network complexities.

For example, collision-free addressing allows virtual organization participants to join and leave the group without any risk of a conflict arising from their network address. The popular use of network address translation (NAT) devices and network address port translation (NAPT) devices has greatly complicated, possibly compromised, any direct use of IP addresses as part of an endpoint's unique denomination. For example, protocols that transmit IP addresses within the packet payload are at high risk of failing whenever there are NAT/NAPT intervening devices.

There have been documented cases of hosts joining a Grid while transmitting from locations that have NATs nested three or four levels deep. Therefore, the use of IPv4 addresses for participant's identification would become utterly meaningless. Also, the dualism between IPv4 and IPv6 addresses would be awkward to manage at the Grid infrastructure level. Therefore, it is mandatory to standardize rules to provide consistent and unambiguous addressing and address resolution in spite of arbitrary incidence of NAT/NAPT-like devices. Grid networking initiatives are devising solutions to address this issue.

3.2.9 SCALABILITY

Grid environments are highly scalable – they can be distributed across large geographic regions, enabling the reach of specialized capabilities to extend across the world, they can be extended to new types of resources, and resources can be increased without limits. The ability to seamlessly access resources is a major force multiplier for a Grid and its collaborative problem-solving potential. As yet another variation of the Metcalfe law, the value of a Grid appears to grow with the number of access ramps onto core networks, which are capable of connecting other users, devices, or Grids.

However, there remain obstacles – related to Grid network access, performance, dependability, trust, and other variables – that can directly prevent seamless interaction and impede the growth of a particular Grid instantiation. Some of these are complex problems that current network architecture initiatives are beginning to address.

3.2.10 HIGH PERFORMANCE

Another motivation for the development of Grids was a need to provide, with a reasonable investment, capabilities for extremely high-performance services, for example by aggregating multiple distributed processors and by using parallel computation. Similarly, Grids have been based on extremely high-performance networks. However, Grids have usually not been based on optimal high-performance networks. For example, traditional Grid networks have been especially problematic for latency-intolerant applications and for large-scale data flows.

The need to resolve this issue is driving much of Grid networking architectural development today. As a context for this problem, it may be useful to note Figure 3.1, which depicts three types of data flows. The first represents the general class of network users, approximately one billion people who use the current Internet, whose datastreams are many hundreds of millions but very small. The second represents large-scale enterprises and projects whose datastreams are 100 to 1000 times larger. The third represents a number of projects that have data flows millions of times larger than those of the second group.

Today, the majority of these projects are large-scale science and engineering projects. However, these requirements will one day migrate to other areas of the economy. Many projects are anticipated that will require the communication of sizeable and sustained bursts of data that need to transit reliably through a network. Today the need to communicate terabytes is not uncommon, and soon the communication of petabytes will be required. Today, no common network service can provide for these needs with reasonable cost and efficiency. Therefore, a number of research and development projects have been established specifically to address this issue.

One key issue is allowing extremely large-scale and very small-scale traffic to co-exist without interference on the same infrastructure. Some network researchers have used a metaphor of "mice and elephants" to describe this type of network traffic. Mice are short packets used to carry control information or small data amounts. Elephants are full-length packets which carry as much data at once as the network permits, end to end, so as to improve the overall throughput. Often, Grids require the rapid migration of vast "herds of elephants." Within a Grid, the set of network endpoints, which either sources or sinks the data burst, may be relatively small and static. Therefore, only a subset of those endpoints would have sufficient resources to handle one of these large bursts properly.

This type of common provisioning leads to major network bottlenecks. As Grids gain in popularity and as many different Grids are superimposed over the same common network infrastructure, the problem increases. Aggregates of migration patterns begin to overwhelm local resources. Also, when the large data flows are supported on the same network as the smaller flows, often both are highly disrupted. Frequently, the smaller flows are completely overwhelmed while the larger flows exhibit severely degraded performance. This issue is discussed in some detail in Chapters 9 and 10.

It is notable that in a shared network infrastructure Grid traffic must adhere to that infrastructure's rules, which define fair behavior. For example, the Internet Engineering Task Force (IETF), the Internet standards organization, addresses issues related to data transport protocols, and attempts to define their architecture so that they produce a fair utilization for the widest number of uses of this large-scale, shared commodity network. At this time, some of the transport protocol variants reviewed in Chapter 8 and 9 are not suited for use in the general Internet. Researchers and practitioners who are developing protocol variants often bring those concepts to the IETF as an Internet draft to attempt to make them standards.

3.2.11 SECURITY

Because Grids are shared distributed environments, security and defense against external attacks has always been a key concern, and it has been the focus of multiple crucial architectural development initiatives. To ensure Grid security, all resources are surrounded by security envelopes. In general, networks are also surrounded by many levels of security, related to access, physical resource protection, traffic flow reliability, defense against external attacks, information confidentiality, and integrity. Increasingly, powerful new methods are being used for ensuring network security.

One architectural approach to network security that is being developed by the IETF is the authentication, authorization, and accounting (AAA) standard. Authentication is a method for validating the credentials of a request for resources by the entity (an individual, a software agent, a device, a signal). Authorization is a process that matches the particular entity and request against a set of policies that govern resource use, to determine whether the subject is entitled to a particular resource. Accounting describes the measurement of network resource utilization for troubleshooting, billing, planning, and other measures.

Currently, however, many of the mechanisms being used for network security are not integrated into general Grid security architecture, especially some of the more advanced mechanisms. More recent initiatives are beginning to integrate network security methods with traditional Grid security techniques. These combined methods supplement and complement common end-to-end, mutually agreed security stipulations between two communicating partners. If AAA policies are correctly formulated, it is possible to enhance the potential for multiple capabilities being implemented on a common infrastructure.

3.2.12 PERVASIVENESS

One aspect of scalability is a capability for pervasiveness. Grids can be highly pervasive, extended to many types of edge environments and devices. Currently, Grids are being extended to multiple forms of edge devices, such as sensors, RFID devices, cameras, and visualization devices. As a result, Grid networks are being developed to enable new types of edge capabilities, especially related to wireless communications.

3.2.13 CUSTOMIZATION

A primary feature of Grids is that they can be customized to address specialized requirements. Consequently, Grid networks also have inherent capabilities for customization. For example, many distributed environments have an inherent dominant paradigm of recipient(s) pulling data from nodes to their location at their convenience. In contrast, Grids have an inherent paradigm of nodes pushing data through the network to one or more recipients. Grid operation directly contrasts with peer-to peer technologies, environments in which every participant is expected to pull data at different times. In a Grid complex environment, however, the "push" versus "pull" characterization may be elusive, with both styles being featured at the same time. For example, a Grid may enable a node to push an experiment's data onto another Grid node. The latter node may detect that it lacks some of the data working set and therefore begins pulling historical data out of a file server.

In keeping with the "push" paradigm, it is desirable to ensure that Grids make use of one-to-many communication semantics while taking advantage of complexity relief and efficiencies in the dissemination of Grid data to multiple, geographically dispersed recipients. Reliability and security can augment the basic multicast semantics in Grid environments.

However, few general networks have multicast enabled. New methods for Grid networking multicast are being created, such as optical multicast, offering new design trade-offs to Grid architects. In addition, there have been efforts to place reliability and security functions at layer 3/4 or directly within the distributed infrastructure at the end systems (e.g., as featured in application-level multicast).

Another example of a specialized service is anycast, a style of one-to-one communication wherein one of the endpoints is lazy evaluated and resolved to a real, actionable endpoint address only when some circumstances (e.g., the actual workload of endpoint candidates) become known. With anycast, applications become capable of dispatching underspecified tasks (e.g., "this task must be sent to whichever location can afford 1000 CPUs at this moment") to a Grid infrastructure. In turn, Grid infrastructures are empowered to resolve an anycast style of addressing into a traditional one. They do so to reflect local and specialized knowledge of resource availability, thus yielding some adaptive load-balancing schema. Like multicast, anycast can be implemented at different levels, within the network layer (layer 3, packet forwarding and routing) and within an application's own protocol (layer 7).

3.3 TRANSLATING REQUIREMENTS TO ARCHITECTURE

Many of the standards groups presented in Chapter 4 are formalizing the requirements described here into formal architectural designs. One of the first attempts to create a topography of the key issues related to designing middleware for Grid networks and related environments was activities related to a workshop sponsored by the National Science Foundation, which resulted in an IETF request for comments (RFC).

3.3.1 IETF RFC 2768

IETF RFC 2768 is an attempt to describe the range of issues related to middleware, the services and other resources that exist functionally between applications and network infrastructure, including Grid networks [3]. This topic has not generally been part of the domain of the IETF's scope of responsibility [4]. However, the RFC was motivated by a need to attempt to investigate closely the requirements of collaboration-based and high-performance distributed computing, which are driving advanced network design and development. No consensus existed then, or exists now, on the exact demarcation point that distinguishes different functionality among middleware domains, e.g., edge, core, access. This initiative noted that various approaches, frequently diverging, were being undertaken by the research community, standards organizations, commercial providers, equipment makers, and software developers.

3.3.2 SERVICE-ORIENTED ARCHITECTURE

A service-oriented architecture (SOA) is one that provides for casual, as opposed to close, binding among software components to support distributed services. SOA is ideally suited for distributed implementation in a large-scale environment, across various discontinuities, whether these originate from physical, security, or administrative demarcation lines. An SOA results in loosely coupled distributed systems wherein the constituent entities – "services" – are independently designed, maintained, and interwoven with other services. These traits are similar to many of the requirements for Grid software that have been mentioned in the earlier sections, especially the ones describing autonomy and SLA orientation.

An SOA dictates that the relationship between services is solely based on a set of formalized data definitions and message exchanges. A service has no a priori knowledge of how and where companion services are implemented. Specifically, services do not rely upon a uniform system or a homogeneous runtime technology.

With an SOA, services can be composed to represent a workflow – e.g., a whole business process or a complex science experiment.

Web Services are a popular framework with which to reduce an SOA to practice. Web Services are rapidly gaining acceptance in the industry as they build upon a vendor-neutral suite of standards of broad applicability. While the Web Services momentum has clearly originated at the end systems, there have been clear signs of Web Services technology gaining acceptance with network vendors and network operators (Chapter 4 describes standardization efforts in this area). Similarly, workflow languages and toolkits have emerged to assist the process of composing Web Services.

3.3.3 A MULTITIER ARCHITECTURE FOR GRIDS

Grid software exploits multiple resource types, while meeting defining requirements for each resource type. Requirements for the network resource were analyzed in the Sections 3.2.1 to 3.2.12.

The breadth and extensibility of multiple heterogeneous resource types motivate the creation of the multitiered architecture shown in Figure 3.3. The first tier contributes a virtualization layer. The virtualization function is specific to each resource type and wraps around each resource instantiation given a resource type. For ease of programming, the ensuing logical representation for a resource is typically first supported by companion off-the-shelf software constructs. The upper tiers

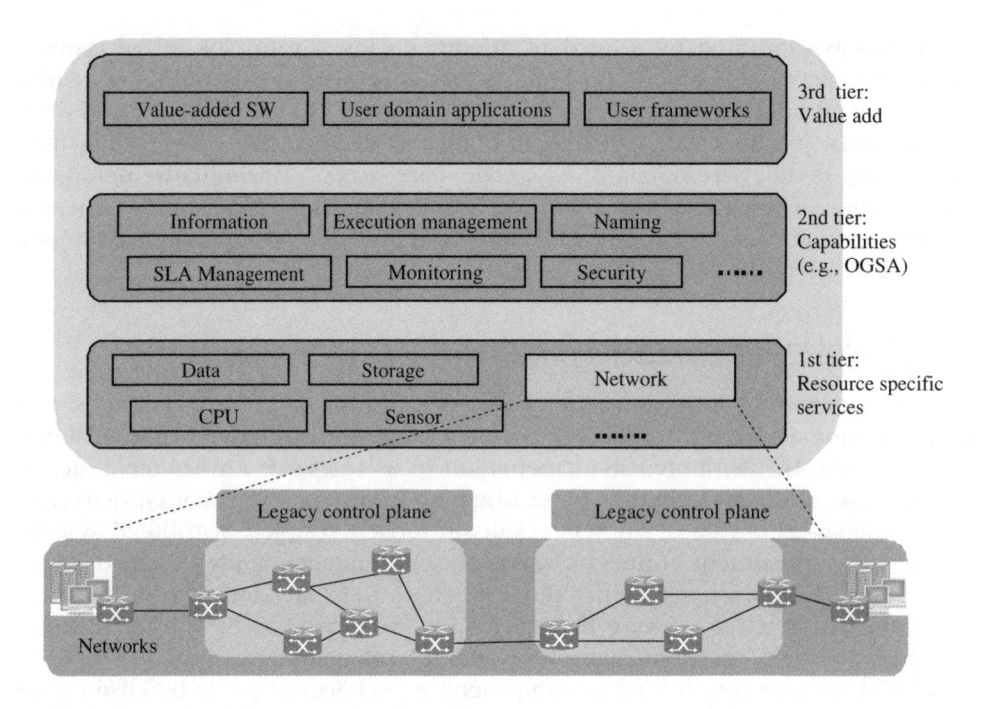

Figure 3.3. Multitiered architecture for Grid software. Graphical representation adapted from ref. 3 and Admela Jukan's contribution to ref. 4.*

must handle the logical representation of the resource and refrain from direct access to any specific mechanism for resource lifecycle management (e.g., to configure, provision, monitor the resource). For portability and complexity management, it is important to provide the upper tiers with only a minimalist view of the resource, yet without overlooking any of its core capabilities.

Although the first tier may still perceive individual resources as silos, the second tier provides a foundation for horizontal integration among resources (silos). Within this tier, the SOA property to compose autonomous services is most relevant. Conforming to SOA principles, a service is capable of engaging with other service(s) at either the same tier or at the bottom tier, in a peer-to-peer fashion. The ensuing pool of services featured in the second tier is a departure from strict software layering techniques, which have shown severe limits in reflecting complex synapses across entities.

The Global Grid Forum's Open Grid Services Architecture (OGSA) [3] is a blueprint with which to structure services that belong to the second tier and exhibits

---

* Copyright © Global Grid Forum (2003). All Rights Reserved. This document and translations of it may be copied and furnished to others, and derivative works that comment on or otherwise explain it or assist in it implementation may be prepared, copied, published and distributed, in whole or in part, without restriction of any kind, provided that the above copyright notice and this paragraph are included on all such copies and derivative works.

3.3 Translating Requirements to Architecture

multivendor interoperability. In OGSA, services fall into eight distinct realms, according to the definitions in ref. 5:

- *Infrastructure services* enable communication between disparate resources (computer, storage, applications, etc.), removing barriers associated with shared utilization.
- *Resource management services* enable the monitoring, reservation, deployment, and configuration of Grid resources based on quality of service requirements.
- *Data services* enable the movement of data where it is needed – managing replicated copies, query execution and updates, and transforming data into new formats if required.
- *Context services* describe the required resources and use policies for each customer that utilizes the Grid – enabling resource optimization based on service requirements.
- *Information services* provide efficient production of, and access to, information about the Grid and its resources, including status and availability of a particular resource.
- *Self-management services* support the attainment of stated levels of service with as much automation as possible, to reduce the costs and complexity of managing the system.
- *Security services* enforce security policies within a virtual organization, promoting safe resource sharing and appropriate authentication and authorization of users.
- *Execution management services* enable both simple and more complex workflow actions to be executed, including placement, provisioning, and management of the task life cycle.

Web Services are the implementation framework of choice for OGSA-compliant implementations.

The third and topmost tier provides the ultimate view of a business process or a science experiment. At this tier, it is expected that a virtual organization's own policies are defined and implemented, while the underlying second-tier services are considered the programming model to access resources. For instance, a virtual organization may provide a specific scheduling policy for Grid resources that best matches its style of operation. Unlike the second tier, the requirements for interoperability at this tier can be limited to a virtual organization's constituency.

3.3.4 INTRODUCING GRID NETWORK SERVICES

Grid network services are SOA-compliant services, intended to isolate and control abstracted network resources (e.g., SLAs, route determination, performance data, topology), while meeting the requirements in Section 3.2 and masking the complexity, churn, and heterogeneity that are inherent to the network resource. Grid network services originate in the first tier in Figure 3.3 and reach into the second tier. With Grid network services, the network resource becomes a first-class abstracted resource for use in OGSA and OGSA-like second tier constructs. The

Global Grid Forum's High Performance Networking Research Group (GHPN-RG) has first introduced the notion of Grid network services [6].

Notably, the network resource is known to be a shareable and quality of service (QoS)-capable asset. Furthermore, a network resource is made of segments subject to different ownership/administration and controlled by legacy functions, which supply stability, policy enforcement, and interoperability. This process is depicted in Figure 3.3 – beginning with the first tier and moving downward, with two independent network clouds forming a whole end-to-end network extent.

For instance, a class of Grid network services may be capable of negotiating an end-to-end SLA, with an actual activation time and release time set at specific hours of the day. Within networks that have such agility, this Grid network service will hand off reservation statements in the network's own language. Within networks which feature a static allocation model, this Grid network services will proceed with requesting a conservative 24×7 implementation of the SLA.

3.3.4.1 Service query and response

Another class of Grid network services may be defined to reply to local service queries. In response to such a query, this Grid network service would return a reference to the service that matches the query attributes and which is nearest to the querying service, given the actual round trip time experienced over the network (and irrespective of physical coordinates or network hops).

By composing these services and services related to other resources, it becomes possible to realize a service of greater complexity, such as a delay-tolerant data mover. This data mover could take a (typically large) dataset as input together with a "deliver by" date as an indication of delay tolerance. Deadline permitting, the data mover would then proceed to tap off-peak network resources. In some circumstances, for instance, it would store the data at intermediate locations while other network hops are momentarily busy or off-line. To store the data along the way, such a data mover would, in a timely way, broker anonymous storage at proper junction points along the end-to-end path [7].

For a totally different network experience, a global load-balancing service might be launched to assist the migration of computation between data centers while maintaining an upper bound on the downtime perceived by some interactive external clients. This service would seek available CPU resources, combined with a short-lived, end-to-end network SLA whose metrics are highly deterministic across the Metropolitan-Area Network (MAN) or Wide-Area Network (WAN) [8].

The two examples referenced here thrive on the "collective" outcome of using multiple heterogeneous resources at once, which is a hallmark of OGSA-like second-tier constructs.

3.3.4.2 Stateful attributes of Grid network services

A Grid network service's northbound interface adheres to the syntax of the Web Services specification and the specific usage rules such as the ones defined by the GGF. Special consideration must be given to the stateful nature of a Grid network service. Simply stated, this service must be capable of accruing and handling internal state to reflect the state of an SLA ("demand side") and/or the network resource ("supply side"). The Web Services community has made overtures to stateful services (e.g., Web Services Resource Framework, WSRF [9]), thus extending their original, stateless design center. Specifications such as WSRF thus bear high relevance to the network resource as well as other resources, as described in greater detail in Chapter 7.

In a typical Grid network service implementation, the southbound interface tracks a network's reality. This type of interface is intended to intercept legacy control planes (e.g., Open Shortest Path First (OSPF), Border Gateway Protocol (BGP), Generalized Mulltiprotocol Label Switching (GMPLS)), management interfaces (e.g., SNMP, TL1), service-oriented wire protocols (e.g., UNI), or virtualization layers (e.g., WBEM) provided by the network equipment provider. As described in Chapter 4, these bindings build upon the standardization efforts of several organizations. As such, they are outside the scope of new Grid standardization efforts.

Although the standardization of interfaces for Grid network services is still in its infancy, there is an increasingly large body of projects and experimentation, as described in research literature, workshops, and public demonstrations. Examples include: user-controlled lightpaths (UCLPs) [10], dense wavelength division multiplexing (DWDM)-RAM [11], the DRAC service plane [12], the VIOLA control plane [13], the network resource management system [14], and Optical dynamic intelligent network (ODIN) [15].

ACKNOWLEDGMENT

The analysis in Section 3.2.1 and Figure 3.1 were develped by Cees de Laat.

REFERENCES

[1] C. de Laat, E. Radius, and S. Wallace (2003) "The Rationale of the Current Optical Networking Initiatives," iGrid2002 special issue, *Future Generation Computer Systems*, 19, 999–1008.

[2] H. Zimmerman (1980) "OSI Reference Model – The ISO Model of Architecture for Open Systems Interconnection," *IEEE Transactions on Communications*, 28, 425–432.

[3] B. Aiken, J. Strassner, B. Carpenter, I. Foster, c. Lynch, J. Mambretti, R. Moore, and B. Teitelbaum (2000) *Network Policy and Services: A Report of a Workshop on Middleware*, IETF RFC 2768, February 2000.

[4] B. Carpenter (ed.) (1996) *Architectural Principles of the Internet*, RFC 1958, June 1996.

[5] I. Foster, H. Kishimoto, A. Savva, D. Berry, A. Djaoui, A. Grimshaw, B. Horn, F. Maciel, F. Siebenlist, R. Subramaniam, J. Treadwell, and J. Von Reich (2005) *Open Grid Services Architecture Version 1.0*, Grid Forum Document, No. 30, January 2005.

[6] G. Clapp, T. Ferrari, D.B. Hoang, A. Jukan, M.J. Leese, P. Mealor, and F. Travostino, *Grid Network Services*, Grid Working Document, revisions available for download at https://forge.gridforum.org/projects/ghpn-rg.

[7] F. Dijkstra and C. de Laat (2004) "Optical Exchanges", GRIDNETS Conference Proceedings, October 2004, http://www.broadnets.org/2004/workshop-papers/gridnets/DijkstraF.pdf.

[8] P. Daspit, L. Gommans, C. Jog, C. de Laat, J. Mambretti, I. Monga, B. van Oudenaarde, S. Raghunath, F. Travostino, and P. Wang (2006) "Seamless Live Migration of Virtual Machines over the MAN/WAN", *Future Generations Computer Systems* (in press).

[9] The Web Services Resource Framework (WSRF) Technical Committee, Organization for the Advancement of Structured Information Standards, http://www.oasis-open.org/committees/tc_home.php?wg_abbrev=wsrf.

[10] *User Controlled LightPaths*, http://www.canarie.ca/canet4/uclp/.

[11] T. Lavian, D. Hoang, J. Mambretti, S. Figueira, S. Naiksatam, N. Kaushil, I. Monga, R. Durairaj, D. Cutrell, S. Merrill, H. Cohen, P. Daspit, and F. Travostino (2004) *A Platform for Large-Scale grid Data Service on Dynamic High-Performance Networks*, First International Workshop on Networks for Grid Applications (GridNets 2004), San Jose, CA, October 29, 2004.

[12] L. Gommans, B. van Oudenaarde, F. Dijkstra, C. de Laat, T. Lavian, I. Monga, A. Taal, F. Travostino, and A. Wan (2006) "Applications Drive Secure Lightpath Creation across Heterogeneous Domains," special issue with feature topic on Optical Control Planes for Grid Networks: Opportunities, Challenges and the Vision. *IEEE Communications Magazine* 44(3).

[13] The VIOLA project, http://www.imk.fraunhofer.de/sixcms/detail.php?template=&id=2552&_SubHP=&_Folge=&abteilungsid=&_temp=KOMP.

[14] M. Hayashi (2005) "Network Resource Management System for grid Network Service," presented at GGF15 Grid High Performance Networks Research Group, Boston, October 5, 2005, https://forge.gridforum.org/docman2/ViewProperties.php?group_id=53&category_id=1204 &document_content_id=4956.

[15] J. Mambretti, D. Lillethun, J. Lange, and J. Weinberger (2006) "Optical Dynamic Intelligent Network Services (ODIN): An Experimental Control Plane Architecture for High Performance Distributed Environments Based on Dynamic Lightpath Provisioning," special issue with feature topic on Optical Control Planes for Grid Networks: Opportunities, Challenges and the Vision. *IEEE Communications Magazine*, 44(3), 92–99.

Chapter 4

Relevant Emerging Network Architecture from Standards Bodies

Franco Travostino

4.1 INTRODUCTION

The previous chapters describe Grid attributes, which motivate architectural design goals. Grid architecture is being developed in accordance with recognized standards. Standards, which are important for any technology requiring interoperability, are central to Grid networks. Without standards, Grids cannot be based on hardware and software from multiple providers, and they cannot interoperate with required resources – networks, devices (e.g., visualization devices, sensors, specialized industrial components, whether designed for shared or private use), and services.

Interoperability restrictions are unacceptable. Such restrictions can severely compromise the evolution of the Grid paradigm and its wide adoption across different infrastructure and communities. Grid networks build upon a unique ensemble of emerging Grid standards and standards developed outside the Grid community, including networking standards and core Web Services standards. Within each of these diverse realms there are established standards, emerging standards, and challenging areas for which standards may provide a solution.

Established standards cover many broad areas, and they are as diverse as communication methods, such as SOAP [1], various types of process components, and physical infrastructure, such as 10-Gbit Ethernet [2]. Grid architecture has challenged many of these standards. In some cases, Grid experimentation first exposed and documented interoperability challenges.

Emerging standards include recently approved standards and work in progress, for example those that are in the form of draft documents. This section focuses on the emerging standards from many standards organization and their relationship to Grid networks. Because the scope of the material requires categorization, the organizations are presented here generally following the OSI seven-layer model, in descending order.

4.2 GLOBAL GRID FORUM (GGF)

The Global Grid Forum (GGF) is the organization that serves as a focal point for Grid-related standards. It advances standards directly and indirectly, for example through liaison with other standards bodies that are best qualified to address a particular interoperability topic. The GGF first met in 2001 after the merger of three independent organizations, those in North America (originally called "Grid Forum"), in Asia Pacific, and in Europe – the European Grid Forum (eGrid). GGF operates as a community as opposed to a hierarchical organization. The GGF provides individuals with equal opportunities for joining its open processes (as opposed to having a highly formal structure, for example designating seats, alternate seats, and voting rights weighted on a corporate basis, etc.).

The GGF's governance is based on a set of established processes, including those implemented by the GGF Chair, the GGF Advisory Committee (GFAC), the GGF Steering Group (GFSG), the Area Directors, and the Chairs for working groups and research groups. The GGF internal structure and processes have been modeled on those used by the Internet Engineering Task Force (IETF).

GGF's members, as well as its constituency, are extremely diverse. Its unique volunteers include scientists (e.g., high-energy physicists, astronomers, seismologists), academics, industry representatives (e.g., from areas such as R&D, service provisioning, information technology deployment), federal agency institute researchers, government planners, and market analysts.

The list here emphasizes GGF's activities that are directly relevant to networking issues. These activities are organized within the topics and documents highlighted:

- The Open Grid Services Architecture (OGSA) [3] articulates eight categories of services, through which applications obtain shared, secure, dependable access to available resources. These primary services are: infrastructure services, resource management services, data services, context services, information services, self-management services, security services, execution management services.
- WS-Agreement [4] is a design pattern that defines how the producer of a service and the consumer of that service can reach an agreement. With domain-specific derivations, it can be applied to the stipulation of a Service Level Agreement (SLA) with a network provider.
- GridFTP [5] is an enhanced version of FTP [6] addressing typical Grid operating requirements. Its features include: segment-oriented restartable transfers, data

striped over multiple TCP connections, asynchronous application programming interface.
- The Hierarchy for Network Monitoring Characteristics [7] document describes which network metrics (e.g., bandwidth, latency, jitter) are relevant to Grid applications. It introduces standard terminology and a classification hierarchy for such metrics.
- GGF's whole security standards development efforts are relevant because they are concerned with authentication and authorization properties as they apply to multiple resources (including the network) within decentralized, possibly dynamic federations. Starting with GGF13, new working groups have begun to emerge around the specific areas of Grids using firewalls and Virtual Private Networks (VPNs).
- In several areas, GGF's first step is to produce informational documents that capture representative use cases and highlight problems. The Grid High Performance Networking Research Group (GHPN-RG) has compiled a list of issues and opportunities in supporting Grids with network resources [9]. This group has charted the opportunities resulting from optical networks and the most promising research endeavors related to this topic [8]. Furthermore, the GHPN-RG has produced a set of use cases [10] capturing the interaction between applications and Grid network services (commonly referred to as "network middleware"). The analysis of use cases is seen as a starting point toward defining architecture and interfaces of Grid network services. The Data Transport Research Group (DT-RG) has classified [11] transport protocols (i.e., layer 4 of the OSI stack) that are in use within Grid communities and differ from the IETF-sanctioned TCP.
- Another objective of GGF's activities is bringing communities of interest together to identify which impediments versus catalysts are expected along the various Grid adoption paths. For example, among others, the Telecom Community Group [12] is a forum devoted to those network service providers who are interested in enabling Grid traffic, adopting Grids for internal use, and/or supplying Grid-managed services. The Telecom Community Group is chartered to cover the whole spectrum of business drivers and technical enablers, and thus is directed at developing a comprehensive understanding of the push/pull forces at play in the marketplace. The outcomes can impact Grid networks roadmaps, and specifically can formulate expectations related to scalability and dependability from the perspective of provider-provisioned services.

The OGSA and WS-Agreement activities prompt GGF to have active liaisons with the Web Services and virtualization communities of appropriate standards bodies (OASIS (see Section 4.4) and DMTF (see Section 4.9)). Documents related to these GGF/IETF liaison activities include refs 9 and 11.

4.3 ENTERPRISE GRID ALLIANCE (EGA)

The Enterprise Grid Alliance (EGA) is a consortium of Grid vendors and customers who are concerned with the many facets of Grid adoption within an enterprise. EGA operates in an open, independent, vendor-neutral fashion. EGA has published a reference model for enterprise Grids [13].

4.4 ORGANIZATION FOR THE ADVANCEMENT OF STRUCTURED INFORMATION STANDARDS (OASIS)

Established in 1993, the Organization for the Advancement of Structured Information Standards (OASIS) is a not-for-profit, international consortium devoted to develop and foster adoption of e-business standards. OASIS has adopted an atypical membership-oriented organization – one in which member companies gain different access to the process while leadership is based on individual merit and is not a result of financial contributions, corporate standing, or special appointment. With a constituency largely drawn from the industry, the process has been quite effective in reaching consensus.

Of special relevance to Grid communities is that OASIS has undertaken the development of a an extension to new Web Services architecture called Web Services Resource Framework (WSRF) [14]. WSRF complements the Web Services production with the capability to handle stateful services within the Web Services framework. With WSRF, OASIS continues the work that used to the focus of the GGF's OGSI development efforts [15], while basing these activities on more of a mainstream set of standard development activities within the larger Web Services ecosystem. Similarly, the specifications known as WS-Notification are extensions that implement asynchronous messaging with publisher/subscriber semantics [16]. With regard to networks and Grids, WSRF and WS-Notification are seen as enabling technology to represent legacy network resources (which are known to be stateful) as Web Services, and to elevate network resources to become Grid-managed resources.

Grids have been described as having a workflow-oriented style of operation. OASIS's work [17] on Business Process Execution Language for Web Services (BPEL4WS) supplies a key foundation to workflow handling within Web Services. With BPEL4WS, a programmer describes a distributed process occurring across the web, with different entities acting upon it in time and space according to the specified flow. The programmer's description of a workflow is thought to be general enough to apply to supply chain processes as well as Grid constructs. Conforming to the overall Web Services philosophy, the BPELWS program treats the constituent Web Services as "black boxes", without any inference on their internal structuring.

The Web Services Reliability 1.1 specification [18] has a high degree of significance in networked environments. With Web Services, reliability is often oversimplified as an obvious consequence of using the TCP protocol as the transport (OSI layer 4) protocol of choice. While TCP is indeed commonly used in Web Services messaging, it is key to recall that Web Services admit SOAP intermediaries that terminate a transport protocol session (be it TCP or other) and initiate a new one (be it TCP or other). The Web Services Reliability 1.1 specification supports end-to-end reliability via SOAP header extensions that are independent of the underlying transport protocol and their composition over the whole end-to-end extent.

4.5 WORLD WIDE WEB CONSORTIUM (W3C)

The W3C develops interoperable technologies [19] – specifications, guidelines, software, and tools – that further the vision of world-wide web related to standards such as Web Services and Semantic Web. W3C develops specifications such as XML [20], SOAP, and Web Services Description Language (WSDL), which are the foundation for Grid toolkits (as well as other applications of Web Services).

With direct implication to networks, this organization has discussed alternate XML notations, in which human readability is traded for performances. The SOAP and XML technologies have in fact been widely criticized for the large amount of data packets issued and their verbosity. Specifically, the XML Binary Characterization [21] Working Group has gathered information on use cases and the feasibility, compatibility challenges inherent to a Binary XML roll-out.

4.6 THE IPSPHERE FORUM

In IPSphere [22], network vendors and providers have converged to define an intercarrier automation tool for policy, provisioning, and billing settlement. The effort – which started as the "Infranet Initiative" by Juniper Inc. – is seeking methods that can resolve providers' long-standing difficulty of obtaining financial returns from QoS service classes stretching across multiple providers. The outcome of IPsphere can be thought of as adding a business layer around the IP stack and the many mechanisms to implement QoS. For this goal, the SOA "schema and contract" style of interoperability may be a possible implementation method. The IPsphere vision would allow a virtual organization, for instance, to build its Grid upon a premium highly deterministic network service across multiple providers for a monthly subscription fee.

4.7 MPI FORUM

In 1994, the Message Passing Interface (MPI) Forum [23] released the first version of the Message Passing Interface (MPI) specification. The MPI's goal is to give developers a standard way to build message-passing programs, with portability and efficiency being two paramount targets. MPI-2 [24] is the latest specification by the Forum. There exist freely available, portable implementations of the MPI specification series, like the MPICH [25] toolkit from Argonne National Labs. Several Grid applications are built upon MPI constructs. MPI and Web Services are at the opposite ends of the message-passing spectrum. MPI embodies the high-performance end while Web Services aptly represent the large-scale interoperability end.

4.8 INTERNET ENGINEERING TASK FORCE (IETF)

The Internet Engineering Task Force (IETF) is the primary standards body for the Internet, although that designation has no "official" legal sanction, e.g., from a government charter or agreement endorsed by international treaty. The IETF is focused on many concerns related to the standardization of network protocols that are part of layer 3 of the OSI stack and which are generally defined as "Internet technologies." However, the IETF does address protocols and issues that relate to other layers, and has active cooperative relationships with other standards bodies that directly address issues at those other layers. The IETF standards are published as a series of Request For Comment (RFC) documents. Best suggested practices are published as "Best Common Practices' documents. Determinations are advanced through general agreements and open discussion rather than votes – based on the principle of "rough consensus and running code."

The IETF is supported by the Internet Society (ISOC), a nonprofit, membership organization that fosters the expansion of the Internet and, among other activities, provides financial and legal support to the IETF (and its companion Internet Research Task Force).

However, the IETF is a group of individuals operating without a formal incorporated membership structure (there are no officers or directors – individuals join mailing lists and pay registration fees for the plenary meetings). It is an international open forum ("open to any interested individual"), with members from multiple communities: academic and government researchers, Internet service providers, equipment designers and developers, and others. The IETF's governance is provided by the IETF Chair, the Internet Architecture Board (IAB), the Internet Engineering Steering Group (IESG), the Area Directors, and the Chairs for working groups and research groups. There are approximately just over 100 working groups.

The increasingly ubiquitous use of IP-related technologies has resulted in the IETF extending its scope to selected sub-IP areas (e.g., GMPLS [26], Common Control And Measurement Plane (CCAMP) [36]) where IP technologies are still applicable (even though this may be layer 2 rather than layer 3). A key insight into IETF operations is that the IETF does not distinguish between IP-based protocols that are meant for the Internet and those that are used within an intranet only. This posture is rooted on the tenet that "if a system can be attached to Internet, one day it will". As such, the reasons of portability, scalability, and security are known to dominate over the performance ones. The following activities (either recently terminated or still under way) are of special interest to Grid communities:

- Remote Direct Data Placement (RDDP) [27] speaks to the opportunity to minimize the costs of data copies and reduce CPU processing when receiving data at sustained high rates.
- High-speed TCP [28] introduces experimental enhancements to the TCP congestion control algorithm. It has direct applicability to the connections with high-bandwidth-delay products that Grid advanced communities (especially in e-Science) typically experiment with.
- The IPv6-related working groups continue to assist in the delicate transition from an IPv4-only world to a hybrid IPv4–IPv6 one. IPv6 is integral part of the vision of nodes joining and leaving the virtual organization of a Grid without addressing collisions and dualisms over native versus NAT-ed IPv4 addresses.
- The Layer 3 Virtual Private Networks Working Group [29] addresses provider-provisioned VPN solutions taking place at layer 3. In such VPNs, the provider

edge device (regardless of whether it is physically deployed at the provider edge or the customer edge) decides how to route the VPN traffic based upon the IP and/or MPLS headers produced at the customer edge.

- The Layer 2 Virtual Private Networks Working Group [30] concerns itself with provider-provisioned solutions wherein layer 2 networks (e.g., a LAN) are seamlessly extended across an IP network or an MPLS-enabled IP network.
- The Layer 1 Virtual Private Network Working Group [31] is chartered to develop specifications of protocols that realize the architecture and requirements set out in the specification of the International Telecommunication Unit Telecommunication Standardization Sector (ITU-T), within a GMPLS-enabled transport service provider network.
- The Multiprotocol Label Switching Working Group [32] is concerned with the various ramifications of label switching as it applies to a number of link-level technologies, with the manifold objectives of performances, traffic engineering, and isolation (e.g. VPNs), all of which appeal to Grid extensions over the network.
- Multicast-related working groups [e.g., 33–35] provide key extensions to IP multicast in the areas of membership, routing, and security. IP multicast holds potential for effective and efficient dissemination of data to members of a virtual organization.
- The Common Control and Measurement Plane (CCAMP) Working Group [36] is working on standards related to GMPLS and the Link Monitoring Protocol (LMP). These signaling protocols are the foundation for control plane technologies that are capable to produce dynamic behaviors (in provisioning, restoration, etc.). This group is due to coordinate its activities with the ITU-T.
- The working groups in the security area provide key building blocks to the Grid security roadmap. The Internet Protocol Security (IPsec) Working Group [37] and the Transport Layer Security (TLS) Working Group [38] develop extensions to the IPsec and TLS protocol suites featured in client-based and clientless VPNs. The Public Key Infrastructure Working Group [39] continues to populate the Public Key Infrastructure (PKI) ecosystem based on X.509 certificates. The Generic Security Services API Next Generation Working Group [40] evolves the generic security services API, which has been shown to play a key role in Globus and other Grid toolkits.

4.9 DISTRIBUTED MANAGEMENT TASK FORCE (DMTF)

The Distributed Management Task Force (DMTF) is a membership-oriented industry organization chartered to provide common management infrastructure components for instrumentation, control, and communication, while keeping it platform independent and technology neutral. To this extent, DMTF has introduced the Web Based Enterprise Management suite (WBEM) suite [41]. Of particular relevance, the popular Common Information Model (CIM) [42] provides a common and structured definition of management information for systems, networks, applications, and services. Initially aimed at IP-capable network elements and networks, the CIM approach has proven to be a powerful technology of broad applicability. For instance, it is being extended to storage elements [43] and servers [44]. With the proper supporting schemas, CIM can be applied to services and Grid-managed services. Within GGF, the OGSA group is investigating the use of CIM as a virtualization foundation technology for its SOA. With its Utility Computing Working Group [45], the DMTF has taken the noteworthy step of providing those advancements in management technologies – security, dependability, dynamic extensibility – that are motivated by the utility computing vision.

4.10 INTERNATIONAL TELECOMMUNICATION UNION (ITU-T)

The International Telecommunication Union (ITU) governs a large body of standards and normative procedures. Established in 1865, the ITU is now a specialized agency of United Nations. The ITU is organized into three sectors: Telecommunication Standardization (ITU-T), Radio-communication (ITU-R), and Telecommunication Development (ITU-D). The ITU-T was formerly known as the CCITT. The ITU-T is a membership-oriented organization, with members at the state level (i.e., sovereign nations) and sector level (corporations and institutes). Bearing high relevance to Grid users, the ITU-T is known for having defined the Synchronous Digital Hierarchy (SDH) and many physical layer standards (e.g., G.652, G.709). The following work items in ITU-T present Grid communities with new opportunities:

- The Automatically Switched Optical Network (ASON) [46] documents specify a meshed optical network that allows dynamic establishment and repair of layer 1 sessions.
- The Generic Framing Procedure (GFP) [47], the Virtual Concatenation (VCAT) [48], and the Link Capacity Adjustment Schema (LCAS) [49] specifications address complementary facets of the adaptation between data traffic (e.g., Ethernet) and network infrastructure that was designed specifically for voice, e.g., the Synchronous Optical Network (SONET)/Synchronous Digital Hierarchy (SDH). These specifications are often grouped under the "next-generation SONET" denomination. Specifically, G.7041 [47] is a joint specification by the American National Standards Institute (ANSI) and ITU-T which indicates how Ethernet frames are encapsulated in the octet-synchronous SONET/SDH (as well as other non-SONET/SDH framing techniques for DWDM). G.707 [48] specifies how to map a datastream into two or more SONET/SDH circuits matching the capacity of the original datastream. G.7042 [49] specifies how to dynamically provision such logical group of SONET/SDH circuits.
- The L1 VPN draft document [50] describes the scenario of a customer receiving a point-to-multipoint layer 1 service from a provider. This work-in-progress holds great potential when matched to the Grid's notion of virtual organization and the merits of circuit-oriented communication for high-performance data-intensive operations;
- Study Group 13 leads, among activities, the architecture-level processes related to the Next-Generation Network (NGN) [51]. NGN is a packet-based network capable

of providing multiple services, including traditional telecommunication services, to static and nomadic users. Service provisioning is independent of underlying transport-related technologies. The notions of Grids and virtual organizations can benefit from such service demarcation and its adoption by telecom providers, resulting in new services catering to the Grid audiences.

- Still under Study Group 13, the Open Communications Architecture Forum (OCAF) [52] is concerned with the specification of common off-the-shelf component categories resulting in a Carrier Grade Open Environment (CGOE) that is capable of accelerating the deployment of NGN infrastructure and services.
- Study Group 11's mandate includes the definition of the signaling requirements and protocols that are needed to reduce the NGN vision to practice.
- Study Group 4's mandate includes the management of telecommunication services, networks, and equipment, as they evolve toward a NGN reality.
- Study Group 16's mandate includes specifications that reach out to the end-user, by way of multimedia end-systems, with significance to collaborative Grid environments such as AccessGrid.

4.11 OPTICAL INTERNETWORKING FORUM (OIF)

The Optical Internetworking Forum is an industry group chartered to accelerate the adoption of interoperable optical networking technologies. The OIF's constituency is membership oriented and features equipment manufacturers, telecom service providers, and end-users. OIF members include representatives from both the data and optical network areas.

The User Network Interfaces (UNI) Signaling 1.0 specification [53] and the External Network Network Interfaces (E-NNI) Signaling specification [54] are two implementation agreements that yield service-oriented interfaces with which a consumer (or a network) demands and receives service from a provider. When used in conjunction with an agile network such as the ITU-T ASON, the consumer can negotiate the dynamic establishment of a SONET/SDH framed service or an Ethernet service or a layer 1 service. Early experimentation with high-capacity networks for Grids has shown the appeal of Grid network services "dialing" a connection on-demand by way of an OIF UNI service interface.

Furthermore, the OIF has organized interoperability demonstrations [55]. With major participation by providers and network vendors, such events have proven very useful to publicize and improve networks that are predicated upon agile layer 2/1 services, and thus capable of satisfying large capacity Grid users among other communities.

4.12 INFINIBAND TRADE ASSOCIATION (IBTA)

Founded in 1999, the Infiniband Trade Association [56] has produced the specification for Infiniband – a technology for serial I/O and switched fabric interconnect capable to deliver high bandwidth and low latency to clusters of servers within a data center. Infiniband has demonstrated a signaling rate of 30 Gbps, while disclosing that efforts are under way to advance the rate to 120 Gbps. Infiniband is envisioned to play a role in the intra-datacenter portion of a Grid network.

4.13 INSTITUTE OF ELECTRICAL AND ELECTRONICS ENGINEERS (IEEE)

The Institute of Electrical and Electronics Engineers (IEEE) provides scientific, education, and professional services to a large world-wide community. Also, the IEEE operates a standard development program, with participants having either corporate or individual membership. The ballot process is a key event in the development and maturing of standards at the IEEE.

The scope of IEEE's standardization effort is extremely large, and it continues to expand.

Its scope comprises LAN and MAN standards including Ethernet, token ring, wireless LAN, bridged/virtual LANs, mainly within layer 1 and layer 2 realms. It can be argued that all of the layer 2 and layer 1 technologies standardized by the IEEE are applicable to Grid deployments over networks. The recent documents of three working groups are worth emphasizing, as they bear the special significance of enabling technologies to Grids. Later chapters will describe other IEEE standards activities

- The 802.3ae Working Group has produced the 802.3ae-2002 standards for 10 Gbps Ethernet [2]. These have evolved the popular, cost-effective Ethernet technology to the rates seen in the WAN. This standard features two distinct PHYs, the LAN-PHY for nominal 10 Gbps rates and the WAN-PHY for data rate compatible with the payload rate of OC-192c/SDH VC-4-64c. While WAN compatibility has proven enormously useful in large-scale Grid experimentation, the conversion between LAN-PHYs and WAN-PHYs has often created operational challenges.

- The 802.1ad project ("Provider Bridges" [57]) and the 802.1ah project ("Backbone Provider Bridges" [58]) represent a concerted effort to overcome the hard limit on the service definitions in a single VLAN tag (defined in 802.1Q). 802.1ad – often nicknamed "Q in Q" – describes how to stack virtual LAN (vLAN) tags for increased customer separation and differentiated treatment throughout a service provider cloud. 802.1ah – often nicknamed "Mac in Mac"– encapsulates an Ethernet frame into a whole service provider Media Access Control (MAC) header, thus providing a complementary way to scale across large provider clouds.

- The 802.16 working group [59] is concerned with broadband wireless access standards, which is often labeled "WiMAX." WiMAX has been considered a potential means to overcome last-mile connectivity limitations of cable and DSL, yet without compromising throughput (hence the broadband denomination). In turn, WiMAX may allow Grid manifestations to connect users without drops in throughput or pronounced downstream/upstream asymmetry.

REFERENCES

[1] SOAP, The XML Protocol Working Group, the World Wide Web Consortium, http://www.w3.org/2000/xp/Group/.
[2] P802.3ae 10Gb/s Ethernet Task Force, Institute of Electrical and Electronics Engineers, http://grouper.ieee.org/groups/802/3/ae/index.html.
[3] I. Foster, H. Kishimoto, A. Savva, D. Berry, A. Djaoui, A. Grimshaw, B. Horn, F. Maciel, F. Siebenlist, R. Subramaniam, J. Treadwell, and J. Von Reich (2005) "Open grid Services Architecture Version 1.0," Grid Forum Document, 30, January.
[4] A. Andrieux, K. Czajkowski, A. Dan, K. Keahey, H. Ludwig, J. Pruyne, J. Rofrano, S. Tuecke, and M. Xu, "Web Services Agreement Specification (WS-Agreement)," Grid Working Document, revisions available for download at https://forge.gridforum.org/projects/graap-wg.
[5] W. Allcock (2003) "GridFTP: Protocol Extensions to FTP for the Grid," Grid Forum Document, 20, April.
[6] J. Postel and J. Reynolds (1985) "The File Transfer Protocol," RFC 959, Network Information Center, October 1985.
[7] B. Lowekamp, B. Tierney, L. Cottrell, R. Hughes-Jones, T. Kielmann, aand M. Swany (2004) "A Hierarchy of Network Performance Characteristics for Grid Applications and Services," Grid Forum Document No. 23, May.
[8] D. Simeonidou, R. Nejabati, B. St. Arnaud, M. Beck, P. Clarke, D. B. Hoang, D. Hutchison, G. Karmous-Edwards, T. Lavian, J. Leigh, J. Mambretti, V. Sander, J. Strand, and F. Travostino (2004) "Optical Network Infrastructure for Grid," Grid Forum Document, No. 36, August.
[9] V. Sander (ed.) (2004) "Networking Issues for Grid Infrastructure," Grid Forum Document, No. 37, November.
[10] T. Ferrari (ed.), "Grid Network Services Use Cases," Grid Working Document, revisions available for download at https://forge.gridforum.org/projects/ghpn-rg.
[11] M. Goutelle, Y. Gu, E. He, S. Hegde, R. Kettimuthu, J. Leigh, P. Vicat-Blanc Primet, M. Welzl, and C. Xiong (2005) "Survey of Transport Protocols Other Than Standard TCP," Grid Forum Document, No. 55, November 2005.
[12] The Telecom Community Group, Global Grid Forum, https://forge.gridforum.org/projects/telco-cg.
[13] Enterprise grid Alliance's Reference Model 1.0, Enterprise Grid Alliance, http://www.gridalliance.org/en/WorkGroups/ReferenceModel.asp.
[14] The Web Services Resource Framework (WSRF) Technical Committee, Organization for the Advancement of Structured Information Standards, http://www.oasis-open.org/committees/tc_home.php?wg_abbrev=wsrf.
[15] S. Tuecke, K. Czajkowski, I. Foster, J. Frey, S. Graham, C. Kesselman, T. Maquire, T. Sandholm, D. Snelling, and P. Vanderbilt (2003) "Open grid Services Infrastructure," Grid Forum Document No. 15, June.
[16] Web Services Notification Technical Committee, Organization for the Advancement of Structured Information Standards, http://www.oasis-open.org/committees/tc_home.php?wg_abbrev=wsn.
[17] The Web Services Business Process Execution Language (BPEL) Technical Committee, Organization for the Advancement of Structured Information Standards, http://www.oasis-open.org/committees/wsbpel/charter.php.
[18] The Web Services Reliable Messaging (WSRM) Technical Committee, Organization for the Advancement of Structured Information Standards, http://www.oasis-open.org/committees/tc_home.php?wg_abbrev=wsrm.
[19] The World Wide Web Consortium, http://www.w3.org/.

[20] eXtensible Markup Language, The World Wide Web Consortium, http://www.w3.org/TR/REC-xml/.
[21] The XML Binary Characterization Working Group, World Wide Web Consortium, http://www.w3.org/XML/Binary/.
[22] The IPsphere Forum, http://www.ipsphereforum.org/home.
[23] The Message Passing Interface Forum, http://www.mpi-forum.org/.
[24] MPI-2: Extensions to the Message-Passing Interface, The Message Passing Interface Forum, July 1997, http://www.mpi-forum.org/docs/mpi-20-html/mpi2-report.html.
[25] MPICH-A Portable Implementation of MPI, Argonne National Laboratory and Mississippi State University, http://www-unix.mcs.anl.gov/mpi/mpich/.
[26] E. Mannie (ed.) (2004) "Generalized Multi-Protocol Label Switching (GMPLS) Architecture, RFC 3945," RFC 3945, Network Information Center, October 2004.
[27] The Remote Direct Data Placement Working Group, Internet Engineering Task Force, http://www.ietf.org/html.charters/rddp-charter.html.
[28] S. Floyd (2003) "HighSpeed TCP for Large Congestion Windows" RFC 3649, Network Information Center, December 2003.
[29] The Layer 3 Virtual Private Networks Working Group, Internet Engineering Task Force, http://www.ietf.org/html.charters/l3vpn-charter.html.
[30] The Layer 2 Virtual Private Networks Working Group, Internet Engineering Task Force, http://www.ietf.org/html.charters/l2vpn-charter.html.
[31] The Layer 1 Virtual Private Networks Working Group, Internet Engineering Task Force, http://www.ietf.org/html.charters/l1vpn-charter.html.
[32] The Multiprotocol Label Switching Working Group, Internet Engineering Task Force, http://www.ietf.org/html.charters/mpls-charter.html.
[33] The Multicast and Anycast Group Membership (MAGMA) Working Group, Internet Engineering Task Force, http://www.ietf.org/html.charters/magma-charter.html.
[34] The Protocol Independent Multicast (PIM) Working Group, Internet Engineering Task Force, http://www.ietf.org/html.charters/pim-charter.html.
[35] The Multicast Security Working Group, Internet Engineering Task Force, http://www.ietf.org/html.charters/msec-charter.html.
[36] The Common Control and Measurement Plane Working Group, Internet Engineering Task Force, http://www.ietf.org/html.charters/ccamp-charter.html.
[37] The Internet Protocol Security Working Group, Internet Engineering Task Force, http://www.ietf.org/html.charters/ipsec-charter.html.
[38] The Transport Layer Security Working Group, Internet Engineering Task Force, http://www.ietf.org/html.charters/tls-charter.html.
[39] The Public Key Infrastructure Working Group, Internet Engineering Task Force, http://www.ietf.org/html.charters/pkix-charter.html.
[40] The Generic Security Services API Next Generation Working Group, Internet Engineering Task Force, http://www.ietf.org/html.charters/kitten-charter.html.
[41] Web-based Enterprise Management, The Distributed Management Task Force (DMTF), http://www.dmtf.org/about/faq/wbem/.
[42] The Common Information Model (CIM) Schema Version 2.9.0, The Distributed Management Task Force, http://www.dmtf.org/standards/cim/cim_schema_v29.
[43] The Storage Management Initiative, Storage Networking Industry Association, http://www.snia.org/smi/home.
[44] The Server Management Working Group, Distributed Management Task Force, http://www.dmtf.org/about/committees/ServerManagement_WGCharter.pdf.
[45] The Utility Computing Working Group, Distributed Management Task Force, http://www.dmtf.org/about/committees/UtilityComputingWGCharter.pdf.

References

[46] Recommendations on Automatically Switched Transport Network (ASTN) and Automatically Switched Optical Network (ASON), International Telecommunication Union http://www.itu.int/ITU-T/2001–2004/com15/otn/astn-control.html.

[47] ITU-T (2001) "Generic Framing Procedure (GFP)," Recommendation G.7041/Y.1303, October 2001.

[48] ITU-T (2000) "Network Node Interface for the Synchronous Digital Hierarchy," Recommendation G.707, October 2000.

[49] ITU-T (2004) "Link Capacity Adjustment Scheme (LCAS) for Virtual Concatenated Signals," Recommendation G.7042/Y.1305, February 2004.

[50] "Layer 1 Virtual Private Network Generic Requirements and Architecture Elements," Study Group 13, International Telecommunication Union, August 2003.

[51] The Focus Group on Next Generation Networks, International Telecommunication Union, www.itu.int/ITU-T/ngn/fgngn/index.html.

[52] The Open Communications Architecture Forum Focus Group, International Telecommunication Union, www.itu.int/ITU-T/ocaf/.

[53] "The User Network Interface (UNI) 1.0 Signalling Specification, Release 2," an Implementation Agreement Created and Approved by the Optical Internetworking Forum, February 2004, http://www.oiforum.com/public/documents/OIF-UNI-01.0-R2-Common.pdf.

[54] "Intra-Carrier E-NNI Signalling Specification", an Implementation Agreement Created and Approved by the Optical Internetworking Forum, February 2004, http://www.oiforum.com/public/documents/OIF-E-NNI-Sig-01.0-rev1.pdf.

[55] OIF Worldwide Interoperability Demonstration 2005, Optical Internetworking Forum, http://www.oiforum.com/public/supercomm_2005v1.html.

[56] The Infiniband Trade Association, http://www.infinibandta.org/home.

[57] 802.1ad Provider Bridges, Institute of Electrical and Electronics Engineers, http://www.ieee802.org/1/pages/802.1ad.html.

[58] 802.1ah Backbone Provider Bridges, Institute of Electrical and Electronics Engineers, http://www.ieee802.org/1/pages/802.1ah.html.

[59] 802.16 Working Group on Broadband Wireless Access, Institute of Electrical and Electronics Engineers, http://grouper.ieee.org/groups/802/16/.

Chapter 5

Grid Network Services and Implications for Network Service Design

Joe Mambretti, Bill St. Arnaud, Tom DeFanti, Maxine Brown, and Kees Neggers

5.1 INTRODUCTION

The initial chapters of this book introduce Grid attributes and discuss how those attributes are inherent in Grid architectural design. Those chapters describe the benefits of designing multiple resources within a Grid services framework as addressable modules to allow for versatile functionality. This approach can provide for both a suite of directly usable capabilities and also options for customization so that infrastructure resources can be accessed and adjusted to match the precise requirements of applications and services.

These chapters also note that until recently this multifaceted flexibility has not been extended to Grid networks. However, new methods and architectural standards are being created that are beginning to integrate network services into Grid environments and to allow for more versatility among network services. Chapter 3 explains that the SOA used for general Grid resources are also being used to abstract network capabilities from underlying infrastructure. This architecture can be expressed in

---

Grid Networks: Enabling Grids with Advanced Communication Technology Franco Travostino, Joe Mambretti, Gigi Karmous-Edwards © 2006 John Wiley & Sons, Ltd

intermediate software that can provide for significantly more capability, flexibility, and adjustability than is possible on today's networks.

This chapter presents additional topics related to the basic requirements and architectural design of Grid network services. The design of Grid network services architecture currently is still at its initial stages. The development of this architecture is being influenced by multiple considerations, including those related to technology innovation, operational requirements, resource utilization efficiencies, and the need to create fundamentally new capabilities. This chapter discusses some of the considerations related to that emerging design, including functional requirements, network process components, and network services integration.

5.2 TRADITIONAL COMMUNICATIONS SERVICES ARCHITECTURE

Traditional architectures for communication services, network infrastructure, and exchange facilities have been based on designs that were created to optimize network resources for the delivery of analog-based services, based on a foundation of core transport services. Such network infrastructure supported only a limited range of precisely defined services with a small, fixed set of attributes. The services have been fairly static because they have been based on an inflexible infrastructure, which usually required changes through physical provisioning. Such networks have also been managed through centralized, layered systems.

Traditional communications models assume that services will be deployed on a fixed hierarchical stack of layered resources, within an opaque carrier cloud, through which "managed services" are provided. Providing new services, enhancing or expanding existing services, and customizing services is difficult, costly, and restrictive. Dedicated channel services, such as VPNs, are generally allowed only within single domains. Private, autonomous interconnections across domains are not possible. The quality of the services on these channels and their general attributes are not flexible and cannot be addressed or adjusted by external signaling.

Today's Internet is deployed primarily as an overlay network on this legacy infrastructure. The Internet has made possible a level of abstraction that has led to a significantly more versatile communications services environment, and Grid network services are being designed to enhance the flexibility of that environment.

5.3 GRID ARCHITECTURE AS A SERVICE PLATFORM

In contrast to traditional telecommunication services, Grid environments can be designed to provide an almost unlimited number of services. A Grid is a flexible infrastructure that can be used to provide a single defined service, a set of defined services, or a range of capabilities or functions, from which it is possible for external entities to create their own services. In addition, within those environments, processes can request that basic infrastructure and topologies be changed dynamically – even as a continuous process.

Just as the term "Grid" is analogous to the electric power system, a Grid service has been described as being somewhat analogous to the services provided by electrical utilities. Multiple devices can attach to the end of an electrical power grid, and they can use the basic services provided by that common infrastructure for different functions. However, the electrical power grid, like almost all previous infrastructure, has been designed, developed, and implemented specifically to provide a single defined service or a limited set of services.

Previous chapters note that the majority of Grid services development initiatives have been oriented to applications, system processes, and computer and storage infrastructure – not to network services. Although network services have always been an essential part of Grid environments, they have been implemented as static, undifferentiated, and nondeterministic packet-routed services – rarely as reconfigurable, controllable, definable, deterministic services.

Recently, research and development projects have been adapting Grid concepts to network resources, especially to techniques for services abstraction and virtualization. These methods are allowing network resources to be "full participants" within Grid environments – accessible, reconfigurable resources that can be fully integrated with other Grid resources.

For example, with the advent of Grid Web Services described in Chapter 3, the constituent components of a network from the physical to the application layer can be represented as an abstraction layer that can fully interact with other Grid services on a peer-to-peer basis, rather than traditional hierarchical linkages in a stack as is now common with typical telecommunication applications. This approach represents a major new direction in network services provisioning, a fundamentally new way to create and implement such services. It does not merely provide a path to additional access to network services and methods of manipulating lower level resources functionality, it also provides an extensive toolkit that can be used to create complete suites of new networks services.

5.3.1 GRID NETWORK SERVICES ARCHITECTURE

The Grid standards development community has adopted a general framework for a SOA based on emerging industry standards, described in Chapter 4. This architectural framework enables the efficient design and creation of Grid-based services by providing mechanisms to create and implement Grid service processes, comprising multiple modular processes that can be gathered and implemented into a new functioning service whose sum is greater than the parts. Such standards-based techniques can be used to create multiple extensible integrated Grid-based services, which can be easily expanded and enhanced over the services' lifetime. In addition, this architecture enables these modular services to be directly integrated to create new types of services.

This architecture provides for several key components, which are described in Chapter 7. The higher level of service, and the highest level of services abstraction, consists of capabilities or functions that are made available through advertisements through a standard, open communication process. These high-level processes interact with intermediate software components between that top layer and core facilities and resources. The core facilities and resources can consist of almost any information technology object, including any one of a wide array of network services and other resources.

This infrastructure is currently being developed, and as it is implemented it is becoming clear that this new services approach will manifest itself in many forms. In some cases, organizations will focus on providing only end-delivered services, and rely on using Grid services provided by other organizations. In other cases, organizations will focus on providing mid-level services to those types of organizations, while perhaps relying on Grid infrastructure providers for core resources. Other organizations may provide only basic infrastructure resources. However, this new model enables any organization to access and provide capabilities at any level.

As a type of Grid service, individual network resources can become modular objects that could be exposed to any legitimate Grid process as an available, usable service. In general, these resources will probably be advertised to mid-level services rather than to edge processes, although that capability also remains an option. As part of a Grid services process or service workflow procedure, network resources can be directly integrated with any other type of Grid service, including those that are not network related. Consequently, multiple network resource objects, advertised as services, can be gathered, integrated, and utilized in virtually almost unlimited numbers of ways. They can be combined with other types of Grid objects in ad hoc integrated collections in order to create specialized communication services on demand. All resource elements become equal peers that can be directly addressable by Grid processes.

Grid services-oriented architecture provides capabilities for external processes, on a peer-to-peer basis, to provision, manage, and control customized network services directly – without any artificial restrictions imposed from centralized networking management authorities, from server-based centralized controls, or from hierarchical layering. This design allows multiple disparate distributed resources to be utilized as equal peers, which can be advertised as available services that can be directly discovered, addressed, and used.

This architecture allows different types of services, including highly specialized services, to co-exist within the same core, or foundation, infrastructure, even end-to-end across multiple domains. This approach significantly increases capabilities for creating and deploying new and enhanced services, while also ensuring cost effectiveness through infrastructure sharing. This approach can incorporate traditional distributed management and control planes, e.g., as exposed resources within a services-oriented architecture, or it can completely eliminate traditional control and management functions.

5.4 NETWORK SERVICES ARCHITECTURE: AN OVERVIEW

5.4.1 SERVICES ARCHITECTURE BENEFITS

Grid network services architecture provides multiple benefits. It supports a wider range of communication services, and it allows those services to have more attributes than traditional communication services. This architecture can be implemented

5.4 Network Services Architecture: An Overview

to expand services offerings, because basic individual resource elements can be combined in almost limitless ways. With the implementation of stateful services, or using workflow languages that maintain state, multiple network resources can be treated as individual components that can be used in any form or combination as required by external services and applications, thereby precisely matching application needs to available resources.

A major advantage of this architecture is that it is more flexible and adaptive than traditional networks. This flexibility can be used to make communication services and networks more "intelligent," for example by enabling an applications web service to be bound to a network web service, thereby enabling the combined service to be more "context aware." Using this model, applications can even be directly integrated into network services, such that there is no distinction between the application and the network service. This architecture also provides integrated capabilities at all traditional network layers, not just individual layers, eliminating dependencies on hierarchical protocol stacks. Also, it provides for enhanced network scalability, including across domains, and for expandability, for example by allowing new services and technologies to be easily integrated into the network infrastructure.

Processes external to the network can use these core component as resources in multiple varieties of configurations. Applications, users, infrastructure processes, and integrated services, all can be integrated with network service and resources in highly novel configurations. These external processes can even directly address core network resources, such as lightpaths and optical elements, which to date have not been accessible through traditional services. This approach does not simply provide access to lower level functionalities, but it also enables full integration of higher level services with those functionalities, in part by removing traditional concepts of hierarchical layers.

As Chapter 3 indicates, the concept of layers and planes has been a useful abstraction to classify sets of common network functions. The OSI layer model [1] depicted in Figure 3.2 is an artifact, designed to address such tasks as the limitations of buffering and of managing different types of telecommunication transport services. However, the Grid network services approach departs from the traditional vertical model of services provided through separate OSI network layers. The concept of a "stack" of layers from the physical through to the application largely disappears in the world of Grid services. This concept is also gaining acceptance by communications standards bodies, as noted in Chapter 14, including the ITU, which produced a future directions document indicating that standard model may not be carried forward into future designs [2]. Although this architectural direction may engender some complexity for provisioning, it will also result in multiple benefits.

Similarly, traditional network services incorporate concepts of management planes, control planes, and data planes, which are architectures that define specific, standardized sets of compartmentalized functionality. Because Grid network services architecture includes basic definitions of the set of Grid network services functions, it would be possible to extend this approach to also incorporate a concept of a "Grid network services plane." However, while convenient, this notion of a "plane" would obscure a fundamental premise behind the Grid network services architecture, which is being designed such that it is not limited to a set of functions within a traditionally defined "plane"; instead it provides a superset of all of these functionalities, incorporating all traditional functions within a broad standard shared-use set of capabilities.

Another advantage of implementing Grid network resources within a SOA is that it provides for a transition path from traditional communications infrastructure. The enhanced levels of abstraction and virtualization provided through Grid network services architecture can be used as a migration path from limited legacy infrastructure toward one that can offer a much wider and more powerful set of capabilities, from centrally managed processes with hierarchical controls to highly distributed processes.

5.5 GRID NETWORK SERVICES IMPLICATIONS

Within a Grid network services environment, it is possible to accept either a predefined default service or to highly customize individualized network services. Network services, core components, specialized resources such as layer 3 services with customized attributes, dedicated layer 2 channels, reconfigurable cross-connections, and even lightpaths and individual physical network elements, such as ports, can be identified and partitioned into novel integrated services. These services can be provided with secure access mechanisms that enable organizations, individuals, communities, or applications to discover, interlink, and utilize these resources. For example, using this architecture, end-users and applications can provision end-to-end services, temporarily or permanently, at any individual level or at multiple levels.

Because of these attributes, this architecture allows Grid network services to extend from the level of the communications infrastructure directly into the internal processes of other resources, such as computers, storage devices, or instruments. Using techniques based on this architecture, the network can also be extended directly into individual applications, allowing those applications to be closely integrated with network resources. Such integration techniques can be used to create novel communications-based services.

The approach described here provides multiple advantages for Grid environments. However, even when used separately from Grid environments, this approach can be used to provide significantly more functionality, flexibility, and cost efficiency for digital communications services, and it can provide those benefits with much less complexity. These advantages are key objectives in the design of next generation digital communication services, and new architecture that provides for service level abstracts are important methods for achieving those goals.

5.6 GRID NETWORK SERVICES AND NETWORK SERVICES

Among the most important advantages of Grid network services architecture is the ability to match application requirements to communication services to a degree that has not been possible previously. This capability can be realized through network services-oriented APIs that incorporate signaling among Web Services. At a basic level, such a signal could request any number of standard services, either connectionless and connection oriented, e.g., TCP/IP communications, multicast, layer 2 paths, VPNs, or any other common service.

Grid network Web Services can allow for specialized signaling that can be used for instantiating new service types, in accordance with the general approach of Grid architecture. For example, such signaling can enable requests for particular highly defined services through interactions between applications and interfaces to the required network resources. Instead of signaling for standard best effort services, this signal could be a request for a service with a precisely defined level of quality assurance. Through this type of signaling, it is possible to integrate Grid applications with deterministic networking services.

5.6.1 DETERMINISTIC NETWORKING AND DIFFERENTIATED SERVICES

5.6.1.1 Defining and customizing services

Today, almost all Internet services are best effort and nondeterministic. Few capabilities exist for external adjustments for individual service attributes. Specialized, high-quality services have been expensive to implement, highly limited in scalability, and difficult to manage. Particularly problematic is specialized, inter-domain services provisioning. The Internet primarily consists of an overlay network supported by a core network consisting of static, undifferentiated electronic switched paths at the network edge and static optical channels within the network core. Because the current Internet is an overlay network, operating on top of a fixed hierarchical physical infrastructure with minimal interaction between the layer that routes packets (layer 3) and other layers, basic topologies usually cannot be changed dynamically to enhance layer 3 performance, for example by using complementary services from other layers. Consequently, differentiated services have not been widely implemented. They have usually been implemented within LANs or within specialized enterprise networks.

Grid network services architecture can be used to provide determinism in networks. High-level signaling, in conjunction with intermediate software components, can provide for optimized matches between multiple application requirements, which can be expressed as specified deterministic data flows and available network resources. These processes can be based on specialized communications (either in-band or out-of-band) comprising requests for network services signaled into the network, information on the network resources and status signaled by network elements, various performance monitoring and analysis reports, and other data. This architecture allows both link state and stateless protocol implementation, and provides for information propagation channels among core network elements.

5.6.1.2 Quality of service and differentiated services

The need for differentiated services has been recognized since the earliest days of data networks. There have been attempts to create differentiated services at each traditional service level. Many earlier projects focused on signaling for specific Quality of Service (QoS) levels. A number of these initiatives have been formalized through standards bodies, such as the IETF DiffServ efforts, described in Chapters 6 and 8.

Other projects attempted at QoS provisioning at layers 2 and 1. However, because of management, provisioning logistics and cost considerations, these services have not been widely implemented. Currently, almost all Grid services are being supported by undifferentiated, nondeterministic, best effort IP services.

5.6.1.3 Grid network services

Through standard Grid abstraction techniques, individual users or applications (either ad hoc or through scheduling) are able to directly discover, claim, and control network services, including basic network resources. Such services can be standard, such as IP or transport (TCP or User Datagram Protocol (UDP)) or specialized (Stream Control Transmission Protocol, SCTP) [3], or they can be based on layers below layer 3, such as layer 2 paths and light paths. These capabilities can be utilized across multiple domains locally, regionally, nationally, and internationally. Furthermore, they can dynamically change the attributes and configurations of those resources, even at the application level. Grid applications have been demonstrated that can discover and signal for specific types of network services, including by dynamically configuring and reconfiguring lightpaths locally, within metropolitan areas, nationally, and internationally.

Another powerful capability of this network services architecture is that it can provide for a unique capability that allows for a scalable, reliable, comprehensive *integration* of data flows, with various service parameters at all traditional service layers, i.e., layers 1, 2, 3, and 4 and above. Different types of services required by applications with various specified parameters (e.g., stringent security, low latency, minimal jitter, extra redundancy, minimal latency) can be blended dynamically as needed.

This architecture can provide for the incorporation of integrated services at all levels, each with options for various service parameters, layer 3 services (e.g., high-performance IPv4, IPv6, unicast, and multicast), layer 2 services, including large-scale point-to-point layer 2 services, and layer 1 wavelength-based transport, including end-to-end lightpaths, with options for single dedicated wavelengths, multiple wavelengths, and subwavelengths. Dynamically provisioned lightpaths have been demonstrated as a powerful capability whether integrated with layer 3 and layer 2 services or as direct layer 1-based dedicated channels.

5.7 GRID NETWORK SERVICE COMPONENTS

A Grid network service architecture includes various processes that are common to other Grid services, including functions for resource discovery, scheduling, policy-based access control, services management, and performance monitoring. In addition, the architecture includes components that are related specifically to network communication services.

5.7.1 NETWORK SERVICE ADVERTISEMENTS AND OGSA

A key theme for Grid environments is an ability to orchestrate diverse distributed resources. Several standards bodies are designing architecture that can be used for Grid resource orchestration. Many of these emerging standards are described in Chapter 4. Grid network services are being developed within the same framework as other Grid services, e.g., the Open Grid Services Architecture (OGSA), which is being created by the Global Grid Forum (GGF) [4]. The work of the GGF complements that of the OASIS standards group (Organization for the Advancement of Structured Information Standards) [5]. Also, W3C is developing the Web Services Definition Language (WSDL) and the Web Services Resource Framework (WSRF) [6]. These standardized software tools provide a means by which various Grid processes can be abstracted such that they can be integrated with other processes. The GGF has endorsed this architecture as a means of framing Grid service offerings.

5.7.2 WEB SERVICES

In a related effort, OASIS is developing the Web Services Business Process Execution Language (WSBPEL or BPEL4WS). The WSBPEL initiative is designing a standard business process execution language that can be used as a technical foundation for innumerable commercial activities. At this time, there is a debate in the Web Services OGSA community about the best way to support state. The current OGSA approach is to create stateful Web Services. An alternative approach is to keep all Web Services stateless and maintain state within the BPEL. The latter approach is more consistent with recursive object-oriented design.

Although oriented toward business transaction processing and common information exchange, this standard is being developed so that it can be used for virtually any process. The architecture is sufficiently generalized that it can be used for an almost unlimited number of common system processes and protocols, including those related to resource discovery and use, access, interface control, and initiating executable processes.

This model assumes that through a SOA based on WSRF, multiple, highly distributed network resources will be visible through service advertisements. Over time, increasing numbers of these network services and related resources will be exposed as Web Services, e.g., using web tags to describe those services. Using these tools, a Web Services "wrapper" can be placed around a resource, which can then be advertised as a component for potential use by other services within Grid environments. Eventually, some of these resources may contain such Web Services components as an integral part of their basic structure.

5.7.3 WEB SERVICES DEFINITION LANGUAGE (WSDL)

However, if they are to be widely advertised and discovered, a standard mechanism is required, such as a standards-based registry service devoted to supporting Web Services as defined by the W3C standards. The international advanced networking community has established a process, in part through an international organizational partnership, to create WSDL schema that will design supersets of User-to-Network Interface (UNI) functionality, including multiple WSRF stateful elements. The initial instantiations of this model have been designed and implemented, and are being used as early prototypes. The international advanced networking research community is currently creating common XML schema for optical network services, provisioning, and management.

5.7.4 UNIVERSAL DESCRIPTION, DISCOVERY, AND INTEGRATION (UDDI)

As with other types of Web Services, discovery mechanisms can be simple or complex. Efforts have been undertaken that can present Web Services to different communities, at different levels, with different perspectives, for multiple end objectives. The OASIS organization is developing a mechanism called Universal Description, Discovery, and Integration (UDDI), a protocol that is part of the interrelated standards for its Web Services stack. UDDI defines a standard method for publishing and for discovering the network-based software components of a SOA (www.uddi.org). Although this standard is currently commercial process oriented, it can be extended to incorporate other types of processes.

5.7.5 WEB SERVICES-INSPECTION LANGUAGE (WSIL)

A related emerging standard is the Web Services-Inspection Language (WSIL), which specifies an XML format, or "grammar," that can help inspect a site for available services and rules that indicate how the information discovered through that process should be made available for use. A WS-Inspection document provides a method for aggregating references to service description documents in a variety of formats preexisting within a repository created for this purpose. Through this process, inspection documents are made available at a point-of-offering for the service. They can also be made available through references that can be placed within content media, such as HTML. Currently, public search portals are becoming a preferred approach for advertising and consuming Web Services. Keyword searching on a service description or Uniform Resource Identifier (URI) may be as effective as building UDDI or WSIL linkages.

5.7.6 NETWORK SERVICE DESIGN AND DEVELOPMENT TOOLS

The SOA approach described here presents limitless opportunities for communication services design, development, and implementation. Within this environment, services creation can be undertaken by multiple communities and even individuals – research laboratories, community organizations, commercial firms, government agencies, etc. To undertake these development and implementation tasks, such organizations may wish to use common sets of tools and methods. Concepts for these tools and methods are beginning to emerge, including notions of programming languages for network services creation. As these languages are being designed, it is important to consider other developments related to general Grid services.

For example, as noted in Chapter 3, currently, the GGF is developing a Job Submission Description Language (JSDL) [7]. This document specifies the semantics and structure of JSDL, used for computational jobs submitted within Grid environments, and it includes normative XML schema. At this time, this language does not incorporate considerations of network services. However, its overall structure provides a model that can be extended or supplemented to include mechanisms for requesting network services, either through an extension of JSDL or through a related set of specifications.

Currently, commercial Web Services development software tools are available that can be used to create services-oriented communication systems, by constructing specific, customized Grid communication functionality from network resources advertised as services.

5.8 NEW TECHNIQUES FOR GRID NETWORK SERVICES PROVISIONING

5.8.1 FLEXIBLE COMMUNICATION SERVICES PROVISIONING

The architecture described here implies a need for a fundamentally new model for communication services provisioning, one that is highly distributed in all of its aspects. This distributed environment does not resemble the traditional carrier networks, or even a traditional network. As noted, the type of communications services provisioning described here is significantly different from the traditional model of acquiring communications services from a centrally managed authority, delivered through an opaque carrier cloud. Instead, it is based on a wide-area communications facility that provides a collection of advertised resources that can be externally discovered, accessed, and managed. This distributed facility supports a flexible, programmable environment that can be highly customized by external processes. A major benefit of this approach is that it can provide an unlimited number of services – each with different sets of attributes.

5.8.2 PARTITIONABLE NETWORK ENVIRONMENTS

These attributes sharply distinguish this new network environment from traditional communication services and infrastructure. The model presented here is one that provides not merely dedicated services and resources to external processes, e.g., a dedicated VPN or tunnel, but also a full range of capabilities for managing, controlling, and monitoring those resources, even by individual applications. This environment can be partitioned so that each partition can also have its own management and control function, which also can be highly customized for individual requirements. Therefore, packages of distinct capabilities can be integrated into customized end-delivered service suites, which can either expose these capabilities or render them totally transparent.

For example, although this technique can incorporate functions for scheduling and reservations, these capabilities are not required. Therefore, a particular partition does not have to incorporate this capability. There are many types of applications and services that have irregular demands over time and unknown advance requirements. For such applications and services, it is not practical to try to predetermine exact measures of demand and resource utilization. To address such irregular resource demands, one approach could be attempting to implement sophisticated methods for optimization and predication. However, to date these mechanisms have proven to be unreliable and problematic to implement. Another approach that is often used is to overprovision with static resources, a technique that can generate high costs.

An alternative would be to use Grid network services to provide a flexible environment that would automatically and constantly adjust to meet on-going changing demands. Furthermore, this environment, although based on shared infrastructure, could be partitioned so that within each partitioned area subenvironments could be established and customized to meet the needs of applications and services. Within each partition or subpartition a complete set of tools would be provided to enable local customization, including capabilities for adjusting deep within the underlying infrastructure fabric.

A related important point is that Web Services also allow communities or individual users to create integrated network environments, within which they can create integrated heterogeneous network resources from various network service providers. They can create a virtualized homogeneous network entity within which the resource integrator can create new network resource-based Web Services, such as VPNs, QoS partitioning, etc. These services would be independent of the underlying service provided by the original service providers. At this point, Web Services functions do more than provide a means to "lease" a subset of another entities resources. The Web Services/SOA model allows the creation of new services for which the sum is much greater than the individual parts.

5.8.3 SERVICES PROVISIONING AND SIGNALING

One challenge in implementing Grid network services has been the lack of a standard signaling mechanism for network resources by external processes. Such a signaling mechanism is a critical component in providing distributed services within and across domains. Although SOA eliminates most requirements for specialized signaling, some circumstances may exist that requires innovative intelligent network processes, based on IP communications and signaling, both in-band and out-of-band, to accomplish functions that traditionally have been provided only by management and control processes. Such functions include those for general optical resource management, traffic engineering, access control, resource reservation and allocations, infrastructure configuration and reconfiguration, addressing, routing (including wavelength routing), resource discovery (including topology discovery), protection mechanisms through problem predication, fault detection, and restoration techniques.

5.9 EXAMPLES OF GRID NETWORK SERVICES PROTOTYPES

This chapter describes a number of considerations related to Grid network services architecture, along with some of the primary issues related to that architecture. The next sections provide a few examples of prototype implementations based on those concepts, as further illustrations of those concepts. As noted, incorporation of network services into a Grid environment involves several components, e.g., a high-level advertisement, mid-level software components that act as intermediaries between edge processes, such as applications, and core resources that are utilized through these intermediate processes. Examples are provided that are related to signaling methods for layer 3, layer 2 and layer 1. Other examples are provided in later chapters.

5.9.1 A LAYER 3 GRID NETWORK SERVICES PROTOTYPE

Early attempts to integrate Grid environments and specific network behaviors were primarily focused on APIs that linked the Grid services to layer 3 services. For example, some of these prototypes were implemented to ensure specified quality of services, for example by using the IETF differentiated services (DiffServ) standard, which is described in Chapters 6 and 10. Using this approach, Grid processes were directly integrated with DiffServ router interfaces to ensure that application requirements could be fulfilled by network resources. Other mechanisms interrogated routers to determine available resources, manipulated them to allocate bandwidth, and provided for resource scheduling through advance reservations.

For example, an early experimental architecture that was created to link Grid services to specific layer 3 packet services that could be manipulated was a module that is part of the Globus toolkit – the General-purpose Architecture for Reservation and Allocation (GARA) [8]. The Globus toolkit is open source software services and libraries that are used within many Grid environments [9]. GARA was created to govern admission control, scheduling, and configuration for Grid resources, including network resources. GARA has been used in experimental implementations to interlink Grid applications with DiffServ-compliant routers as well as for layer 3 resource allocation, monitoring, and other functions. GARA was used to implement layer 3 QoS services on local, wide-area, and national testbeds.

5.9.2 APIS AND SIGNALING FOR DYNAMIC PATH PROVISIONING

Other research initiatives experimented with integrating large-scale science applications on Grid layer 2 and optical metropolitan area, national and international testbeds. Within a context of OGSA intermediate software, these experiments enabled science applications to provision their own layer 2 and layer 1 paths. To accomplish this type of direct dynamic path provisioning, several mechanisms that address the requirements of dynamic network APIs and external process signaling were created, particularly for explicit dynamic vLAN and optical path provisioning.

An example of the type of signaling protocol that proved useful for these experiments and could be utilized in a customizable communications environment is the Simple Path Control (SPC) protocol, which is presented in an IETF experimental method draft [10]. This protocol can be used within an API, or as a separate signal, to establish ad hoc paths at multiple service levels within a network.

This protocol does not replace existing communication signaling mechanisms; it is intended as a complementary mechanism to allow for signaling for network resources from external processes, including applications. SPC can be integrated with existing signaling methods.

This protocol can be used to communicate messages that allow ad hoc paths to be created, deleted, and monitored. SPC defines a message that can be sent to a compatible server process that can establish paths among network elements. SPC can also be used to interrogate the network about current basic state information. When a request is received, the compatible server process identifies the appropriate path through a controlled network topology and configures the path.

Specific paths do not have to be known to requesting clients. The SPC protocol can be integrated with optimization algorithms when determining and selecting path options. This integration allows decisions to be based on any number of path attribute criteria, e.g., related to priority, security, availability, optimal performance, and others. SPC can be used as an extension of other protocols, such as those for policy-based access control and for scheduling. For communications transport, it can use any standard IETF protocol.

5.9.3 A LAYER 2 GRID NETWORK SERVICES PROTOTYPE

Another experimental architecture that was designed and developed to support large-scale Grid applications is the Dynamic Ethernet Intelligent Transit Interface (DEITI). This experimental architecture was created to allow for the extension of Grid services-enabled optical resource provisioning methods to other mechanisms used for provisioning dynamic vLANs, specifically 10-Gbit Ethernet vLANs [11]. This experimental prototype has been used successfully for several years on optical testbeds to extend lightpaths to edge devices within Grid environments using dynamic layer 2 path provisioning. However, it can also be used as separately within a layer 2 environment, based on IEEE standards (e.g., 802.1p, 802.1q, and 802.17). A key standard is 802.1q, the standard for virtual bridged local area networks, which is an architecture that allows traffic from multiple subnets to be supported by a single physical circuit. This specification defines a standard for explicit frame tagging, which is essential for path identification. This explicit frame tagging process is implemented externally so that it can be used both at the network edge and in the core. This standard is further described in Chapter 11.

Goals for this architecture are to provide a means, within a Grid services context, for traffic segmentation to ensure QoS, to enable enhanced, assured services based on network resource allocations for large-scale flows, and to provide for dynamic layer 2 provisioning. This architecture uses the SPC protocol for signaling.

5.9.4 SERVICES-ORIENTED ARCHITECTURE FOR GRIDS BASED ON DYNAMIC LIGHTPATH PROVISIONING

Experimental architecture for dynamic lightpath provisioning is beginning to emerge, based on Grid services architecture. One experimental service architecture being developed for dynamic optical networking is the Optical Dynamic Intelligent Network (ODIN) service architecture. Another example, which provides the most complete set of capabilities for distributed communication infrastructure partitioning at the optical level, is the User-Controlled LightPath architecture (UCLP).

5.9.5 OPTICAL DYNAMIC INTELLIGENT NETWORK SERVICES (ODIN)

The experimental Optical Dynamic Intelligent Network services (ODIN) architecture was designed specifically to allow large-scale, resource-intensive dynamic processes within highly distributed environments, such as Grids, to manage core resources within networks, primarily lightpaths [12]. It has generally been implemented within an OGSA context, using standard software components from that model. The initial implementations were based on OGSI. It has also been integrated with other network-related components such as an access policy module based on the IETF AAA standard and a scheduler based on a parallel computation scheduler adapted for network resource allocations. This architecture was designed to enable Grid applications to be closely integrated, through specialized signaling and utilizing standard control and management plane functions, with low-level network resources, including lightpaths and vLANs. This service architecture uses the SPC protocol for signaling, with which it establishes a session path that receives and fulfills requests.

The session becomes a bridge that directly links applications with low-level network functions. It contains mechanisms for topology discovery and for reconfiguring that topology, within a single domain or across multiple domains. It can be used to allow core network resources to be directly integrated into applications, so that they can dynamically provision their own optical channels, or lightpaths, vLANs, or essentially any layer 1 or layer 2 path. Through this mechanism, an external process can transport traffic over precisely defined paths. When these resources are no longer required, they are released.

Through its signaling mechanism, this services architecture creates a means by which there is a continuous dialog between edge processes and basic network middleware and underlying physical fabric. This iterative process ensures that resources are matched to application (or end-delivered service), requirements, and it provides unique capabilities for network resource provisioning under changing conditions. This architecture also allows for network resources, including topologies, to be customized (configured and reconfigured by external processes), and allows those processes to be matched with resources that have specific sets of attributes. This services process can be implemented within centralized server processes, or it can be highly distributed.

5.9.6 USER-CONTROLLED LIGHTPATH PROVISIONING

Another example of this SOA model for networks is the "User-Controlled LightPath" (UCLP) architecture [13]. UCLP is also an instantiation of this SOA, based on OGSA and using Globus toolkit 3 and Java/Jini services. This architecture provides for creating individual objects from core network resources so they can be used as elements from which higher level structures can be created. For example, a lightpath can be an object that can be placed in and manipulated within a Grid environment.

Significantly, UCLP does not merely constitute a means to provide on-demand lightpaths to users. The UCLP architecture enables distributed optically based facilities to be partitioned and subpartitioned into sets of management and engineering functions as well as network resources. UCLP allows users to integrate various heterogeneous network resources. These partitions can then be allocated to external processes that can shape networking environments in accordance with their needs. The designation of the approach as "user controlled" is a key declaration that provides a sharp demark from the traditional approach to communications infrastructure. The "user" in this sense can be any legitimate request external to the network infrastructure. These requests can ask for any combination of options related to discovery, acquisition of resources, provisioning, management, engineering, reconfigurations, and even protection and restoration.

This architecture does not require a central control or management plane, although if required it can integrate those functions. Similarly, it does not require advanced reservation or scheduling mechanisms, although they are also options.

UCLP allows end-users to self-provision and dynamically reconfigure optical (layer 1) networks within a single domain or across multiple independent management domains. Integrating network resources from multiple domains is different than setting up a lightpath across several domains. UCLP can do both. UCLP even allows users to suballocate resources, for example create subpartitions, e.g., for optical VPNs and provide control and management of these VPNs to other users. A key feature of this architecture is that it allows services and networks to be dynamically reconfigured at any time. No prior authorization from network managers is required. Access policies and security implementations are integrated into the infrastructure environment.

Consequently, this technique is complementary to the ad hoc provisioning methods of Grid services, allowing processes within Grid environments to create – as required – application-specific IP networks, as a subset of a wider optical networking infrastructure. This capability is particularly important for many science disciplines that require optimized topology and configurations for their particular application requirements. It is particularly effective for supporting large-scale, long-duration, data-intensive flows.

UCLP can be used for authenticated intra-domain and inter-domain provisioning. For example, it can be used with another procedure, OBGP [14], to establish paths between domains. OBGP is an example of the use of UCLP for inter-domain applications. Autonomous System (AS) path information in a Border Gateway Protocol (BGP) route can be obtained to create an identifier that can be used to discover authoritative servers, which can be the source of information on potential optical paths across domains. Such servers can be access policy servers, specialized lightpath provisioning servers, or other basic network service nodes.

5.10 DISTRIBUTED FACILITIES FOR SERVICES ORIENTED NETWORKING

The services-oriented network architecture model described in this chapter will require core infrastructure and facilities that are fundamentally different from those used by standard telecommunications organizations. Implementing a services architecture requires a new type of large-scale, distributed infrastructure, based on services exchange facilities that are much more flexible than those typical telecommunications central offices and exchange points. One major difference is that these facilities will deliver not only standard communication services but also multiple types of highly advanced services, including Grid services. They will be composed of resources that can be controlled and managed by external communities. These types of services are currently being designed and developed in prototype by research communities [15].

The foundation for these services will be a globally distributed communications infrastructure, based on flexible, large-scale facilities, which can support multiple, customizable networks and communication services. The international advanced networking community is designing next-generation communications infrastructure that is based on these design concepts [16]. They are transitioning from the traditional concept of a creating a network infrastructure to a notion of creating a large-scale distributed "facility," within which multiple networks and services can be created – such as the Global Lambda Integrated Facility (GLIF) [17]. A number of such facilities that are currently being designed will have highly distributed management and control functions, within the SOA context. Potential implementation models for these types of facilities are further described in Chapter 14.

5.10.1 PROVISIONING GRID NETWORK SERVICES

Provisioning Grid network services within highly distributed environments as fully integrated resources is a nontraditional process comprising multiple elements. This chapter presents some of the concepts behind Grid network services, which are motivating the creation of a new Grid network services architecture. Chapter 6 continues this discussion with a description of how these concepts relate to traditional network services with several OSI layers. Chapter 6 also describes several experiments and methods that explored mechanisms that can provide for flexible models for service provisioning within layer 3, layer 2, and layer 1 environments based on adjustable resources, such as through implementations of DiffServ, QoS for layer 2 services, and defined lightpaths.

Chapter 6 also notes the challenges of implementing service provisioning for those capabilities. Some of these challenges can be attributed to the lack of a complete Grid services middleware suite specifically addressing network resource elements. Chapter 7 presents an overview of these middleware services in the context of Grid network services.

REFERENCES

[1] H. Zimmerman (1980) "OSI Reference Model – The ISO Model of Architecture for Open Systems Interconnection," *IEEE Transactions on Communications*, 28, 425–432.
[2] "General Principles and General Reference Model for Next Generation Networks," ITU-T Y.2011, October 2004.
[3] R. Stewart, Q. Xie, K. Morneault, C. Sharp, H. Schwarzbauer, T. Taylor, I. Rytina, M. Kalla, L. Zhang, and V. Paxson (2000) "Stream Control Transmission Protocol," RFC 2960, October 2000.

[4] www.ggf.org.
[5] http://www.oasis-open.org.
[6] Web Services Resource Framework (WSRF) Technical Committee, Organization for the Advancement of Structured Information Standards, http://www.oasis-open.org.
[7] A. Anjomshoaa, F. Brisard, M. Drescher, D. Fellows, A. Ly, S. McGough, D. Pusipher, and A. Savva (2005) "Job Submission Description Language (JSDL)," Global Grid Forum, November 2005.
[8] www.icair.org/spc.
[9] A. Roy and V. Sander (2003) "GARA: A Uniform Quality of Service Architecture," *Resource Management: State of the Art and Future Trends*, Kluwer Academic Publishers, pp. 135–144.
[10] www.globus.org.
[11] www.icair.org.
[12] J. Mambretti, D. Lillethun, J. Lange, and J. Weinberger (2006) "Optical Dynamic Intelligent Network Services (ODIN): An Experimental Control Plane Architecture for High Performance Distributed Environments Based on Dynamic Lightpath Provisioning," special issue with feature topic on Optical Control Planes for Grid Networks: Opportunities, Challenges and the Vision. *IEEE Communications Magazine*, 44(3), pp. 92–99.
[13] User Controlled Lightpaths, http://www.canarie.ca/canet4/uclp/.
[14] http://obgp.canet4.net/.
[15] T. DeFanti, M. Brown, J. Leigh, O. Yu, E. He, J. Mambretti, D. Lillethun, and J. Weinberger (2003) "Optical Switching Middleware For the OptIPuter," special issue on Photonic IP Network Technologies for Next-Generation Broadband Access. *IEICE Transactions on Communications E86-B*, 8, 2263–2272.
[16] T. DeFanti, C. De Laat, J. Mambretti, and B. St. Arnaud (2003) "TransLight: A Global Scale Lambda grid for E-Science," special issue on "Blueprint for the Future of High Performance Networking." *Communications of the ACM*, 46(11), 34–41.
[17] www.glif.is.

Chapter 6

Grid Network Services: Building on Multiservice Networks

Joe Mambretti

6.1 INTRODUCTION

The Grid network community is developing methods that will enhance distributed environments by enabling a closer integration of communication services and other resources within the Grid environment. At the same time, the more general network research and development community continues to create new architectures, methods, and technologies that will enhance the quality of services at all traditional network layers. Almost all of these initiatives are, like Grid development efforts, providing core capabilities with higher levels of abstraction. Most current Grid development efforts are focused on leveraging these efforts to combine the best innovations of networking research and development with those of the Grid community.

This chapter provides a brief introduction to several basic network service concepts, primarily standard network services at layers 1 through 4, as a prelude to discussions that will be the focus of much of the rest of this book. Later chapters describe architectural and technical details of the services at each of these layers.

6.2 GRID NETWORK SERVICES AND TRADITIONAL NETWORK SERVICES

Grid network services are those that manage, control, or integrate some aspect of communication service or other network-related resource, such as a characteristic of the service (quality of service class), policy access processes, individual network elements within a service wrapper, network information such as topology, network performance metrics, and other elements. A Grid network service can be created from multiple elements. It can have an interface that is exposed to a particular API, or it can have several interfaces that are exposed to multiple other network resources. It can have a schema of attributes. It can incorporate other network services as static elements or as dynamic elements accessed through scheduling or through an ad hoc procedure. It can be localized or highly distributed.

Within a Grid context, these types of architectural considerations are usually part of a horizontal design, as opposed to a hierarchical stack. In contrast, the traditional OSI model (Figure 3.2) is an example of a stacked vertical architecture, in which the functionality of one layer has dependencies on those of the immediately adjacent layers but not on others. To access lower layer functionality, it is necessary to transverse through all of the interceding layers. Traditional network services are defined within the context of the individual layers of the OSI model.

Grid network architects are attempting to use methods for abstracting capabilities to provide a horizontal services model as an overlay on this traditional vertical architecture. One goal in this approach is to take advantage of the QoS capabilities at each layer, as an aggregate service rather than as a service at a specific layer. However, another goal is to ensure that Grid environments can take advantage of service qualities inherent in these individual layers, which are being continually improved through new techniques and technologies. Over the past few years, a basic goal of Grid networking designers has been to provide appropriate matches through service integration between Grid application requirements and high-quality communication services provided by individual OSI layers.

6.2.1 THE GRID AND NETWORK QUALITY OF SERVICE

As Grid architecture continues to evolve, network services are also being constantly improved. The general direction of these developments is complementary to evolving designs for Grid network services, especially with regard to attempts to provide for a greater degree of services abstraction. The standards organizations described in Chapter 4 are continually developing new methods to improve communications services at all layers. Approaches exist for ensuring service quality at each network layer is managed in accordance with specifications that are defined by these standards organizations. Each layer also provides for multiple options for ensuring service guarantees. These options can be used to define a service concept, which can be implemented within the architecture defined at that layer. Service definitions for all layers are specified in terms of quality levels, for example the quality of a particular service within a class of services, as measured by general performance objectives. Often these objectives are measured by standard metrics related to specific conditions. Although the approaches used to ensure quality at each layer are somewhat different, all are based on similar concepts.

In general, these services are supported by a suite of components, defined within the framework of commonly recognized, defined architectural standards. The services have an associated set of metrics to measure levels of service quality and ensure that all traffic is provided by default with at least a basic level of quality guarantees.

Another component consists of means for signaling for any specialized services, governed by policies to ensure that such signals are appropriate, e.g., authenticated, authorized, and audited. To distinguish among types of traffic, a method of marking it with identifiers is required to map it to one or several levels of service classes. Mechanisms have been created for identifying individual traffic elements, e.g., packets or frames, with markers that allow for mapping to quality levels, governed by policy-driven provisioning mechanisms. This type of mechanism can also be used to provide traffic with specialized service-quality attributes. Other mechanisms detect these markers, determining what resources are available to accomplish services fulfillment, and provide for assuring the requested quality of service. In addition, a means must exist to monitor the service to ensure the fulfillment of its requirements over time, and, if problems arise, to intercede to guarantee service quality.

The following sections provide an overview of traditional network services and indicate how their development relates to the design of Grid network services. Many Grid network service design projects have been undertaken in close coordination with those intended to provide for enhanced Internet quality of services, and these development efforts are following similar paths. However, in addition, other Grid development efforts are exploring integrating Grid environments with layer 2 and layer 1 services. The following sections introduce basic concepts related to network services at layers 1–4 and provide preliminary information about how those concepts relate to Grid network services as a resource. Later chapters will expand on each of these topics.

6.3 NETWORK SERVICE CONCEPTS AND THE END-TO-END PRINCIPLE

Network services do not exist in the abstract. They are produced by the infrastructure that supports them. To understand how services are designed, it is useful to examine the design of the infrastructure on which those services are based. An important concept in system design is often described as the "end-to-end principle," which was set forth in a paper published in 1981, and again in a revised version in 1984 [1]. One objective for this paper was to present guidelines for "the placement of functions among modules of a distributed computer system." The paper presents the argument that functions placed at low levels of a system may be of less benefit and have higher associated costs than those placed at higher levels. One of the systems referenced as an example to explain this argument was the first packet-routed network, the ARPANET, specifically its process mechanisms such as delivery guarantees.

Later, this argument was frequently summarized as a core premise, basically that, when designing a distributed system, the core should be as simple as possible and the intelligence should be placed at the edge. This premise became a fundamental principle for Internet architects and remains a key design principle. Often this concept is used to compare fundamental Internet design principles with traditional carrier networks. Internet designers provide for a simple, powerful network core, while placing intelligence within edge devices. In contrast, traditional communications providers, assuming simple edge devices, have tended to place complexity in the core of the network.

6.3.1 NETWORK QUALITY OF SERVICE AND APPLICATIONS QUALITY OF SERVICE

To provide for quantitative descriptions of network services, metrics have been devised to define specific "quality of service" designations. However, it is important to note, especially within a Grid context, that within application communities network QoS is measured by different metrics than those used by networking communities. Required application QoS parameters almost never directly translate into general infrastructure objective performance measures, and they certainly do not do so for network services. This observation has been made since the inception of data networks, and it is noted in the original end-to-end design publications [1].

For example, application QoS is usually measured using parameter metrics important to objectives that directly relate to the end-delivery of specific characteristics. These can include interactive real-time responsiveness, control signal sensitivity, and capabilities for mixing large and small flows with equal performance results.

In contrast, network performance quality is usually measured using parameter metrics that often relate to other considerations. These considerations include network design, configuration, engineering, optimal resource allocation, reliability, traffic and equipment manageability, characteristics of general overall traffic flow, such as average and peak throughput ratios, jitter and delay, general "fairness" of resource utilization, end-to-end delay, and error rates, for example as indicated by lost packets, or duplicated, out-of-order or corrupt packets. The type of performance that may be optimal for overall network quality of service provisioning may be suboptimal for specific application behavior.

Also, provisioning for network quality of service implementation and optimization can directly and negatively influence application quality of service implementation and optimization, and the latter can be problematic for the former. The dynamic interaction between these two processes can be complex. These types of challenges are especially important for Grid environments because of the close integration between applications, services, and infrastructure resources.

6.4 GRID ARCHITECTURE AND THE SIMPLICITY PRINCIPLE

The spectacular success of the Internet is a strong argument for the strength of its basic architectural premise, in particular the argument that simplicity of design provides for a more successful, scalable result. Basic Internet architectural documents articulate a number of key principles that were inherent in early design decisions but not formalized until several RFCs were developed later.

This end-to-end principle is basic to Internet design; for example, it is reiterated in *Architectural Principles of the Internet* (IETF RFC 1958) [2] and in *Some Internet Architectural Guidelines and Philosophy* (RFC 3439), which notes that "the end-to-end principle leads directly to the Simplicity Principle" [3]. These, and related publications, note that complexity in network design is usually motivated by a need for infrastructure robustness and reliability as opposed to the QoS delivered. Consequently, there is a direct correlation between simplicity of design and QoS. These considerations give rise to the conceptual notion of the Internet design as an "hourglass," consisting of the IP-enabled edge devices at either end and the minimalist IP layer in the middle. (RFC 3439) [3].

Similarly, Grid architecture is based on this minimalist hourglass design concept. Basic Grid architecture and Internet architecture are highly complementary. Grid architecture is based on the same design premise. The Grid services-oriented architecture places a wide range of functionality at the system edge. It is notable that the end-to-end argument was first advanced in the context of a design for a distributed computing system and was never intended to apply only to data networks. Current efforts in architectural design for Grid network services are continuing this tradition of enhancing edge functionality based on powerful, but simple, core facilities.

6.4.1 NETWORK DESIGN AND STATE INFORMATION

A key premise to the end-to-end principle is that design simplicity is enhanced by minimizing the need to maintain state information. Building on the earlier end-to-end design principles, RFC 1958 notes that "an end-to-end protocol design should not rely on the maintenance of state (i.e., information about the state of the end-to-end communication) inside the network." Instead, state "should be maintained only in the endpoints, in such a way that the state can only be destroyed when the endpoint itself breaks." This concept is known as "fate-sharing" [4]. One conclusion that results from this observation is that "datagrams are better than classical virtual circuits."

Therefore, the network should be designed solely to transmit datagrams optimally. All other functions should be undertaken at the edges of the network. This point is a particularly important one that will be discussed further in other sections, especially those related to layer 2 and layer 1 services within a Grid environment. This section will note that the goal of providing infrastructure to transmit datagrams optimally is not mutually exclusive with incorporating circuit-oriented technologies within Grid environments. The abstraction layers enabled by basic Grid architecture support the optimization of any resource.

Also, as RFC 3439 emphasizes, this approach does not mean that the Internet "will not contain and maintain state," because much basic state is contained and maintained in the network core. Core state is orthogonal to the end-to-end principle. The primary point is that edge should not be dependent on core state. This concept is another important point that will be discussed in other sections related to Grid network services and stateful elements.

6.4.2 INTERNET BEST EFFORT SERVICES

One of the key architectural innovations that has propelled the rapid growth of the Internet has been its simple, scalable architecture. The basic model of the Internet – a common service provided to all data traffic within shared infrastructure resource – has proven to be a powerful, successful concept. From its initiation as a network service, the Internet has supported data traffic with one basic level of service, a best effort service, which treats all packets equally. This technique has been described as "fair," a term that indicates that no particular traffic stream is given priority over any other stream with regard to recourse allocation. In the context of a best effort service, every packet is treated identically in the provisioning of the available network resources, insofar as this idealized goal is possible through network design and engineering. Although providing for absolute fairness is an unrealizable goal, this egalitarian approach has been highly successful. This type of services approach provides for efficiency, optimal resource utilization, and exceptional scalability.

On the other hand, from the earliest days of the Internet, the limitations of this approach have also been recognized [5]. By definition, this type of service cannot guarantee measurable quality to any specific traffic stream. Therefore, while general traffic is extremely well served, special traffic may not be, and certainly there are no quality guarantees for such traffic. The most frequently used example for a need for this type of differentiation is a comparison between email traffic, which can tolerate network delay, and digital media, which can be degraded by such latency.

With this single-service approach, one of the few options to provide for QoS is to increase network resources, perhaps through overprovisioning. Also, it has always been clear that different types of applications require network services with different characteristics. Using a single service, applications cannot be developed that require multiservice implementations. Router congestion can result in nonoptimal traffic support, and the performance of large-scale flows can seriously deteriorate when mixed with multiple smaller scale flows. Similarly, smaller flows also can be negatively affected by large-scale flows. A single modestly sized flow can seriously degrade the performance of very large numbers of smaller flows. In addition, latency intolerant applications can be seriously compromised.

Therefore, the need for high-quality, scalable, reliable, and interoperable options in addition to the default Internet services has been recognized since the early days of the Internet. A need has been recognized to distinguish, or differentiate, among the types of various traffic streams, especially as increasing numbers of individual users, applications, and communities adopted the Internet. With a method for such differentiation, streams that require higher quality can be allocated more resources than those that do not.

Providing consistent quality of services for network traffic is an on-going challenge because many elements contribute to delivered results, including infrastructure, configurations, specific protocols, protocol stacks and tunings, kernel tunings, and numerous parameters. Addressing the attributes of all of these elements is important in ensuring QoS. Also, adjusting these elements often influences the behavior of others in unpredictable ways. Many basic determinates of quality relate to resource allocations within the network infrastructure.

6.5 GRIDS AND INTERNET TRANSPORT LAYER SERVICES

Layer 4 of the OSI model, the transport layer, is the data delivery service for network traffic. The two key Internet protocols for transport are the Transmission Control Protocol (TCP), which provides for assured, reliable end-to-end transit, and the User Datagram Protocol (UDP), which does not provide for transport guarantees. TCP provides for flow control and error detection. TCP packets are transmitted encapsulated within IP packets. The TCP/IP protocol suite has been the basic standard for the Internet for many years, and it is now being examined as the growth of Internet infrastructure and services provide challenges to common implementation.

To address reliability, quality of service, and performance challenges related to the Internet, multiple initiatives have been established to investigate the basic architecture of classic TCP and UDP, and to investigate the potential for alternatives. A significant body of research literature exists related to these issues.

Because of the importance of these efforts, a number of Grid network service research and development communities are also actively engaged in exploring new transport architecture related to TCP, UDP, and related stacks. Recently, the GGF undertook a survey of the primary research efforts in the area and published it as a reference document [6]. These topics are discussed in Chapters 8 and 9.

6.6 IETF DIFFERENTIATED SERVICES

Beginning in the early 1990s, this technical challenge has been addressed through several approaches to basic Internet architectural design. Some of these innovative techniques began to migrate to formalization through standards committees. For example, at that time, the Resource Reservation Protocol (RSVP) [7], which has been developed principally through the IETF, was created to allow traffic that required special consideration to signal across the network to prepare resources in advance of data flows. Essentially, the Internet is a stateless architecture, and this approach was a small departure from that basic principle. In order to create an end-to-end path, initial signals have to travel from source to destination and back to create a path across the network fabric.

In the mid-1990s, another approach to QoS was created, the differentiated services (DiffServ) architecture, which the IETF began to formalize in 1999 [8–10]. The IEFT DiffServ architecture model provides a framework that specifies methods for defining and implementing a spectrum of network services defined by detailed quality measures (RFC 2474) [11]. This approach is especially useful if resources supporting high-priority flows constitute only a small portion of the aggregate traffic and those resources are only a small portion of the total available.

This architecture preserves basic IETF principles such as end-to-end continuity, keeping the network core simple, and placing complexity at the network edge. For example, using this model, edge routers perform the basic differentiation function, while core routers can focus on forwarding processes. This architecture is based on several key concepts, including mechanisms to define edge processes that are associated with specific types of network traffic behavior and mechanisms related to the implementation of those behaviors through standard router configurations and traffic engineering.

6.6.1 DIFFSERV MECHANISMS

The IETF developed the DiffServ concept in order to create a framework for describing and implementing a basic, predicable QoS. DiffServ uses concepts of traffic classification associated with forwarding behavior that implements services classes. DiffServ also provides for on-going monitoring and adjustments, e.g., through packet drops, scheduling, and other measurable parameters. DiffServ provides a specification for individually marking packets, a function undertaken by edge routers, on a packet-by-packet basis, so that they can be identified and classified within specific traffic categories, such as high, low, or normal priority. Per Hop Behavior (PHB) designations are provided by the designation of values, within a field of the packet header [12]. The DiffServ field, an octet IP header, is used to designate flows within categories. The initial six bits (allowing for 64 classes) are designated as the codepoint, or Differentiated Services Codepoint (DSCP), that indicates the PHB. The other bits are reserved.

Because the packets are marked, traffic behavior requirements can be related to individual packets, or classes of packets based on the DiffServ field values. These markings specify a particular type of PHB, which is provided by network routers as the traffic flows through the network. Forwarding behavior can be implemented within DiffServ routers based on those values. Using PHB, specific traffic treatments, such as traffic conditioning, can be linked to specific packets and flows within designated classifications across a network fabric. For example, common traffic can be segmented from other high-priority traffic that may require special treatment, such as faster forwarding.

This technique addresses the need to identify different types of traffic, within an appropriate policy context. Routers can be configured to treat different packets or packet classes in accordance with preset policy, related to previously defined parameters, e.g., priorities. Policies implemented within DiffServ routers can ensure that high-priority traffic receives required resource allocations and that low-priority traffic does not receive more resource allocations than is required. Because of the association of resources with specific flows, these policy implementations are sometimes termed "policing profiles." This technique can be used to limit the ingress flows of designated traffic in order to govern submission to specific services to prevent them from being oversubscribed.

Part of this architectural approach includes the options for using queues that are each associated with a particular class of traffic quality. For example, high-priority traffic can be sent to special queues that forward their packets faster than alternative queues. Traffic behavior can be treated as micro-streams or as aggregated streams.

The IETF has also created standards for marking traffic for particular types of PHBs, such as Expedited Forwarding (EF) (RFC 2598), which can be used to create a type of "virtual leased line" across autonomous networks [13,14]. In other words, it can be used to set up an end-to-end path across a DiffServ-compliant network for designated priority traffic that has high-quality attributes, such as guaranteed bandwidth levels and minimal latency and jitter. Another example of a PHB is assured forwarding, which is better than best effort but not as good as the EF premium service. This topic is discussed further in Chapter 10.

6.6.2 GRIDS AND QUALITY OF SERVICE NETWORK SERVICES

Many of the designs that have been developed to ensure specific network behaviors within Grid environments have been based on APIs that interlink Grid services to layer 3 services. One early architectural design was the General-purpose Architecture for Reservation and Allocation (GARA, originally, Globus Architecture for Reservation and Allocation), which was created in accordance with Globus architectural principles. [15,16]. This architecture provides a common mechanism for dynamically discovering and then reserving heterogeneous resources with the assumption that those resources can be independently controlled and administered. GARA was developed so that it could be broadly applied to any resource, including networks, and it was intended to be used for multiple heterogeneous resources simultaneously.

GARA defines mechanisms for passing application requirements to resource managers in order to allow those managers to discover resources to fulfill those requests and to provide for adjustments in those allocations if sufficient resources are not available or if additional resources are required as part of an on-going process. GARA is extensible to other network layers and is not specifically oriented to services at a specific layer.

6.7 GARA AND DIFFSERV

Several early research experiments used GARA as an interface for manageable Grid layer 3 network services, based on RSVP and Integrated Services (IntServ). Later experiments were based primarily based on DiffServ. In this type of implementation, messages passed from applications to resource managers were linked to processes that could provide for reservations, interrogate DiffServ-compliant routers to ascertain the level of available network resources, ensure the appropriate allocation of bandwidth, and ensure that on-going QoS performance was met through process monitoring.

6.8 GRIDS AND NONROUTED NETWORKS

The advanced networking research community has been investigating the potential for complementing the general layer 4/layer 3 services in Grid environments with methods based on complementary nonrouted services. Such services range from flow switching to IP-based virtual paths, to dynamic layer 2 vLANs. For example, today a number of research efforts are directed at resolving these issues through flow routing, as well as variants often labeled as "IP switching." An early version of this approach was "tag switching," which later evolved into the IETF Multiprotocol Label Switching (MPLS) architecture. Any of these techniques can be implemented within Grid environments to supplement layer 3 routing. Because these techniques are IP based and implemented in routers, they are sometimes labeled "layer 2.5" techniques. These architectures and techniques are described in Chapter 10.

6.8.1 LAYER 2.5 SERVICES AND QUALITY STANDARDS

Flow routing is based on the premise that new techniques for traffic stream switching could be formulated from extensions of standard layer 3 routing mechanisms for traffic performance examination and adjustment. This approach extends those capabilities to achieve a form of switching performance, without having to depend on traditional switching techniques, such as vLAN layer 2 switching.

Flow routing is based on a process that parses source and destination information for packets. When patterns are detected for certain flows, an identifier is placed on the flow. This identifier is then used for transmitting packets to their final destinations, or as close to those final destinations as possible, without the additional processing required for complete packet information parsing. Other implementations of this approach give each packet an identifier that is used to send it through the network in an optimal way as opposed to having each network router completely examine each packet. This identifier can be used to segment traffic by priority and to provide specialized treatments to certain traffic categories.

The IETF standard architecture MPLS, which is widely available in commercial routers, has become a popular mechanism for segmenting specialized traffic streams, especially for providing dedicated resources to large-scale streams.

6.8.2 GRIDS AND LAYER 2.5 SERVICES

Flow-based treatments such as MPLS can be useful for Grid environments. They can be used to provision paths as ad hoc resource allocations. They can be especially useful for providing high-quality service support for high-performance, low-latency traffic. Although some of these attributes may be obtained through other techniques, such as dedicated layer 2 paths, e.g., vLANs, MPLS can be used for Grid environments that require large-scale peer-to-peer networking, which can be difficult to achieve with layer 2 paths alone. This topic is discussed further in Chapter 11.

6.9 LAYER 2 SERVICES AND QUALITY STANDARDS

Other research and development initiatives are exploring new methods for supplementing Internet services with those that provide for more deterministic control over nonrouted paths, for example using new techniques for high-performance layer 2 switching and for dynamically allocated layer 1 lightpaths.

The IEEE has developed a specification for layer 2 services quality that is described in the 802.1p standard. This standard sets forth the required attributes and measures for services quality, based on service availability, out-of-order frames, frame duplication, frame loss, priority and other parameters. This standard is part of the IEEE

802.1D standard, which defines attributes of Media Access Control (MAC) bridges, such as traffic classification.

The IEEE's efforts to enhance Ethernet architecture to support a wider range of higher quality and more directly manageable services are assisting in transitioning Ethernet from a LAN/enterprise architectural standard to a wide-area standard. This development is being supported by parallel SONET/SDH standards development activities, which allow Ethernet to be mapped onto SONET/SDH channels with a high degree of flexibility. In part, these capabilities are being made possible by new SONET/SDH-related standards.

SONET/SDH, a common layer 2 transport protocol used in carrier networks, has been implemented primarily with Add-Drop Multiplexers (ADMs) that support only standard services. More recently, standard SONET/SDH has been enhanced with options that can provide for additional services capabilities, including many layer 2 services such as encapsulated Ethernet-over-SONET. These include the Generic Framing Procedure (GFP), which is a standard for mapping multiple packet-based (frame-based) protocols onto SONET/SDH transport circuits [17]. Another related standard is Virtual Concatenation (VCAT, G.709), which is used in conjunction with GFP. VCAT is a standard that allows packet (frame)-based traffic to be enclosed within Time Division Multiplexing (TDM) frames such that they can be aggregated and mapped to appropriately sized SONET/SDH channels (G.707) [18]. The Link Capacity Adjustment Scheme (LCAS) is another key standard that allows for allocations and management of bandwidth-on-demand channels within a SONET/SDH environments (G.7042) [19]. These topics are further discussed in Chapter 11.

6.9.1 GRIDS AND LAYER 2 QUALITY OF SERVICE

Layer 2 services are key resources for Grid environments. They can be used to create multiple ad hoc paths that can be individually addressed by Grid processes and which can specify attributes related to network connectivity, e.g., point to point, point to multipoint, or multipoint to multipoint, as well as variable levels of quality, security implementations, performance levels, and other characteristics.

Ethernet is a common standard used in Grid networks that has quickly migrated from the LAN environment to WANs. Some experimental Grid environments have been based exclusively on layer 2 services. Currently, several initiatives have been established that integrate flexible, ad hoc provisioning of Ethernet services within large-scale distributed environments. These capabilities are being extended across wide areas by utilizing the new SONET/SDH standards noted in the previous section.

Other Grid research initiatives are creating techniques that provide for layer 2 large-scale transport by mapping an Ethernet frame directly on to ad hoc, dynamically provisioned DWDM lightpaths.

6.10 LAYER 1 SERVICES AND QUALITY STANDARDS

Multiple research projects are developing prototype methods that can provide for high-quality data services based on agile advanced optical technologies, especially dynamically allocated lightpaths and optical VPNs, which are controlled and managed by layer 2 and layer 3 methods. Some of these concepts are being formalized within standards groups, such as the ITU-T and IETF, which have established several initiatives focused on standardizing architecture for optical VPNs. For example, the ITU-T Study Group 13 is developing architecture for an layer 1 VPN. This group's Y.1311 specification provides for a generic VPN architecture and general services requirements, and Y.1312 provides for layer 1 VPN service requirements.

In addition, other activities are exploring the potential to create new methods for ad hoc provisioning of DWDM-based lightpaths. These methods are based not on the requirements of traditional service providers but on requirements that are emerging from multiple Grid applications and services communities. Many of these projects are based on Grids that are interconnected with directly addressable lightpaths. The utility of these methods for large-scale data-intensive science projects, using regional, national, and international infrastructure, has been demonstrated for several years. These demonstrations have included application-driven lightpath provisioning [20].

6.10.1 GRIDS AND LAYER 1 QUALITY OF SERVICE

Currently, various research and development projects, primarily oriented to the requirements of large-scale data-intensive science, are examining new mechanisms for determining how next-generation Grids can benefit from advances in basic optical networking technologies. These projects are examining options for supplementing packet-routed services with those based on dynamic lightpath provisioning.

For example, one such project is the OptIPuter, a specialized internationally distributed environment comprising multiple integrated resources (optical networking, high-performance computers, mass storage, and visualization) [21]. This environment is based on a foundation of agile optical networking channels. It is an infrastructure that closely integrates traditional layer 7 resources, such as computational resources, with parallel optical networks, using IP, but it also exploits the potential of a distributed environment in which high-performance optical channels, not computers, constitute the core resource.

The optical foundation of this environment consists of dynamic allocated lightpaths (lambdas) that can be signaled by individual applications. Within this environment, these lightpaths function not as a traditional network but as a reconfigurable backplane to a large-scale, distributed, high-performance virtual computer.

The design of this environment (sometimes termed a "Lambda Grid") is being driven by large-scale global science applications. A goal of this architecture is to provide scientists who are generating extremely large volumes of data (terabytes and petabytes) with a capability for interactively investigating that data with advanced tools for visualization, analysis, and correlations from multiple sites world-wide. Recently, multiple large-scale research projects have demonstrated, e.g., at national and international conferences, the utility of these capabilities for global compute- and data-intensive science research [22,23]. These topics are discussed further in Chapter 12.

6.11 THE GRID AND NETWORK SERVICES

Although the Grid was initially established as essentially a "layer 7" service that used default Internet best effort TCP/IP services, the development community understood the potential for expanding those communications capabilities with a wide range of additional service options. Initial experiments and prototypes demonstrated that expanded communications services and enhanced service flexibility could provide support for Grid applications that could not be undertaken with traditional packet router services. Over the last few years, these options have continued to expand as the flexibility inherent in Grid architecture allows for applications and services to take advantage of multiple types of capabilities at every layer of the OSI model. Subsequent chapters explore in more detail these options and what they mean for the designers and developers of Grid environments.

REFERENCES

[1] J. Saltzer, D. Reed, and D. Clark (1981) "End-to-End Arguments in System Design," *Proceedings, 2nd International Conference on Distributed Systems Design*, Paris, France, April 8–10, 1981, pp. 509–512, revised as "End-to-End Arguments in System Design," (1984) *ACM Transactions on Computer Systems*, (4), 277–288.

[2] B. Carpenter (1996) "Architectural Principles of the Internet," RFC 1958.

[3] R. Bush, T. Griffin, and D. Meyer (2002) "Some Internet Architectural Guidelines and Philosophy," RFC 3439.

[4] D. Clark (1988) "The Design Philosophy of the DARPA Internet Protocols," Proceedings of the ACM, SIGCOMM '88, *Computer Communications Review*, 18(4), 106–114.

[5] W. Willinger and J. Doyle (2002) *"Robustness and the Internet: Design and Evolution,"* White Paper, March 2002, Caltech.

[6] M. Goutelle, Y. Gu, E. He, S. Hegde, R. Kettimuthu, J. Leigh, P. Primet, M. Welzl, and C. Xiong (2005) "Survey of transport protocols other than Standard TCP", Grid Forum Document, No. 55, November 2005.

[7] R. Braden, L. Zhang, S. Berson, S. Herzog, and S. Jamin (1997) "Resource ReSerVation Protocol (RSVP)," RFC 2205, September 1997.

[8] S. Blake, D. Black, M. Carlson, E. Davies, Z. Wang and W. Weiss (1998) "An Architecture for Differentiated Services," RFC 2475, December 1998.

[9] K. Nichols, V. Jacobson, and L. Zhang (1999), "A Two-bit Differentiated Services Architecture for the Internet," RFC 2638, July 1999.

[10] K. Kilkki (1999) *Differentiated Services for the Internet*, Macmillan Technical Publishing.

[11] K. Nichols, S. Blake, F. Baker, and D. Black (1998) "Definition of the Differentiated Services Field (DS Field) in the IPv4 and IPv6 Headers," RFC 2474, December 1998.

[12] D. Black, S. Brim, B. Carpenter, and F. Le Faucheur (2001) "Per Hop Behavior Identification Codes," IETF RFC 3140, June 2001.

[13] B. Davie, A. Charny, J.C.R. Bennet, K. Benson, J.Y. Le Boudec, W. Courtney, S. Davari, V. Firoiu, and D. Stiliadis (2002) "An Expedited Forwarding PHB (Per-Hop Behavior)," RFC 3246, March 2002.

[14] A. Charny, J. Bennet, K. Benson, J. Boudec, A. Chiu, W. Courtney, S. Davari, V. Firoiu, C. Kalmanek, and K. Ramakrishnan (2002) Supplemental Information for the New Definition of the EF PHB (Expedited Forwarding Per-Hop Behavior), RFC 3247, March 2002.

[15] I. Foster, C. Kesselman, C. Lee, B. Lindell, K. Nahrstedt, and A. Roy (1999) "A Distributed Resource Management Architecture that Supports Advance Reservations and Co-Allocation," International Workshop on Quality of Service, June 1999, pp. 27–36.
[16] I. Foster, A. Roy, and V. Sander (2000) "A Quality of Service Architecture that Combines Resource Reservation and Application Adaptation," Proceedings of the Eighth International Workshop on Quality of Service (IWQOS 2000), June 2000, pp. 181–188.
[17] "Generic Framing Procedure (GFP)," ITU-T 7071.
[18] "Network Node Interface for the Synchronous Digital Hierarchy (SDH), ITU-T 707.
[19] "Link Capacity Adjustment Scheme (LCAS)," ITU-T 7042.
[20] J. Mambretti, J. Weinberger, J. Chen, E. Bacon, F. Yeh, D. Lillethun, R. Grossman, Y. Gu, and M. Mazzuco (2003) "The Photonic TeraStream: Enabling Next Generation Applications Through Intelligent Optical Networking at iGRID2002," *Journal of Future Computer Systems*, 19(6), 897–908.
[21] www.optiputer.net.
[22] www.igrid2005.org.
[23] Joe Mambretti, Rachel Gold, Fei Yeh, and Jim Chen (2006) "AMROEBA: Computational Astrophysics Modeling Enabled by Dynamic Lambda Switching," special issue, *Future Generation Computer Systems*, 22 (in press, Elsevier).

Chapter 7

Grid Network Middleware

Franco Travostino and Doan Hoang

7.1 INTRODUCTION

Often labeled simply as "middleware," the software infrastructure that supports a Grid is complex. Its architectural complexity increases when the network is elevated to the status of a first-class Grid-managed resource like other Grid resources, such as computation cycles and storage.

This chapter avoids the use of the word "middleware," because it is too vague for the topics discussed. It introduces definitions for Grid network services and Grid network infrastructure. It then profiles representative examples both for Grid infrastructure in general and for Grid network infrastructure specifically, while reviewing their salient characteristics. The chapter continues with a description of Grid network infrastructure components, such as network bindings, virtualization, access control and policy, and scheduling. The chapter concludes with the challenges and opportunities in federating multiple instances of Grid network infrastructures, each representing a different, independently managed network domain.

Clearly, Grid network infrastructures are at an early developmental stage and predate interoperability standards. Although there are no firm foundations in these areas, the chapter attempts to highlight design points, characterize alternate views, and provide references to literature.

Grid Networks: Enabling Grids with Advanced Communication Technology Franco Travostino, Joe Mambretti, Gigi Karmous-Edwards © 2006 John Wiley & Sons, Ltd

7.2 DEFINITIONS

7.2.1 NETWORK SERVICES AND GRID NETWORK SERVICES

Chapter 6 has reviewed the plurality of services that a network implements at layers 1–4 of the OSI stack. Such network services are focused primarily on the movement of data. In addition to these services, other supporting network services exist, such as such as the Domain Name System (DNS) [1], quality of service (e.g., DiffServ) bandwidth brokers, and policy-based access mechanisms, such as the ones contemplated by the IETF Authentication, Authorization, and Accounting (AAA) architecture [2]. These services are implemented as software agents that interface with the network.

A Grid network service is one with a function in alignment with the requirements described in Chapter 3 and with an interface that is compatible with Grid infrastructure software structured in the context of Service-Oriented Architecture (SOA). Examples of Grid network services include a broker for a minimal latency end-to-end path, a bandwidth reservation service, a delay-tolerant data mover service (see Section 3.3.4). The Global Grid Forum's High Performance Networking Research Group (GHPN-RG) has introduced the concept of a Grid network service in a formal document [3].

While network services and their derivatives are not directly exposed to elements of the Grid infrastructure, Grid network services are directly exposed to elements of the Grid infrastructure (such as a Grid meta-scheduler that can be used for all resource types).

7.2.2 GRID INFRASTRUCTURE SOFTWARE

The software for Grid infrastructure provides a common basis for creating and operating virtual environments, including virtual organizations. It provides a common functionality for controlling and managing cross-organizational resources. As such, Grid infrastructure consists of core components that support:

- resource discovery to provide a virtual community capabilities dictionary;
- resource monitoring to ensure availability of resources and their performance;
- data management for reliable transfer of data;
- common security services for handling authentication, authorization, delegation and community authorization;
- common execution environment for distributed service management;
- communication services for messaging among Grid-managed objects.

7.2.3 GRID NETWORK INFRASTRUCTURE SOFTWARE

Early Grid manifestations assumed that the network is unknowable and at the same time capable of yielding adequate performance, as a result of ad hoc support. Applications that could benefit from network information had no standard way to access it. Similarly, the applications and the Grid infrastructure had no ability to inform networks of their needs.

Grid network infrastructure is the ensemble of services that elevate the network to Grid-managed resource. This software typically culminates with advertised Grid network services or Grid network services management tools. To the point, Grid network infrastructure implements:

- communication infrastructure for Grids and similar manifestations of distributed systems;
- data transport services;
- resource management services for managing network resources;
- scheduling services for coordinating users' demand and underlying network resources;
- access control for network resources;
- QoS managing services for supporting Grid applications;
- monitoring services;
- network information services.

While ideally matched to Grid requirements, a Grid network infrastructure does not necessarily have to be designed exclusively for Grid applications. Elements of Grid network infrastructure – e.g., the virtualization platform – can also be used to support functions that are part of horizontal IT integrations (e.g., across departments or institutions) that do not necessarily qualify as Grids.

7.3 GRID INFRASTRUCTURE SOFTWARE

A number of integrated packages of Grid software suites, often termed "toolkits," have been created. For example, the Globus toolkit exemplifies a component of a Grid infrastructure that allows a platform to support a virtual organization and to enable Grid applications. In this section, the Globus toolkit and its components are profiled. While each Grid infrastructure toolkit – Globus [4], Condor [5], Legion [6], Unicore [7], or proprietary solutions – reflects an independent view and specializes its design accordingly, the Globus components are representative of the basic issues of this topic and responses to those issues.

7.3.1 THE GLOBUS TOOLKIT

The Globus Toolkit (GT) [4] is a open community-based, open source set of services and software libraries that was originally developed at a national laboratory. Globus provides standard building blocks and tools for use by application developers and system integrators. It has already gone through four versions in the last few years: the original version in the late 1990s, GT2 in 2000, GT3 in 2003, and GT4 in 2005. GT2 was the basis for many Grid developments worldwide. GT3 was the first full-scale implementation of Grid infrastructure built upon Web Services by way of the GGF's OGSI intermediate layer [8]. GT4 is the first implementation that is compliant with mainstream Web Services as well as Grid services based on WSDL [9] and WSRF [10].

The upcoming releases of Globus are expected to show continued alignment with the OGSA suite of specifications (see Section 3.3.3) that are being defined within the GGF.

Throughout the development of various GT releases, Globus designers have focused on providing tools that implement common interfaces for interacting with heterogeneous system components. Specifically, GT has defined and implemented protocols, APIs, and services that provide common solutions to usability and interoperability problems such as authentication, resources discovery, and resources access. GT addresses these issues through mechanisms that provide security, information discovery, resources management, data management, communication, fault detection, and portability.

GT's resource layer protocols are used for remote invocation of computation, resource discovery and monitoring, and data transport. The Grid Resource Allocation and Management (GRAM) suite provides for the secure, reliable creation and management of processes on remote resources. The Monitoring and Discovery Service (MDS) provides a uniform framework for discovering and accessing configuration and status information of Grid resources such as compute server configuration, network status, and the capabilities and policies of services. GridFTP is an extended version of the FTP application and protocol [14]. Extensions include use of connectivity layer security protocols, partial file access, and management of parallelism for high-speed transfer. GT's security layer is defined by the public key-based Grid Security Infrastructure (GSI) protocols [12,13], which provide for single sign-on authentication, communication protection, and for some support for restricted delegation.

The components of Globus Toolkit version 4 (GT4) are often classified in five categories: execution management, security services, data management, information services, and common runtime. The categories are shown in Figure 7.1 (with ref. 11 for authoritative reference) and delineated here:

- For execution management, the toolkit provides services for resource and allocation management and workspace management and a community scheduler.
- For security, the toolkit provides services for handling authentication, authorization, delegation, and community authorization.
- For data management, the toolkit provides services for reliable data transfer, data access and integration, and data replication.
- For information services, the toolkit provides services for monitoring and discovery services.
- For runtime support, the toolkit provides various Web Services core and libraries and Extensible I/O support.

The GT4 provides a set of predefined services. Currently, it implements nine Web Services interfaces and is still evolving:

(1) job management: Grid Resource Allocation and Management (GRAM);
(2) Reliable File Transfer (RFT);
(3) delegation;
(4) Monitoring and Discovery System-index (MDS-index);
(5) Monitoring and Discovery System-trigger (MDS-trigger);
(6) Monitoring and Discovery System-aggregate (MDS-aggregate);

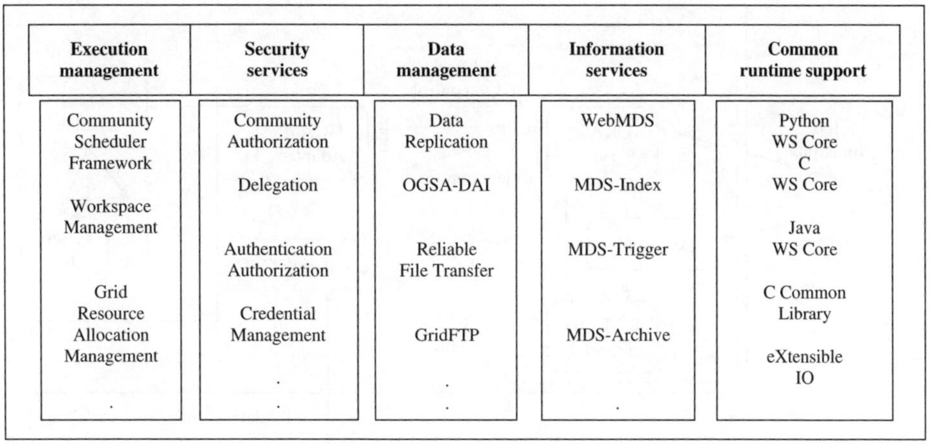

7.3 Grid Infrastructure Software

Execution management	Security services	Data management	Information services	Common runtime support
Community Scheduler Framework	Community Authorization	Data Replication	WebMDS	Python WS Core
Workspace Management	Delegation	OGSA-DAI	MDS-Index	C WS Core
Grid Resource Allocation Management	Authentication Authorization	Reliable File Transfer	MDS-Trigger	Java WS Core
	Credential Management	GridFTP	MDS-Archive	C Common Library
				eXtensible IO
.	.	.	.	

Figure 7.1. Components of the Globus toolkit – version 4 (GT4). Reproduced by permission from Globus Researchers.

(7) Community Authorization (CAS);

(8) Data Access and Integration (OGSA-DAI);

(9) Grid Telecontrol Protocol (GTCP) for online control of instrumentation.

GT4 includes three additional services: GridFTP data transport, Replica Location Service (RLS), and MyProxy online credential repository. Other libraries provide authentication and authorization mechanisms. The Extensible I/O (XIO) library provides access to underlying transport protocols.

GT4 uses the concept of a "container" as the implementation unit within which it is practical to integrate different services. A container implements fundamental message transport protocols, message-level security, and various Web Services interfaces concerning addressing, notification, resource management, and mechanisms for registry administration. Figure 7.2 shows an example of the GT4 container for execution management. The container may host "custom Web Services," whose client interfaces are defined in terms of basic Web Services specifications, "custom WSRF Web Services," whose client interfaces make use of WSRF and related mechanisms, and "GT4 WSRF Web Services" – advanced services provided by GT4, such as GRAM, MDS, and reliable file transfer. A GT4 container also contains administration interfaces for basic administration functions.

Often, Grid application requirements can be satisfied only by using resources simultaneously at multiple sites. Applications use an extensible Resource Specification Language (RSL) to communicate requests for resources to resource brokers, resource co-allocators, and eventually resource managers. Resource meta-schedulers take high-level RSL specifications and transform them into concrete requests that can be passed to a resource co-allocator. A resource co-allocator breaks a composite request into its constituent elements and distributes each component to the appropriate resource manager. Each resource manager in the system is responsible for taking a relevant component of the RSL request and mapping it into the operations within the local, site-specific resource management system.

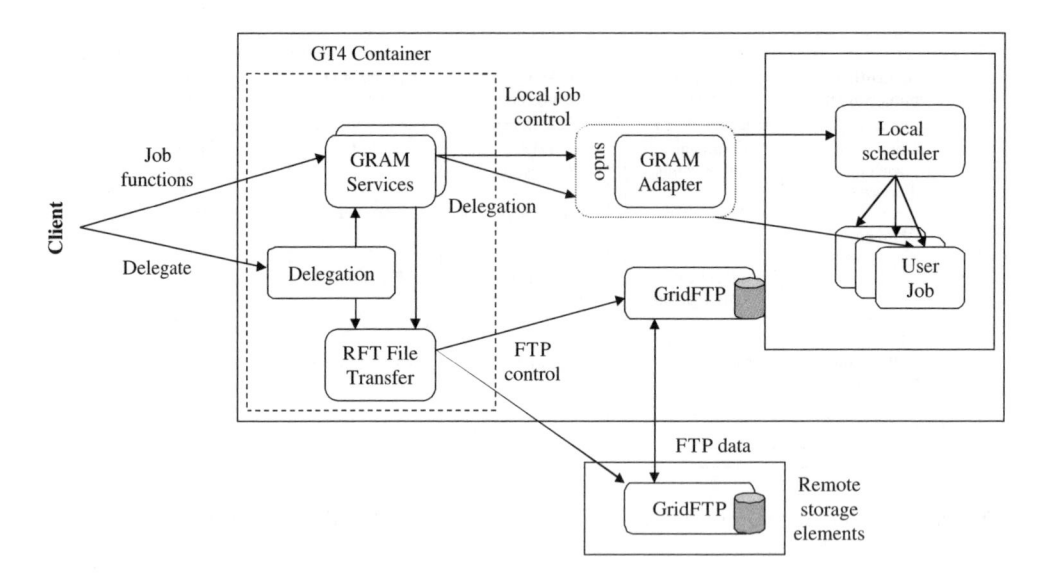

Figure 7.2. GRAM implementation structure [15]. Reproduced by permission from Globus Researchers.

7.3.1.1 Execution management

Execution management tools support the initiation, management, scheduling, and coordinating of remote computations. GT4 provides a suite of Web Services, collectively termed WS-GRAM (Web Services Grid resource allocation and management) for creating, monitoring, and managing jobs on local or remote computing resources.

An execution management session begins when a job order is sent to the remote compute host. At the remote host, the incoming request is subject to multiple levels of security checks. WS-Security mechanisms are used to validate the request and to authenticate the requestor. A delegation service is used to manage delegated credentials. Authorization is performed through an authorization callout. Depending on the configuration, this callout may consult a "Grid-mapfile" access control list, a Security Assertion Markup Language (SAML) [16] server, or other mechanisms. A scheduler-specific GRAM adapter is used to map GRAM requests to appropriate requests on a local scheduler. GRAM is not a resource scheduler, but rather a protocol engine for communicating with a range of different local resource schedulers using a standard message format. The GT4 GRAM implementation includes interfaces to Condor and Load Sharing Facility (LSF) [17] and Portable Batch System (PBS) [18] schedulers, as well as to a "fork scheduler" that simply forks a new process, e.g., a Unix process, for each request.

As a request is processed, a "ManagedJob" entity is created on the compute host for each successful GRAM job submission and a handle (i.e., a WS-Addressing [19] end-point reference, or EPR) for this entity is returned. The handle can be used by the client to query the job's status, kill the job, and/or "attach" to the job to obtain notifications of changes in job status and output produced by the job. The client can also pass this handle to other clients, allowing them to perform the same operations

7.3 Grid Infrastructure Software

if authorized. For accounting and auditing purposes, GRAM deploys various logging techniques to record a history of job submissions and critical system operations.

The GT4 GRAM server is typically deployed in conjunction with delegation and Reliable File Transfer (RFT) servers to address data staging, delegation of proxy credentials, and computation monitoring and management. The RFT service is responsible for data staging operations associated with GRAM. Upon receiving a data staging request, the RFT service initiates a GridFTP transfer between the specified source and destination. In addition to conventional data staging operations, GRAM supports a mechanism for incrementally transferring output file contents out of the site where the computational job is running.

The Community Scheduler Framework (CSF [20]) is a powerful addition to the concepts of execution management, specifically its "collective" aspect of resource handling that might be required for execution. CSF introduces the concept of a meta-scheduler capable of queuing, scheduling, and dispatching jobs to resource managers for different resources.

One such resource manager can represent the network element, and, therefore, provide a noteworthy capability that could be used for Grid network infrastructure.

7.3.1.2 Data management

Data management tools are used for the location, transfer, and management of distributed data. GT4 provides various basic tools, including GridFTP for high-performance and reliable data transport, RFT for managing multiple transfers, RLS for maintaining location information for replicated files, and Database Access and Integration Services (DAIS) [21] implementations for accessing structured and semistructured data.

7.3.1.3 Monitoring and discovery

Monitoring is the process of observing resources or services for such purposes as tracking use as opposed to inventorying the actual supply of available resources and taking appropriate corrective actions related to allocations.

Discovery is the process of finding a suitable resource to perform a task, for example finding and selecting a compute host to run a job that has the correct CPU architecture and the shortest submission queue among multiple, distributed computing resources.

Monitoring and discovery mechanisms find, collect, store, and process information about the configuration and state of services and resources.

Facilities for both monitoring and discovery require the ability to collect information from multiple, perhaps distributed, information sources. The GT4's MDS provides this capability by collating up-to-date state information from registered information sources. The MDS also provides browser-based interfaces, command line tools, and web service interfaces that allow users to query and access the collated information. The basic ideas are as follows:

- Information sources are explicitly registered with an aggregator service.
- Registrations have a lifetime. Outdated entries are deleted automatically when they cease to renew their registrations periodically.

- All registered information is made available via an aggregator-specific Web Services interface.

MDS4 provides three different aggregator services with different interfaces and behaviors (although they are all built upon a common framework). MDS-Index supports Xpath queries on the latest values obtained from the information sources. MDS-trigger performs user-specified actions (such as sending email or generating a log file entry) whenever collected information matches user-defined policy statements. MDS-archiver stores information source values in a persistent database that a client can query for historical information.

GT4's MDS makes use of XML [22] and Web Services to register information sources, to locate, and to access required information. All collected information is maintained in XML form and can be queried through standard mechanisms. MDS4 aggregators use a dynamic soft-state registration of information sources with a periodic refreshing of the information source values. This dynamic updating capability distinguishes MDS from a traditional static registry as accessible via a UDDI [23] interface. By allowing users to access "recent" information without accessing the information sources directly and repeatedly, MDS supports scalable discovery.

7.3.1.4 Security

GT4 provides authentication and authorization capabilities built upon the X.509 [24] standard for certificates. End-entity certificates are used to identify persistent entities such as users and servers. Proxy certificates are used to support the temporary delegation of privileges to other entities.

In GT4, WS-Security [25] involves an authorization framework, a set of transport-level security mechanisms, and a set of message-level security mechanisms. Specifically:

- Message-level security mechanisms implement the WS-Security standard and the WS-SecureConversation specification to provide message protection for GT4's transport messages;.
- Transport-level security mechanisms use the Transport Layer Security (TLS) protocol [26].
- The authorization framework allows for a variety of authorization schemes, including those based on a "Grid-mapfile" access control list, a service-defined access control list, and access to an authorization service via the SAML protocol. For components other than Web Services, GT4 provides similar authentication, delegation, and authorization mechanisms.

7.3.1.5 General-purpose Architecture for Reservation and Allocation (GARA)

The GRAM architecture does not address the issue of advance reservations and heterogeneous resource types. Advance reservations semantics can guarantee that a resource will deliver a requested QoS at the time it is needed, without requiring that the resource be made available beginning at the time that the request is first made.

To address the issue of advanced reservations, the General-purpose Architecture for Reservation and Allocation (GARA) has been proposed [27]. With the separation of reservation from allocation, GARA enables advance reservation of resources, which can be critical to application success if a required resource is in high demand. Also, if reservation is relatively more cost-effective than allocation, lightweight resource reservation strategies can be employed instead of schemes based on either expensive or overly conservative allocations of resources.

7.3.1.6 GridFTP

The GridFTP software for end-systems is a powerful tool for Grid users and applications. In a way, GridFTP sets the end-to-end throughput benchmark for networked Grid solutions for which the network is an unmodifiable, unknowable resource and the transport protocol of choice is standard TCP. GridFTP builds upon the FTP set of commands and protocols standardized by the IETF [14,28,29]. The GridFTP aspects that enable independent implementations of GridFTP client and server software to interwork are standardized within the GGF [30]. Globus' GridFTP is an implementation that conforms to [30].

GridFTP's distinguishing features include:

- restartable transfers;
- parallel data channels;
- partial file transfers;
- reusable data channels;
- striped server mode;
- GSI security on control and data channels.

Of particular relevance to the interface with the network are the striped server feature and the parallel data channel feature, which have been shown to improve throughput. With the former feature, multiple GridFTP server instantiations at either logical or physical nodes can be set to work on the same data file, acting as a single FTP server. With the parallel data channel feature, the data to be transferred is distributed across two or more data channels and therefore across independent TCP flows. With the combined use of striping and parallel data channels GridFTP can achieve nearly 90% utilization of a 30-Gbps link in a memory-to-memory transfer (27 Gbps [31]). When used in a disk-to-disk transfer, it resulted in a 17.5-Gbps throughput given the same 30-Gbps capacity [31].

The use of parallel data channels mapped to independent TCP sessions results in a significantly higher aggregate average throughput than can be achieved with a single TCP session (e.g., FTP) in a network with typical loss probability and Bit Error Ratio (BER). Attempts have been made to quantify a baseline for the delta in throughput, given the three simplifying assumptions that the sender always has data ready to send, the costs of fan-out and fan-in to multiple sessions are negligible, and the end-systems afford unlimited I/O capabilities [32].

GridFTP can call on a large set of TCP ephemeral ports. It would be impracticable (and unsafe) to have all these ports cleared for access at the firewall, a priori. At the GGF, the Firewall Issues Research Group [33] is chartered to characterize the issues with (broadly defined) firewall functions.

The GridFTP features bring new challenges in providing for the best matches between the configurations of client and server to a network, while acknowledging that many tunable parameters are in fact co-dependent. Attempts have been made to provide insights into how to optimally tune GridFTP [34]. For example, rules can be based upon prior art in establishing an analytical model for an individual TCP flow and predicting its throughput given round-trip time and packet loss [34].

In general, GridFTP performance must be evaluated within the context of the multiple parameters that relate to end-to-end performance, including system tuning, disk performance, network congestion, and other considerations. The topic of layer 4 performance is discussed in Chapters 8 and 9.

7.3.1.7 Miscellaneous tools

The eXtensible I/O library (XIO) is an I/O library that is capable of abstracting any bytestream-oriented communication under primitive verbs: open, close, read, write. XIO is extensible in that multiple "drivers" can be attached to interface with a new bytestream-oriented communication platform. Furthermore, XIO's drivers can be composed hierarchically to realize a multistage communication pipeline.

In one noteworthy scenario, the GridFTP functionalities can be encapsulated in a XIO driver. In this style of operation, an application becomes capable of seamlessly opening and reading files that are "behind" a GridFTP server, yet without requiring an operator to manually run a GridFTP session. In turn, the XIO GridFTP driver can use XIO to interface with transport drivers other than standard TCP.

7.4 GRID NETWORK INFRASTRUCTURE SOFTWARE

Section 7.3 has described end-system software that optimally exploits a network that is fixed and does not expose "knobs," "buttons," and "dials" for the provisioning and/or control of the network's behavior. Instead, the end-systems' software must adapt to the network.

This section reviews the role of Grid network infrastructure software that can interact with the network as a resource. When applied to a flexible network, this software allows a network to adapt to applications. This capability is not one that competes with the one described earlier. Rather, it is seen as a synergistic path toward scaling Grid constructs toward levels that can provide both for the requirements of individual applications and for capabilities that have a global reach.

The research platform DWDM-RAM [35–37] has pioneered several key aspects of Grid network infrastructure. Anecdotally, the "DWDM-RAM" project name was selected to signify the integration of an optical network – a dense wavelength division multiplexing network – with user-visible access semantics as simple and intuitive as the ones of shareable RAM. These attributes of simplicity and intuitiveness were achieved because this research platform was designed with four primary goals. First, it encapsulates network resources into a service framework to support the movement of large sets of distributed data. Second, it implements feedback loops between demand (from the application side) and supply (from the network side), in an autonomic fashion. Third, it provides mechanisms to schedule network

7.4 Grid Network Infrastructure Software

resources, while maximizing network utilization and minimizing blocking probability. Fourth, it makes reservation semantics an integral part of the programming model.

This particular Grid network architecture was first reduced to practice on an advanced optical testbed. However, the use of that particular testbed infrastructure is incidental. Its results are directly applicable to any network that applies admission control based on either capacity considerations or policy considerations or both, regardless of its physical layer – whether it is optical, electrical, or wireless. With admission control in place, there is a nonzero probability ("blocking probability") that a user's request for a service of a given quality will be denied. This situation creates the need for a programming model that properly accounts for this type of probability.

Alternative formulations of Grid network infrastructure include User-Controlled Lightpaths (UCLPs) [38], discussed in Chapter 5, the VIOLA platform [39], and the network resource management system [40].

7.4.1 THE DWDM-RAM SYSTEM

The functions of the DWDM-RAM platform are described here. To request the migration of a large dataset, a client application indicates to DWDM-RAM the virtual endpoints that source and sink the data, the duration of the connection, and the time window in which the connection can occur, specified by the starting and ending time of the window. The DWDM-RAM software reports on the feasibility of the requested operation. Upon receiving an affirmative response, DWDM-RAM returns a "ticket" describing the resulting reservation. This ticket includes the actual assigned start and end times, as well as the other parameters of the request. The ticket can be used in subsequent calls to change, cancel, or obtain status on the reservation. The DWDM-RAM software is capable of optimally composing different requests, in both time and space, in order to maximize user satisfaction and minimize blocking phenomena. After all affirmations are completed, it proceeds to allocate the necessary network resources at the agreed upon time, as long as the reservation has not been canceled or altered.

Table 7.1 shows three job requests being issued to the DWDM-RAM system. Each of them indicates some flexibility with regard to actual start times. Figure 7.3 shows how DWDM-RAM exploits this flexibility to optimally schedule the jobs within the context of the state of network resources.

Table 7.1 Request requirements

Job	Job run-time	Window for execution
A	1.5 hours	8 am to 1 pm
B	2 hours	8.30 am to 12 pm
C	1 hour	10 am to 12 pm

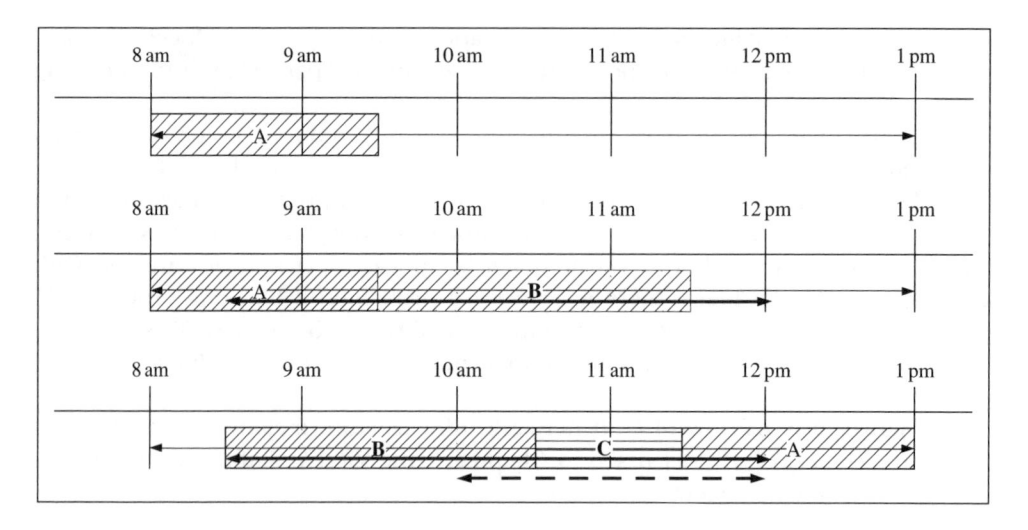

Figure 7.3. The DWDM-RAM Grid network infrastructure is capable of composing job requests in time and space to maximize user satisfaction while minimizing the negative impact of nonzero blocking probability.

The DWDM-RAM architecture (Figure 7.4) is a service-oriented one that closely integrates a set of large-scale data services with those for dynamic allocation of network resources by way of Grid network infrastructure. The architecture is extensible and allows inclusion of algorithms for optimizing and scheduling data transfers, and for allocating and scheduling network resources.

At the macro-level, the DWDM-RAM architecture consists of two layers between an application and the underlying network: the application layer and the resource layer.

The application layer responds to the requirements of the application and realizes a programming model. This layer also shields the application from all aspects of sharing and managing the required resources.

The resource layer provides services that satisfy the resource requirements of the application, as specified or interpreted by the application layer services. This layer contains services that initiate and control sharing of the underlying resources. It is this layer that masks details concerning specific underlying resources and switching technologies (e.g., lambdas from wavelength switching, optical bursts from optical burst switching) to the layer above.

At the application layer, the Data Transfer Service (DTS) provides an interface between an application and Grid network infrastructure. It receives high-level client requests to transfer specific named blocks of data with specific deadline constraints. Then, it verifies the client's authenticity and authorization to perform the requested action. Upon success, it develops an intelligent strategy to schedule an acceptable action plan that balances user demands and resource availability. The action plan involves advance co-reservation of network and storage resources. The application expresses its needs only in terms of high-level tasks and user-perceived deadlines, without knowing how they are processed at the layers below. It is this layer that shields the application from low-level details by translating application-level requests

7.4 Grid Network Infrastructure Software

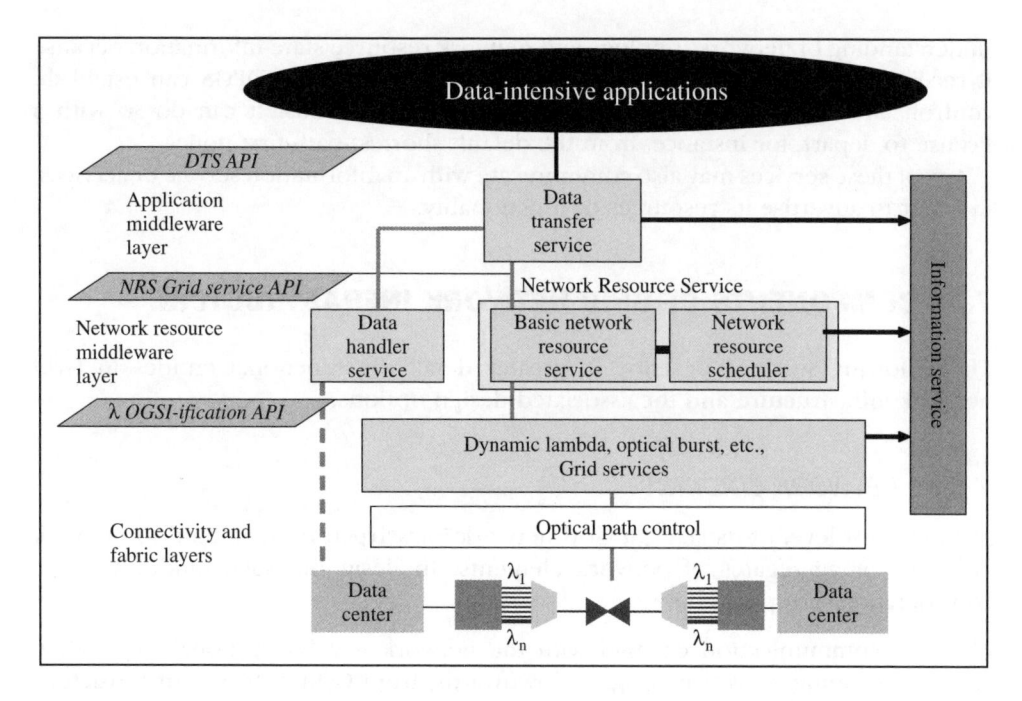

Figure 7.4. The DWDM-RAM architecture for Grid network infrastructure targets data-intensive applications.

into its own tasks, e.g., coordinating and controlling the sharing of a collective set of resources.

The network resource layer consists of three services: the Data Handler Service (DHS), the Network Resource Service (NRS), and the Dynamic Path Grid Service (DPGS). Services provided by this layer initiate and control the actual sharing of resources. The DHS deals with the mechanism for sending and receiving data and performs the actual data transfer when needed by the DTS.

NRS makes use of the DPGS to encapsulate the underlying network resources into an accessible, schedulable Grid service. The NRS queues requests from the DTS and allocates proper network resources according to its schedule. To allow for extensibility and reuse, the NRS can be decomposed into two closely coupled services: a basic NRS and a network resource scheduler. The basic NRS presents an interface to the DTS for making network service requests and handling multiple low-level services offered by different types of underlying networks and switching technologies.

The network resource scheduler is responsible for implementing an effective schedule for network resources sharing. The network resource scheduler can be implemented independently of the basic NRS. This independence provides the NRS with the flexibility to deal with other scheduling schemes as well as other types of dynamic underlying networks.

The DPGS receives resource requirement requests from the NRS and matches those requests with the actual resources, such as path designations. It has complete understanding of network topology and network resource state information because it receives this information from lower level processes. The DPGS can establish, control, and deallocate complete end-to-end network paths. It can do so with a license to depart, for instance, from the default shortest-path-first policy.

Any of these services may also communicate with an information service or services, in order to advertise its resources or functionality.

7.5 COMPONENTS OF GRID NETWORK INFRASTRUCTURE

The following sections describe in greater detail the functional entities in Grid network infrastructure and the associated design options.

7.5.1 NETWORK BINDINGS

At the lowest level of its stack, a Grid network infrastructure must bind to network elements or aggregates of network elements. In designing such bindings, three considerations are paramount:

(1) The communication channel with the network is a bidirectional one. While provisioning actions propagate downwards, from Grid network infrastructure into the network, the monitoring actions result in information that must be propagated upwards.

(2) It is typical that network elements expose many different, often proprietary, mechanisms for provisioning and retrieval of information such as statistics.;

(3) In communicating with the network, the network side of the interfaces is one that in general cannot be altered.

These considerations pose a general requirement that the network bindings be extensible to implement various client sides of provisioning protocols and information retrieval protocols.

The mechanisms to either push provisioning statements or pull information fall in two realms:

(1) control plane bindings;

(2) network management bindings.

The control plane is in many ways the network's "intelligence," for example it undertakes decisions on path establishment and recovery routinely and autonomically, within a short time (e.g., seconds or milliseconds in some cases).

Network management incorporates functions such as configuration, control, traffic engineering, and reporting that allow a network operator to perform appropriate network dimensioning, to oversee network operation, to perform measurements, and to maintain the network [41]. Unlike the control plane, the network management plane has historically been tailored to an operator's interactive sessions, and usually exhibits a coarse timescale of intervention (hours or weeks). Network management bindings exploit preexisting network management facilities and specifically invoke actions that can be executed without total reconfiguration and operator involvement.

Should the network community converge on a common virtualization technique (e.g., WBEM/CIM [42]), the role of network bindings will be greatly simplified, especially with regard to the network management bindings.

Regardless of the mechanisms employed, the bindings need to be secured against eavesdropping and malicious attacks that would compromise the binding and result in the theft of network resources or credentials. The proper defense against these vulnerabilities can be realized with two-way authentication and confidentiality fixtures such as the ones found in the IPsec suite of protocols.

7.5.1.1 Control plane bindings

Control plane bindings interact with functions that are related to directly manipulating infrastructure resources, such as through legacy network control planes (e.g., GMPLS [43], ASTN [44]) by way of:

- service-oriented handshake protocols, e.g., a UNI like the Optical Interworking Forum (OIF) UNI [45] (see Chapter 12);
- direct peering, where the binding is integrated with one or more of the nodes that actively participate in the control plane;
- proprietary interfaces, which network vendors typically implement for integration with Operations Support Systems (OSSs).

The UNI style of protocol is useful for binding when such a binding must cross a demarcation line between two independently managed, mutually suspicious domains. In this case, it is quite unlikely that the target domain will give the requesting domain an access key at the control plane level. This approach would require sharing more knowledge than is required between domains, and contains an intrinsic weakness related to the compartmentalization required to defend against failures or security exploitations. In contrast, an inter-domain service-oriented interface enables code in one domain to express its requirements to another domain without having to know how the requested service specification will be implemented within that recipient domain.

7.5.1.2 Network management bindings

As explained in earlier sections, these bindings can leverage only a small subset of the overall network management functionality. Techniques like Bootstrap Protocol (BOOTP), configuration files, and Graphical–User Interface (GUI) station managers are explicitly not considered for such bindings, in that they do not lend itself to the use required – dynamic, scripted, operator-free utilization.

Command Line Interfaces (CLIs). A CLI is a set of text-based commands and arguments with a syntax that is used for network elements. The CLI is specified by the element manufacturer and it can be proprietary. While most CLI sessions involve an operator typing at a console, CLIs have also been known to be scriptable, with multiple commands batched into a shell-like script.

Transaction Language 1 (TL1). As a special manifestation of CLI, TL1 [46] standardizes a set of ASCII-based commands that an operator or an OSS can use to manage a network element. Although SNMP/Management Information Bases (MIBs) dominate the enterprise, TL1 is a widely implemented management protocol for controlling telecommunications networks and its constituent network elements. It has received multiple certifications, such as OSMINE (operations systems modificaions for the integration of network elements).

The Simple Network Management Protocol (SNMP). SNMP is a protocol to create, read, write, and delete MIB objects. An MIB is a structured, named dataset that is expressed in ASN.1 basic coding rules and adheres to IETF RFC standard specifications whenever the management data concerns standardized behaviors (e.g., TCP tunable parameters and IP statistics). SNMP is a client–server protocol. Management agents (clients) connect to the managed devices and issue requests. Managed devices (servers) return responses. The basic requests are GET and SET, which are used to read and write to an individual MIB object, identified by its label identifier (or object identifier, OID). SNMP has a message called TRAP (sometimes known as a notification) that may be issued by the managed device to report a specific event. The IETF has standardized three versions of SNMP. SNMPv1 [47], the first version, and SNMPv2 [48] do not have a control process that can determine who on the network is allowed to perform SNMP operations and access MIB modules. SNMPv3 [49] includes application-level cryptographic authentication for authentication.

XML data representation combined with new protocols in network management. The nearly ubiquitous use of SNMP/MIBs has pointed out the limited and often cumbersome syntax for the management data definition language. Also, the explosive growth in XML adoption within several industries has generated interest in matching XML with requirements for network management data. XML is a subset of the SMGL specified in the ISO 8879. XML defines data objects known as XML documents and the rules by which applications access these objects. XML provides encoding rules for commands that are used to transfer and update data objects. The strength of XML is in its extensibility. The IETF is standardizing the Netconf as a protocol to transport XML-based data objects [50]. Although the Netconf protocol should be independent of the data definition language, the development of security features closely links Netconf with XML. XML objects can also be carried with XML Remote Procedure Call (RPC) [51] and SOAP [52].

Web Based Enterprise Management (WBEM)/Common Information Model (CIM). Also related to XML is the Distributed Management Task Force's (DMTF) WBEM body of specifications [42], which defines an important information architecture for distributed architecture, the Common Information Model (CIM), as well as XML representation of data as messages and message carriage via HTTP (though other associations are possible). CIM has developed a noteworthy approach to data modeling, one that is directly inspired by object-oriented programming (e.g., abstraction, inheritance). As such, it promotes abstraction, model reuse, and consistent semantics for networking (as well as for other information technology resources, such as storage devices and computers).

It has been said that proper network bindings allow information to be fetched and allow the Grid network infrastructure code to monitor the status of the network nodes and to detect network conditions such as faults, congestion, and network "hotspots" so that appropriate decisions can be made as quickly as possible. The aforementioned techniques are capable of a two-way dialog.

With regard to SNMP/MIBs, IETF standard specifications describe how to structure usage counters that provide basic statistical information about the traffic flows through specific interfaces or devices [53]. SNMP does have limitations related to network monitoring. They stem from the inherent request–response of the SNMP (and many other network management techniques). Repeated operations to fetch MIB data require the network node each time to process a new request and proceed with the resolution of the MIB's OID. In addition, the repetition of request–response cycles results in superfluous network traffic.

Various solutions specializing in network monitoring have emerged. NetFlow, for instance, allows devices to transmit data digests to collection points. The effectiveness of this approach led to the creation of a new standardization activity in the IETF around "IP flow information export" [54]. Chapter 13 reviews several concepts and technologies in the network monitoring space.

7.5.2 VIRTUALIZATION MILIEU

7.5.2.1 Web Services Resource Framework

The Web Services Resource Framework (WSRF) is a recent effort by OASIS to establish a framework for modeling and accessing stateful, persistent resources such as components of network resources through Web Services. Although general Web Services have been and still are stateless, Grids have introduced requirements to manipulate stateful resources, whether these are long-lived computation jobs or service level specifications. Together with the WS-Notification [55] specifications, WSRF supersedes the earlier work by the GGF known under the name of Open Grid Services Infrastructure (OGSI). WSRF is a new important step in bringing the management of stateful resources into the mainsream of the Web Services movement. Although widespread market adoption may dictate further evolutions of the actual framework specification, the overall theme is well delineated.

In Web Services, prior attempts to deal with statefulness had resulted in some ad hoc, idiosyncratic way to reflect a selected resource at the interface level in the WSDL document. The chief contribution of the WSRF work is to standardize a way to add an identifier to the message exchange, such that the recipient of the message can use the identifier to map it to a particular, preexisting context – broadly identified as resource – for the full execution of the request carried in the message as well as in the subsequent messages.

Specifically, a resource that a client needs for access across multiple message exchanges is described with a resource properties document schema, which is an XML document. The WSDL document that describes the service must reference the resource properties document for the definition of that resource as well as of other resources that might be part of that service. The Endpoint Reference (EPR), which is to a web service message what an addressing label is to a mail envelope, becomes

```
<wsa:EndpointReference>
  <wsa:Address>http://www.starlight.net/ReservationService
    </wsa:Address>
  <wsa:ReferenceParameters>
     <rpimpl:SLAType>Premium</rpimpl:SLAType>
     <rpimpl:CustomerTicket>123456</rpimpl:CustomerTicket>
  </wsa:ReferenceParameters>
</wsa:EndpointReference>
```

Figure 7.5. Evolution of an endpoint reference in WSRF.

the designated vehicle to carry the resource identifier also. The EPR is transported as part of the SOAP header. With the EPR used as a consolidated locus for macro-level (i.e., service) and micro-level (i.e., resource) identification, the source application does not have to track any state identification other than the EPR itself. Also, the WSDL document is no longer burdened by any resource identification requirement, and the SOAP body is relieved from carrying resource identification.

Figure 7.5 gives a notional example of an EPR that conforms to WS-Addressing and is ready for use within WSRF. The EPR is used when interacting with a fictional bandwidth reservation service located at a "StarLight" communications exchange. The new style of resource identification is included between reference parameters tags.

The use of a network endpoint in an EPR may be problematic when there is a firewall at the junction point between enclaves with different levels of trust. A firewall will set apart the clients in the local network, which can access the service directly, from the external clients, which might be asked to use an "externalized" EPR to route their messages through an application gateway at the boundary between enclaves.

7.5.2.2 WS-Notification

While extremely powerful, WSRF alone cannot support the complete life cycle of stateful, persistent resources. WSRF deals only with synchronous, query–response interactions. In addition, it is important to have forms of asynchronous messaging within Web Services. the set of specfications know as WS-Notification indeed brings asynchronous messaging and specifically publish/subscribe semantics to Web Services.

In several distributed computing efforts, the practicality of general-purpose publish/subscribe mechanisms has been shown – along with companion techniques that can subset channels (typically referred to as "topics") and manage the channel space for the highest level of scalability.

In WS-Notification, the three roles of subscriber, producer, and consumer are defined. The subscriber posts a subscription request to the producer. Such a request indicates the type of notifications and consumers for which there is interest. The producer will disseminate notifications in the form of one-way messages to all consumers registered for the particular notification type.

7.5.2.3 WS-Agreement

The pattern of a producer and a consumer negotiating and implementing a SLA recurs in many resource domains, whether it is a computation service or a networking service.

The GGF has produced the WS-Agreement specification [56] of a "SLA design pattern". It abstracts the whole set of operations – e.g., creation, monitoring, expiration, termination – that mark the life cycle of a SLA.

Each domain of utilization – e.g., networking – requires a companion specification to WS-Agreement, to extend it with various domain-specific terms for a SLA. Research teams (e.g., ref. 57) have efforts under way to experiment with domain-specific extensions to WS-Agreement for networking.

Once it is associated to domain-specific extension(s), a WS-Agreement can be practically implemented with Web Services software executing at both the consumer side and the provider side.

7.5.2.4 Nomenclatures, hierarchies, and ontologies: taking on semantics

The sections on WSRF and WS-Notification have focused on syntax issues. Once proper syntax rules are in place, an additional challenge is designing various data models and definitions. Consider, for instance, three different network monitoring techniques that produce XML digests with "gigabit per second", "bandwidth", and "throughput" tags. It is difficult to relate the three digests in a meaningful, computer-automated way. The same situation occurs with providers advertising a service with different tags and a different service language. Although the remainder of the section focuses on the monitoring use case, these considerations apply to the service language as well.

One approach is to establish a common practice to specify properties like network throughput and pursue the broadest support for such practice in the field. Within the GGF, the Network Monitoring Working Group (NM-WG) [58] has established a goal of providing a way for both network tool and Grid application designers to agree on a nomenclature of measurement observations taken by various systems [59]. It is a NM-WG goal to produce a XML schema matching a nomenclature like the one shown in Figure 7.6 to enable the exchange of network performance observations between different monitoring domains and frameworks. With suitable authentication and authorization it will also be possible to request observations on demand.

Another design approach is to live with diverse nomenclatures and strengthen techniques that allow computers to navigate the often complex relationships among characteristics defined by different nomenclatures. This approach aligns well with the spirit of the Semantic Web [60], a web wherein information is given well-defined meaning, fostering machine-to-machine interactions and better enabling computers to assist people. New projects [61,62] champion the use of languages and toolkits (e.g., Resource Description Framework (RDF) [63] and Web Ontology Language (OWL) [64]) for describing network characteristics in a way that enables automated semantic inferences.

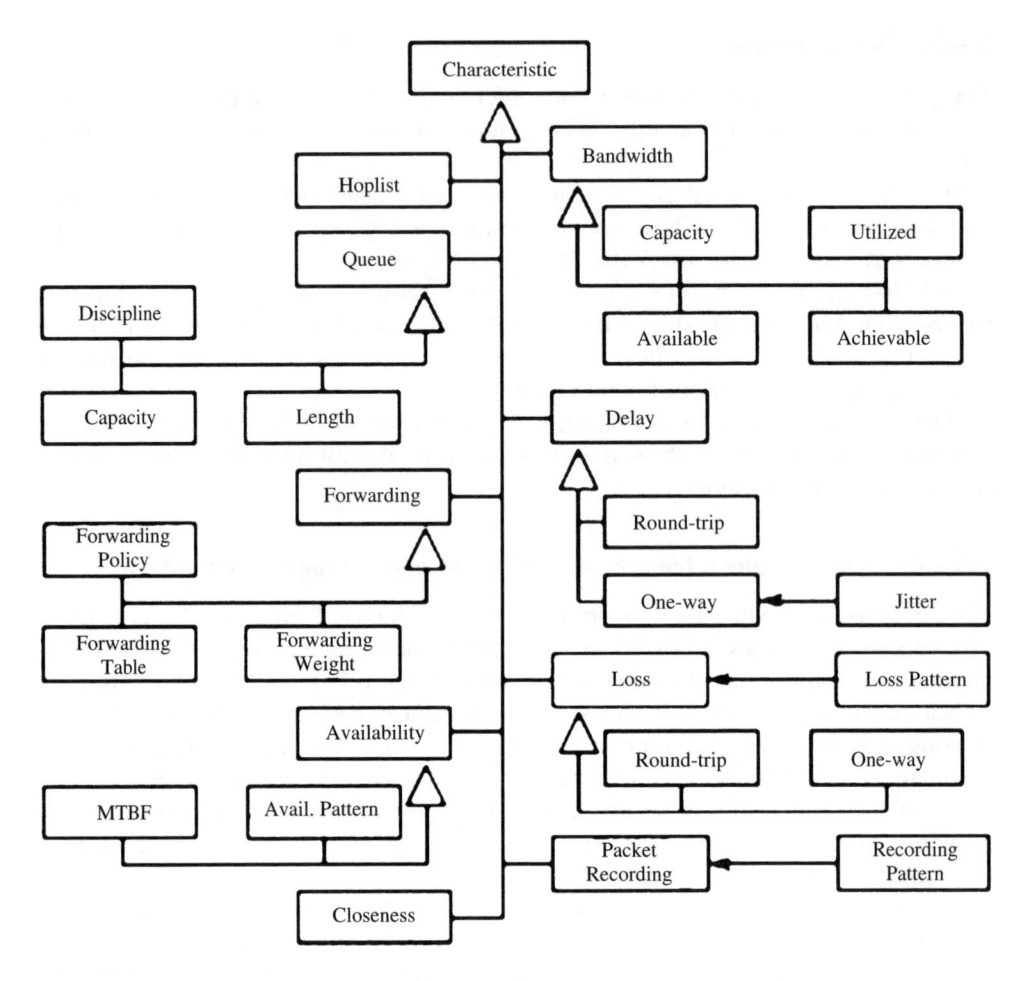

Figure 7.6. GGF NM-WG's nomenclature for network characteristics structured in a hierarchical fashion [59].

7.5.3 PERFORMANCE MONITORING

In its definitions of attributes, Chapter 3 emphasizes the need for determinism, adaptive provisioning schemas, and decentralized control. A Grid network infrastructure must then be associated with facilities that monitor the available supply of network resources as well as the fulfillment of outstanding network SLAs. These facilities can be considered a way of closing a real-time feedback loop between network, Grid infrastructure, and Grid application(s). The continuous acquisition of performance information results in a dataset that Grid network infrastructure can value for future provisioning actions. For greatest impact, the dataset needs to be fed into Grid infrastructure information services (e.g., Globus MDS), which will then best assist the resource collective layers (e.g., the community scheduler framework [20]) concerned with resource co-allocations. Chapter 13 features performance monitoring in greater detail and analyzes how performance monitoring helps the cause of fault detection.

Given the SOA nature of the infrastructure, the performance monitoring functions can either be assembled as a closely coupled component or packaged as a Grid network service in its own right or accessed as a specialized forecasting center (e.g., the Network Weather Service, NWS [65]).

Before a Grid activity with network requirements is launched, Grid network infrastructure can drive active network measurements to estimate bandwidth capacity, using techniques like packet pairs [66]. Alternately, or as a parallel process, they can interrogate a NWS-like service for a traffic forecast. Depending on the scale of operations, the Grid network infrastructure can drive latency measurements or consult a general-purpose, nearest-suited service [67]. The inferred performance information determines the tuning of network-related parameters, provisioning actions, and the performance metrics to be negotiated as part of the network SLA.

While a Grid application is in progress, Grid network infrastructure monitors the fulfillment of the SLA(s). It can do so through passive network measurements techniques, such as SNMP traps, RMONs, NetFlow (reviewed in Section 7.5.1.2). An exception must be generated whenever the measurements fall short of the performance metrics negotiated in the SLA. Grid applications often welcome the news of network capacity becoming available while they are executing (e.g., to spawn more activities). For these, the Grid network infrastructure should continue to perform active network measurements and mine available bandwidth or lower latency links. In this circumstance, the measurements may be specialized to the context of the network resources already engaged in the outstanding Grid activity. It is worth noting that the active measurements are intrusive in nature and could negatively affect the Grid activities in progress.

7.5.4 ACCESS CONTROL AND POLICY

Decisions pertaining access control and policy need to be taken at different levels in a Grid network infrastructure, as shown in Figure 7.7.

The layout in Figure 7.7 is loosely modeled after the taxonomy established by the Telecommunications Management Network (TMN) [68].

Starting from the top, there are access control and policy rules that apply to the ensemble of resources in a virtual organization. The network happens to be one of several resources being sought. Furthermore, the roles that discriminate access within the virtual organization are likely to be different than the native ones (much like an individual can be a professor at an academic institution, a co-principal investigator in a multi-institutional research team, and an expert advisor in a review board).

At the service level, there are access control and policy rules that govern access to a service level specification (e.g., a delay-tolerant data mover, or a time of day path reservation).

At the network level, there are access control and policy rules for network path establishment within a domain.

At the bindings level, there are access control and policy rules to discipline access to a network element or a portion of the same, whether it is through the management

```
┌─────────────────────────────┐
│   Access Control & Policy   │
│     at Collective Level     │
├─────────────────────────────┤
│   Access Control & Policy   │
│      at Service Level       │
├─────────────────────────────┤
│   Access Control & Policy   │
│      at Network Level       │
├─────────────────────────────┤
│   Access Control & Policy   │
│     at Bindings Level       │
├─────────────────────────────┤
│      Network Elements       │
└─────────────────────────────┘
```

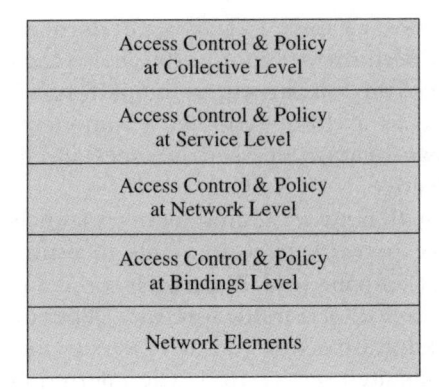

Figure 7.7. Access control and policy considerations are important at multiple places and at different times within a Grid network infrastructure stack, and at each point and time there is a different scope. The scope is larger as one moves upwards in the stack.

plane or the control plane. These rules also thwart exploitations due to malicious agents posing as worthy Grid network infrastructure.

The aforementioned manifestations of access control and policy map well to the layers identified in the DWDM-RAM example of a Grid network infrastructure (Figure 7.3).

The Grid network infrastructures that are capable of spanning across independently managed domains require additional access control and policy features at the service and collective layers. As discussed in the upcoming section on multidomain implications, these features are singled out in a special AAA agent conforming to the vision for Grid network infrastructure.

7.5.5 NETWORK RESOURCE SCHEDULING

Because Grid network infrastructure still is in its infancy, a debate continues on whether there is a need for a scheduler function capable of assigning network resources to Grid applications according to one or more scheduling policies.

At one end, proponents believe that the network is a shareable resource, with both on-demand and advance reservation modes. As such, the access to network resources needs to be scheduled, in much the same way as are threads in a CPU or railway freight trains. The DWMD-RAM example fits well within this camp. Its capability to schedule underconstrained users' requests and actively manage blocking probability has been anticipated in Section 7.4.1. Recently, some groups [69,70] have further explored scheduling policies and their impact on the network, with results directly applicable to systems such as DWDM-RAM. In general, the scheduling capability is very well matched to the workflow-oriented nature of Grid traffic. In other words, a bursty episode is not random event; instead it is an event marked by a number of known steps occurring before and afterwards.

However, some investigators think of the network as an asset that should not be treated as a shared resource. A PC and a car are common individual assets. One viewpoint is that storage, computation, and networking will become so cheap that they do not warrant the operating costs and additional complexity of managing them as a shared resource. The UCLP system [38] is a popular architecture that, in part, reflects this viewpoint.

Scheduling is most effective when it is used to leverage resource requests that afford some flexibility in use or laxity. The laxity of a task using certain resource is the difference between its deadline and the time at which it would finish executing on that resource if it were to start executing at the start time.

When a Grid network infrastructure includes some scheduling intelligence, there are further design points at which a process can decide which scheduling policies should be used with the scheduler, e.g., whether preemption is allowed, whether the programming model allows negotiation, and what the constraints are.

An obvious constraint is network capacity. A more subtle constraint results from the probability of blocking, i.e., the probability that the network layers will be unable to carry out a task because it does not fit in the envelope of available resources, and its execution would impact tasks of equal or greater priority. While blocking probability has been a well-known byproduct of circuit-oriented networks, it can be argued that any network cloud that regulates access through admission control can be abstracted as a resource with nonzero blocking probability. Blocking probability requires additional domain-specific resource rules. In the DWDM networks in the optical domain, for instance, the capability to have an end-to-end path with the same wavelength (or "color") precludes the use of that wavelength over other partially overlapping paths.

Care should be given to properly control the scheduling syndromes that are found in other systems (both digital systems and human-centered systems) exposing shared access and advance reservations semantics. Syndromes include fragmentation (i.e., the time slots that are left are not optimally suited to mainstream requests), overstated reservations or "no-shows" limiting overall utilization, inadequate resource split between on-demand requests and advance reservations, and starvation of requests (e.g., when the system allows scheduling policies other than First In First Out (FIFO)).

Of special significance are efforts such as those that can relate laxity and data transfer size to the global behavior of a network [69,70]. In addition, they shed light on the opportunity to "right size" reservations with regard to optimal operating points for the network.

7.5.6 MULTIDOMAIN CONSIDERATIONS

Independently administered network domains have long been capable of interconnecting and implementing inter-domain mutual agreements. Such agreements are typically reflected in protocols and interfaces such as the Border Gateway Protocol (BGP) [71] (discussed in Chapter 10) or the External Network-to-Network Interface (E-NNI) [72]. While these mechanisms are quite adequate for "best effort" packet services, they do not address well policy considerations and QoS classes, resulting in "impedance mismatches" at the service level between adjacent domains. Initiatives like the IPSphere Forum [73] are a testimonial to the common interest by providers and equipment vendors in a service-oriented network for premium services.

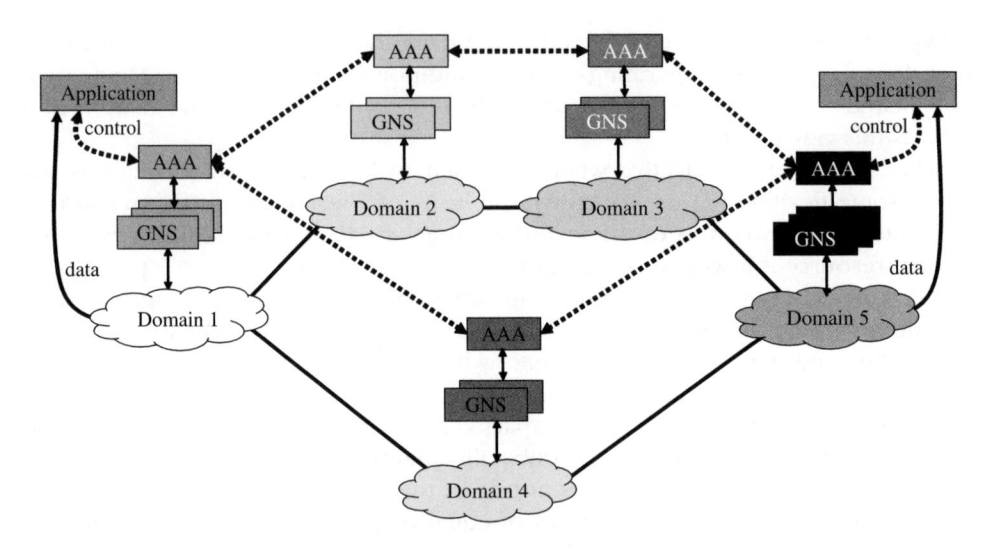

Figure 7.8. The Grid network infrastructures for multiple, independently managed network domains must interconnect with their adjacencies, as shown in this fictional example. They are visualized as a set of Grid Network Services (GNS). A specific GNS is singled out. It is the one that governs Authentication, Authorization, and Accounting (AAA) for a domain's Grid network infrastructure.

Chapter 3 has expressed the need for a discerning Grid user (or application) to exploit rich control of the infrastructure and its QoS classes, such as those described in Chapter 6. To take advantage of these QoS services, the Grid network infrastructure for one domain must be capable of interconnecting with Grid network infrastructure in another domain at the right service level. In the most general case, any two adjacent Grid network infrastructures are independently managed and mutually suspicious, and the most valuable services are those that are most highly protected..

For example, a Grid application in domain 1 (Figure 7.8) requests a very large dataset from a server in domain 5, with a deadline. The request is forwarded to the Grid network infrastructure for domain 1. The Authentication, Authorization and Accounting (AAA) service receives the request and authenticates the user application. Furthermore, it determines whether the request bears the credentials to access the remaining part of Grid network infrastructure for domain 1. Once the request has been validated, the AAA server hands off the request to local Grid network services.

When a Grid network service determines that, for the transfer to occur on time, it is necessary to request a premium QoS class from the network, i.e., guaranteed high and uncontested bandwidth, another Grid network service engages in constructing a source-routed, end-to-end path. It may determine that the path across domains 2 and 3 is the optimal choice, when considering the available resources, the deadline, and the provisioning costs with regard to the specifics of the request and the deadline. It then asks the local AAA to advance the path establishment request to the AAA service in domain 2. This step is repeated until a whole end-to-end nexus is established.

The set of handshakes described in the earlier example is beyond current standards and interoperability agreements. There have been, however, several research efforts addressing these issues. Multidomain efforts are most likely to be conducted first in research-oriented communities like the Global Lambda Integrated Facility (GLIF) [74]. In a research effort at Supercomputing 2004 in Pittsburgh, PA, USA, for instance, a team of several institutions demonstrated the capability to securely negotiate an end-to-end path across four different domains spanning intercontinental distances [75,76]. The experimentation resulted in provisioning times end-to-end in the one minute range. This technique seems useful for large data migrations, and it certainly compares very favorably with the typical turnaround (hours if not days) when operators for multiple organizations are involved.

The following sections articulate some of the key decision points that need to be addressed for interoperability among Grid network infrastructures.

7.5.6.1 Trust model

The inter-domain trust model is a fundamental aspect of the federation of multiple Grid network infrastructures. As it is visualized in the model shown in Figure 7.8, the AAA agent is required to establish a peer relationship with other AAA services representing a transit or a destination domain. Furthermore, there is a trust relationship between an AAA agent and the components of a Grid network infrastructure within each domain.

Once the requesting application is authenticated and authorized by the AAA agent in the source domain, this AAA agent can be set to represent the originating principal and its credentials throughout the whole setup process across domains. The AAA agent is thus given a double role, i.e., to implement access control and policy at the service level for one domain, and to become a requester seeking permission in another domain. The latter domain must mediate request and credentials thereof with the local policy statements, resulting in capabilities to access resources in that domain. Interoperability standards will have to determine how credentials and capabilities are represented across AAA agents.

One research project, which involves the use of "software tokens", is an example of how to communicate widely understood capabilities in an ultra-compact representation (e.g., 128 bits) [77]. A hash-based message authentication code provides the proof of authenticity and integrity of a token. Tokens issued by the proper authority enable acceptance and collection by multiple domains. In addition, tokens can be practically transferred among applications themselves as a currency, for the temporary sublease of services.

Because the token is a compact representation of capabilities, it can optionally be propagated as a recurring watermark directly in the flow of data packets, and it can be validated by edge network devices at full line rate. This particular style of operation binds the actual use of bandwidth with the capabilities. Specifically, it avoids theft of data plane resources by an attacker who attempts to bypass the access control and policy fixtures in the Grid network infrastructure and steals connectivity by interposing at the physical level.

7.5.6.2 Provisioning protocols

There are two basic alternatives to provision a path across the instances of Grid network infrastructure representing the independent domains.

One approach is to commit resources only when all the intervening domains have cleared access control and policy evaluations. While a request travels forward, resources are tested for availability and (optionally) set aside with a "soft" short-lived reservation. On its way back, the request proceeds to committing the resource, whether it is a time-of-day reservation or an instantaneous allocation. The processing protocol must be robust to two or multiple requests producing a race condition over the same resource set.

Another approach is to aggressively commit resources in the forward direction. The subsequent domains are not yet known to clear access control and policy evaluation, and thus the provisioning protocol must be capable of undoing earlier resource commits.

Any time during the life cycle of a service, it might be appropriate to roll back resource commits across all the supporting domains and thus cleanly undo the effects of multidomain provisioning.

The aforementioned provisioning strategies represent recurring traits in networking protocols. They can be practically implemented as Web Services message exchanges. Alternately, they can be implemented by way of off-the-shelf session control protocol suites such as the Session Initiation Protocol (SIP) [78].

7.5.6.3 Path computation

Through its procedures for binding to the network, a Grid network infrastructure becomes capable of actively controlling the pool of inter-domain routes and affects their usage. Given a requested QoS class, for instance, it may elect to route the traffic through a set of domains that supply an uncontested high-capacity circuit and yet do not represent the shortest-distance path between source and destination.

At one end of the spectrum, a Grid network infrastructure may only control the decision over the next hop. Traditionally, this affords the highest degree of adaptivity and scaling while keeping the requirements on distributed topology awareness to a minimum.

At the other end of the spectrum, a Grid network infrastructure may be capable of producing a "flight plan" which comprises hops over two or more subsequent domains, shifting closer to a source-routed paradigm. This model can inject a user's preference in a rather concise way. It does require, however, a wider distribution and synchronization of directories wherein domain capabilities and inter-domain routes are stored.

The latter paradigm has proven useful in so-called hybrid networks with domains that specialize in either packet services or optical services. When bursty traffic is about to start, a Grid application signals its closest Grid network infrastructure to broker a multidomain path entirely in the optical domain, so as to achieve greater throughput efficiencies without disrupting the regular routed traffic. This behavior is often referred to as "cloud bypass." The SURFnet6 advanced network [79], for instance, has a practical realization of a cloud bypass service handled entirely via Grid network infrastructure and without any operator involvement.

7.5.6.4 Distributed service awareness

For inter-domain path computation, performance prediction, and failure recovery, a Grid network service in a domain may need to gain selective awareness of services advertised by other domains. It may need to discover capabilities resident in different domains and influence the composition of services across domains accordingly. In a Grid network, such constituent services pertain to multiple OSI layers (e.g., IP and optical) and exhibit different access semantics (e.g., on-demand versus advance reservation). Lastly, they can be expressed in different service languages, as noted in Section 7.5.2.4.

This process is similar to the one of establishing a supply chain business relationship while discerning among the various attributes of suppliers. As such, the choice of Web Services as the foundation for Grid network services yields quite a fitting platform. Of particular relevance, the Business Process Execution Language for Web Services (BPEL4WS) [80] is a rich mechanism for describing workflow-like chains. The emerging dynamic semantic composition of workflows (e.g., ref. 81) brings new capabilities in discovering and matching service attributes, still within the architectural framework of Web Services.

With regard to publishing service information, the solutions need to be evaluated with regard to the dominant operating models, including richness of the service descriptions, the rate of changes to the descriptions, the requirements for convergence upon changes to the descriptions, and the service population. When a registry design makes the most sense, UDDI and derivatives are a fitting choice of interface into a registry of services. At the opposite end, a fully decentralized design that draws upon Semantic Web techniques has been proposed [61].

REFERENCES

[1] P.V. Mockapetris (1987) "Domain names – concepts and facilities", RFC 1034, November 1987.
[2] C. de Laat, G. Gross, L. Gommans, J. Vollbrecht, and D. Spence (2000) "Generic AAA Architecture", RFC 2903, August 2000.
[3] G. Clapp, T. Ferrari, D.B. Hoang, A. Jukan, M.J. Leese, P. Mealor, and F. Travostino, "Grid Network Services", Grid Working Document, revisions available for download at https://forge.gridforum.org/projects/ghpn-rg.
[4] The Globus toolkit, http://www.globus.org/toolkit/about.html.
[5] The Condor project, http://www.cs.wisc.edu/condor/.
[6] Legion, http://legion.virginia.edu/index.html.
[7] Uniform interface to computing resources (UNICORE), http://www.unicore.org/unicore.htm.
[8] S. Tuecke, K. Czajkowski, I. Foster, J. Frey, S. Graham, C. Kesselman, T. Maguire, T. Sandholm, D. Snelling and P. Vanderbilt (2003) "Open grid Services Infrastructure," Grid Forum Document 15.
[9] Web Services Description Language (WSDL) 1.1, World Wide Web Consortium, http://www.w3.org/TR/wsdl.
[10] The Web Services Resource Framework (WSRF) Technical Committee, Organization for the Advancement of Structured Information Standards, http://www.oasis-open.org/committees/tc_home.php?wg_abbrev=wsrf.

[11] I. Foster (2005) "Globus Toolkit Version 4: Software for Service-Oriented Systems", *IFIP International Conference on Network and Parallel Computing*, LNCS 3779, Springer-Verlag, pp. 2–13.
[12] Foster, I., Kesselman, C., Tsudik, G., and S. Tuecke (1998) "A Security Architecture for Computational Grids," Proceedings of the 5th ACM Conference on Computer and Communications Security.
[13] S. Tuecke, V. Welch, D. Engert, L. Pearlman, and M. Thompson (2004) "Internet X.509 Public Key Infrastructure (PKI) Proxy Certificate Profile," RFC 3820, June 2004.
[14] J. Postel and J.K. Reynolds (1985) "File Transfer Protocol," RFC 959, October 1985.
[15] I. Foster, "GT4 Primer", http://www.globus.org/toolkit/docs/4.0/key/.
[16] The SAML Technical Committee, Organization for the Advancement of Structured Information Standards, http://www.oasis-open.org/committees/tc_home.php?wg_abbrev=security.
[17] Platform, http://www.platform.com/Products/Platform.LSF.Family/.
[18] Portable batch system, http://www.openpbs.org/about.html.
[19] Web services addressing, World Wide Web Consortium, http://www.w3.org/Submission/ws-addressing/.
[20] Community scheduler framework (CSF), https://dev.globus.org/wiki/CSF.
[21] Database Access and Integration Services, a Global Grid Forum Working Group, http://forge.gridforum.org/projects/dais-wg/.
[22] eXtensible Markup Language, The World Wide Web Consortium, http://www.w3.org/TR/REC-xml/.
[23] Universal Description, Discovery, and Integration, http://www.uddi.org/.
[24] ITU recommendation ITU-T X.509 (formerly CCITT X.509) and ISO/IEC/ITU 9594-8.
[25] Web Services Security Technical Committee, Organization for the Advancement of Structured Information Standards, http://www.oasis-open.org/committees/tc_home.php?wg_abbrev=wss.
[26] S. Blake-Wilson, M. Nystrom, D. Hopwood, J. Mikkelsen, and T. Wright (2003) "Transport Layer Security (TLS) Extensions," RFC 3546, June 2003.
[27] Roy, A. and V. Sander (2003) "GARA: A Uniform Quality of Service Architecture", *Resource Management: State of the Art and Future Trends*, Kluwer Academic Publishers, pp. 135–144.
[28] M. Horowitz and S. Lunt, "FTP Security Extensions," RFC 2228, October 1997.
[29] P. Hethmon and R. Elz (1998) "Feature negotiation mechanism for the File Transfer Protocol", RFC 2389, August 1998.
[30] W. Allcock (2003) "GridFTP: Protocol Extensions to FTP for the Grid," Grid Forum Document No. 20, April.
[31] W. Allcock (2005) "GridFTP and Reliable File Transfer Service," GlobusWORLD 2005, Boston, MA, February 2005.
[32] V. Sander (ed.) (2004) "Networking Issues for Grid Infrastructure," Grid Forum Document No. 37, November.
[33] The Firewall Issues Research Group home page, Global Grid Forum, https://forge.gridforum.org/projects/fi-rg/.
[34] T. Itou, H. Ohsaki, and M. Imase (2005) "On Parameter Tuning of Data Transfer Protocol GridFTP for Wide-Area grid Computing," 2nd IEEE/Create-Net Workshop on Networks for Grid Applications (GridNets 2005), Boston, MA, October 2005.
[35] D.B. Hoang, T. Lavian, S. Figueira, J. Mambretti, I. Monga, S. Naiksatam, H. Cohen, D. Cutrell, and F. Travostino (2004) "DWDM-RAM: An Architecture for Data Intensive Services Enabled by Next Generation Dynamic Optical Networks," IEEE GLOBECOM, Dallas, November 2004.

References

[36] T. Lavian, D. Hoang, J. Mambretti, S. Figueira, S. Naiksatam, N. Kaushil, I. Monga, R. Durairaj, D. Cutrell, S. Merrill, H. Cohen, P. Daspit, and F. Travostino (2004) "A Platform for Large-Scale Grid Data Service on Dynamic High-Performance Networks," 1st International Workshop on Networks for grid Applications (GridNets 2004), San Jose, CA, October 29, 2004.

[37] S. Figueira, S. Naiksatam, H. Cohen, D. Cutrell, D. Gutierrez, D. B. Hoang, T. Lavian, J. Mambretti, S. Merrill, and F. Travostino (2004) "DWDM-RAM: Enabling grid Services with Dynamic Optical Networks," GAN'04 (Workshop on Grids and Networks), held in conjunction with CCGrid 2004 (4th IEEE/ACM International Symposium on Cluster Computing and the Grid), Chicago, IL, April 19–22, 2004.

[38] User Controlled LightPaths, Canarie, http://www.canarie.ca/canet4/uclp/.

[39] The VIOLA project, http://www.viola-testbed.de/.

[40] M. Hayashi (2005) "Network Resource Management System for Grid Network Service," presented at GGF15 grid High Performance Networks Research Group, Boston, October 5, 2005, https://forge.gridforum.org/docman2/ViewProperties.php?group_id=53&category_id=1204&document_content_id=4956.

[41] A. Farrel (2004) *The Internet and its Protocols, A Comparative Approach*, Morgan Kaufmann.

[42] Web-based Enterprise Management, The Distributed Management Task Force (DMTF), http://www.dmtf.org/about/faq/wbem/.

[43] E. Mannie (ed.), "Generalized Multi-Protocol Label Switching (GMPLS) Architecture", RFC 3945, October 2004.

[44] Recommendations on Automatic Switched Transport Network (ASTN) and Automatic Switched Optical Network (ASON), International Telecommunication Union http://www.itu.int/ITU-T/2001-2004/com15/otn/astn-control.html.

[45] "The User Network Interface (UNI) 1.0 Signalling Specification, Release 2", an Implementation Agreement Created and Approved by the Optical Internetworking Forum, February 2004, http://www.oiforum.com/public/documents/OIF-UNI-01.0-R2-Common.pdf.

[46] Transaction Language 1, http://www.tl1.com.

[47] J.D. Case, M. Fedor, M.L. Schoffstall, J.D.P. Hethmon, and R. Elz (1989) "Simple Network Management Protocol (SNMP)," RFC 1098, April 1989.

[48] J. Case, K. McCloghrie, M. Rose, and S. Waldbusser (1993) "Introduction to version 2 of the Internet-standard Network Management Framework," RFC 1441, April 1993.

[49] J. Case, D. Harrington, R. Presuhn, and B. Wijnen (2002) "Message Processing and Dispatching for the Simple Network Management Protocol (SNMP)," RFC 3412, December 2002.

[50] Network Configuration Working Group, Internet Engineering Task Force, http://www.ietf.org/html.charters/netconf-charter.html.

[51] XML-RPC, http://www.xmlrpc.com/spec.

[52] SOAP, The XML Protocol Working Group, The World Wide Web Consortium, http://www.w3.org/2000/xp/Group/.

[53] S. Waldbusser, R. Cole, C. Kalbfleisch, and D. Romascanu (2003) "Introduction to the Remote Monitoring (RMON) Family of MIB Modules; RFC-3577," Internet RFC 3577, August 2003.

[54] The IP Flow Information Export Working Group, Internet Engineering Task Force, http://www.ietf.org/html.charters/ipfix-charter.html.

[55] Web Services Notification Technical Committee, Organization for the Advancement of Structured Information Standards, http://www.oasis-open.org/committees/tc_home.php?wg_abbrev=wsn.

[56] A. Andrieux, K. Czajkowski, A. Dan, K. Keahey, H. Ludwig, J. Pruyne, J. Rofrano, S. Tuecke, and M. Xu, "Web Services Agreement Specification (WS-Agreement)," Grid Working Document, revisions available for download at https://forge.gridforum.org/projects/graap-wg.

[57] The VIOLA project, http://www.imk.fraunhofer.de/sixcms/detail.php?template=&id=2552&_SubHP=&_Folge=&abteilungsid=&_temp=KOMP.
[58] Network Measurements Working Group, Global Grid Forum, https://forge.gridforum.org/projects/nm-wg.
[59] B. Lowekamp, B. Tierney, L. Cottrell, R. Hughes-Jones, T. Kielmann, and M. Swany (2004) "A Hierarchy of Network Performance Characteristics for grid Applications and Services," Grid Forum Document, 23, May.
[60] W3C Semantic Web Activity Statement, November 2001.
[61] F. Travostino (2005) "Using the Semantic Web to Automate the Operation of a Hybrid Internetwork," 2nd IEEE/Create-Net Workshop on Networks for Grid Applications (Grid-Nets 2005), Boston, MA, October 2005.
[62] J. van der Ham, F. Dijkstra, F. Travostino, H. Andree, and C. de Laat (2006) "Using RDF to describe networks," *Future Generation in Computer Science* (in press).
[63] Resource description framework (RDF), http://www.w3.org/RDF/.
[64] OWL Web Ontology Language Reference, http://www.w3.org/TR/owl-ref/.
[65] The Network Weather Service http://nws.cs.ucsb.edu/.
[66] N. Hu and P. Steenkiste (2003) "Evaluation and Characterization of Available Bandwidth Probing Techniques", special issue on "Internet and WWW Measurement, Mapping, and Modeling,"*IEEE JSAC*, 21(6).
[67] B. Wong, A. Slivkins and E. Gün Sirer, "Meridian: A Lightweight Network Location Service without Virtual Coordinates," Proceedings of SIGCOMM Conference, Philadelphia, PA, August 2005.
[68] ITU-T M.3000 Series Recommendations.
[69] S. Naiksatam and S. Figueira, "Opportunistic Bandwidth Scheduling in Lambdagrids Using Explicit Control," 2nd IEEE/Create-Net Workshop on Networks for Grid Applications (GridNets 2005), Boston, MA, October 2005.
[70] U. Farooq, S. Majumdar, and E. Parsons (2005) "Dynamic Scheduling of Lightpaths in Lambda Grids," 2nd IEEE/Create-Net Workshop on Networks for Grid Applications (GridNets 2005), Boston, MA, October 2005.
[71] Y. Rekhter and T. Li (1995) "A Border Gateway Protocol 4 (BGP-4)," RFC 1771, March 1995.
[72] "Intra-Carrier E-NNI Signalling Specification," an implementation agreement created and approved by the Optical Internetworking Forum, February 2004, http://www.oiforum.com/public/documents/OIF-E-NNI-Sig-01.0-rev1.pdf.
[73] The IPsphere Forum, http://www.ipsphereforum.org/home.
[74] Global Lambda Integrated Facility, http://www.glif.is/.
[75] B. Oudenaarde, S.B.S.S. Madhav, and I. Monga, "Grid Network Services: Lessons and proposed solutions from Super Computing 2004 demonstration," individual draft submission to the Global Grid Forum,https://forge.gridforum.org/docman2/ ViewProperties.php? group_id=53&category_id=807&document_content_id=3839
[76] L. Gommans, B. van Oudenaarde, F. Dijkstra, C. de Laat, T. Lavian, I. Monga, A. Taal, F. Travostino, and A. Wan (2006) "Applications Drive Secure Lightpath Creation across Heterogeneous Domains," special issue with feature topic on Optical Control Planes for Grid Networks: Opportunities, Challenges and the Vision, *IEEE Communications Magazine* 44(3).
[77] L. Gommans, F. Travostino, J. Vollbrecht, Cees de Laat, and R. Meijer (2004) "Token-Based Authorization of Connection Oriented Network Resources," 1st International Workshop on Networks for grid Applications (GridNets 2004), San Jose, CA, October 29, 2004.
[78] M. Handley, H. Schulzrinne, E. Schooler, and J. Rosenberg (1999) "SIP: Session Initiation Protocol," RFC 2543, March 1999.
[79] SURFnet, http://www.gigaport.nl/info/home.jsp.

[80] The Web Services Business Process Execution Language (BPEL) Technical Committee, Organization for the Advancement of Structured Information Standards, http://www.oasis-open.org/committees/wsbpel/charter.php.

[81] Rajasekaran, J. Miller, K. Verma, and A. Sheth, Enhancing Web Services Description and Discovery to Facilitate Composition, International Workshop on Semantic Web Services and Web Process Composition, 2004 (Proceedings of SWSWPC 2004).

Chapter 8

Grid Networks and TCP Services, Protocols, and Technologies

Bartek Wydrowski, Sanjay Hegde, Martin Suchara, Ryan Witt, and Steven Low

8.1 INTRODUCTION

Previous chapters have discussed multiple network services that are available at the lower layers of the OSI model. This chapter focuses on layer 4 of the OSI model as it is implemented in the Internet through one of its key protocols. Layer 4 is the transport layer, which provides a data delivery service for the application layer that hides the details of communicating over the network. The Internet provides many different transport layer protocols that offer various services to applications. The basic mechanism for obtaining this service is simple. An application engages a particular transport layer protocol by opening a communications socket from within its program code. This simple mechanism has been a key to the success of the scalability of the Internet.

However, Internet transport is a particularly important topic because, under many traffic conditions that are becoming increasingly common in the Internet, classical methods of transport have proven to be inadequate. It has become clear that the classic protocols on which this service is based cannot be used to address a number of emerging application requirements, particularly many related to data-intensive and latency-insensitive Grid applications. These protocols were initially developed

Grid Networks: Enabling Grids with Advanced Communication Technology Franco Travostino, Joe Mambretti, Gigi Karmous-Edwards © 2006 John Wiley & Sons, Ltd

for the Internet when small amounts of traffic were transmitted over circuits with minimal bandwidth.

Today's Internet has changed dramatically – overall traffic volume continues to grow, extremely large volume streams have become common, latency-sensitive applications have become essential, and new types of applications require complex traffic behaviors, such as those generated by ad hoc peer-to-peer networks. For example, standard Internet transport does not provide efficient support for many high-volume data transport streams required by Grid applications, especially over long distances.

This chapter and the next present an overview of the core issues in Internet transport architecture and protocols, the basic problems that are being encountered with classical approaches, possible alternate solutions, and initial experience with those new concepts. The two Internet protocols most commonly used for transport are the Transmission Control Protocol (TCP), which provides for reliable end-to-end transit, for example through flow control and error detection, and the User Datagram Protocol (UDP), which does not provide for transport guarantees. This chapter focuses on TCP, and the next chapter on UDP.

8.2 BACKGROUND AND THEORETICAL BASIS FOR CURRENT STRUCTURE OF TRANSPORT LAYER PROTOCOLS

To meet the needs of the future Internet, the advanced networking research community has undertaken a number of projects that are investigating the basic concepts behind Internet transport architecture, particularly those that have resulted in the current protocols. They are closely examining the components of this architecture and related protocols, creating variant and alternate designs, and conducting empirical experiments with those designs. Many of the leading-edge experiments are being undertaken within Grid environments. Recently, the GGF undertook a survey of the primary research efforts in the area and published it as a reference document [1].

8.2.1 USER DATAGRAM PROTOCOL (UDP)

Although TCP is the focus of this chapter, it may be useful to provide a brief definition of UDP to allow for a comparison of these two protocols. UDP is often used for the bulk transport of data for which the loss of a few packets is not meaningful. When an application sends a block of data using UDP to another host over the network, this "datagram" is sent immediately to the destination host without establishing any connection. UDP provides no guarantees of delivery of the data, no protection either from datagram duplication or from datagram loss.

Because the underlying network is unreliable, the packets may indeed be lost, duplicated by router congestion or reordered because of a network condition such as path load balancing or the parallelism in routers. UDP is useful for applications that are sufficiently robust to overcome these conditions. Its simplicity is attractive to many developers who do not need service guarantees. UDP is popular for many real-time applications, in which the useful lifetime of a datagram is limited and lost packets are not retransmitted. The next chapter provides a further discussion of UDP.

8.2.2 TRANSMISSION CONTROL PROTOCOL (TCP)

If delivery guarantee is required by an application, TCP provides such assurance, essentially creating a reliable data "pipe" between two hosts communicating over the unreliable general network. Before sending data, TCP establishes a connection between the source and destination hosts. After establishing network connectivity, TCP guarantees that the data sent over the pipe will be delivered, and delivered in a specific order. If data cannot be delivered, for example because of loss of connectivity, the application is notified of the failure.

Each segment of data sent is assigned a *sequence number* that allows the receiving host to detect missing segments and ensure in-order delivery of data. Segments are delivered to the application from the TCP layer only when they have been assembled into a contiguous in-order stream. When the TCP receiver sends back an acknowledgement, packets that contain the sequence number of the highest sequence number are delivered to the application. If a packet is lost, then a timer at the sender's TCP session times out, for example if an acknowledgement for the packet is not received within the expected Round-Trip Time (RTT) that it takes for the packet to go from the sender and the acknowledgment to return from the receiver. Any packet considered lost is retransmitted by TCP.

A critical part of TCP is Congestion Control (CC). CC controls the transmission rate to make use of available capacity of the network, while ensuring that this capacity is shared fairly with other flows and that the network is not overloaded with traffic. The basic function of CC is to prevent congestion collapse from occurring, which is the loss of efficiency of the network when the network is overloaded with traffic. CC also improves the throughput, loss, and delay performance of the network.

8.2.3 WINDOW FLOW CONTROL

TCP controls its transmission rate through a mechanism called window flow control. The window (w) is the maximum number of packets that are allowed transmission through the network concurrently without being acknowledged. The window controls the transmission rate as, during each RTT, w packets are transmitted, making the throughput, x, for the source $x = w/\text{RTT}$ (pkts/s).

To better illustrate this principle, consider $w = 42$, i.e., 42 packets have been sent but not acknowledged. By the time that the acknowledgment for the 42nd packet has arrived, 1 RTT has elapsed since its departure and the acknowledgments for all 42 packets have been received during this time. With the arrival of each ACK, the number of unacknowledged packets is reduced by 1. Therefore, during the 1 RTT, a further 42 packets are sent, making the send rate 42 packets per RTT.

As shown in Figure 8.1, the w packets that are in the network exist either in router and switch buffers or are "in flight" in the transmission channel. The total w unacknowledged packets either exist in the network as data packets in the forward path or they are represented by ACK packets in the backward path.

If a channel, for example a fiber-optic cable, has capacity C pkts/s and a propagation delay D s, then a Bandwidth Propagation Delay Product (BPDP), D^*C, determines the number of packets that can be "in flight" over the fiber. To transmit at the full capacity of the fiber, w needs to be at least equal to the BPDP. If w is larger than the BPDP, the excess packets that are not "in flight" fill up the buffers of the routers and switches on the path, as shown in Figure 8.1. However, if the buffers on the path can hold B packets, and if w exceeds BPDP$+B$, packet loss occurs. If just one source is considered, the goal of CC is therefore to choose a value of w so that it is above BPDP and below BPDP$+$B.

Note that whenever a packet is stored in a buffer, it causes an additional "queuing" delay to be experienced by packets arriving at the buffer after it. If the capacity of the channel is C pkts/s, this delay is $1/C$ seconds per packet. Figure 8.2 illustrates this relationship between w, RTT, and packet loss. (Note that the capacity C is assumed to be constant in Figure 8.2.) In a real network, the capacity is a random variable that has an averaged value of C. The fluctuation is usually small in wireline networks but may be significant in wireless networks.

The goal of adjusting w is to ensure that the path capacity is fully utilized and that this capacity is shared fairly between sources using the path. What separates many of the TCP variants is how w is computed. A key element in computing w is detecting

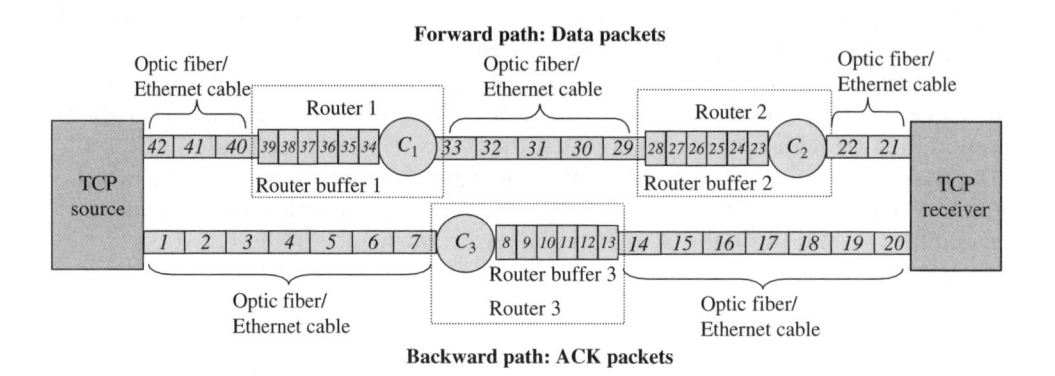

Figure 8.1. Packet propagation and buffering in network.

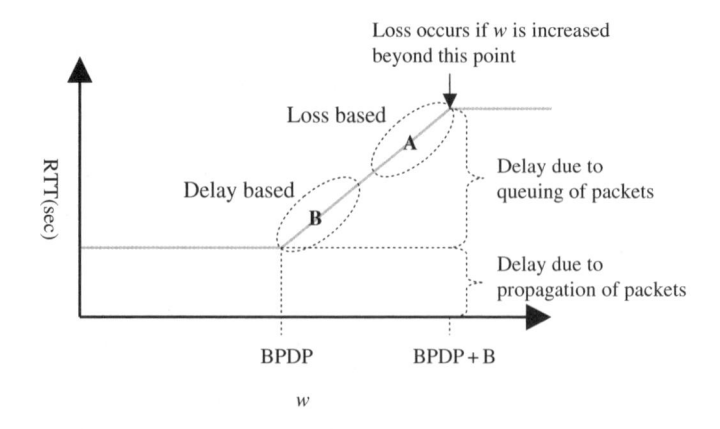

Figure 8.2. Congestion window relationship with delay and loss.

congestion in the network, and the signal used to detected congestion divides TCP CC algorithms into three main classes:

- *Loss based*. These algorithms use packet loss as a signal that the network is congested. The protocols probe the available capacity by increasing w until loss occurs due to overflowing buffers, at which point they decrease w before probing again. These protocols tend to fill up the buffers with their probing action and operate around the BDPD+B region on the graph.
- *Delay based*. These algorithms detect increases in RTT to determine that the buffers are filling up when w is greater than the BPDP. These protocols operate around the BPDP region on the graph and tend to operate with low queuing delay.
- *Explicit signal*. These protocols rely on additional information being communicated by routers and switches to determine the right value for w. The router typically computes a congestion level based on the input traffic rate and capacity, and signals the congestion level to the sources using additional fields in the packet header, such as the Explicit Congestion Notification (ECN) field [2].

Note that any given source is not explicitly aware of any other sources using the same network path. However, each source's use of the network path contributes to the packet loss, or delay experienced by other sources. By measuring the packet loss, delay, or explicit signal, the sources can coordinate among themselves in a distributed way to determine what their fair rate is. Before describing the detailed mechanisms of how different TCP variants adjust the w from these signals, it is important to understand what this "fairness" goal really means.

8.2.4 FAIRNESS

Fairness is a key concept to Internet transport architecture. There is no one absolute concept of fairness. However, two well-accepted and inter-related fairness objectives are maximum utility and max–min fairness:

- *Maximum utility fairness*. A source s can define for itself the value of sending at a particular transmission rate x_s by using the utility function $U(x_s)$. Figure 8.3 gives an intuition for types of utility functions possible for different applications.

    The utility function for a Voice over IP (VoIP) application is described by curve $U_{\text{VoIP}}(x)$. For uncompressed voice traffic, acceptable voice quality requires a transmission rate of about 64 kbps to transmit an 8-kHz 8-bit signal. It is possible to provide an acceptable service with a little less quality than this, but as the graph of $U_{\text{VoIP}}(x)$ shows that the utility of transmitting voice at less than 64 kbps reduces sharply as the transmission rate is decreased, and transmitting beyond 64 kbps does not increase quality significantly.

    $U_{\text{FTP}}(x)$ represents a file transfer FTP application in which the utility function reflects the fact that the application performance increases the faster the file is transferred. The initial steepness of the function describes the need to transfer the file "fast enough" and the decreasing steepness shows the diminishing return for transferring at greater speeds.

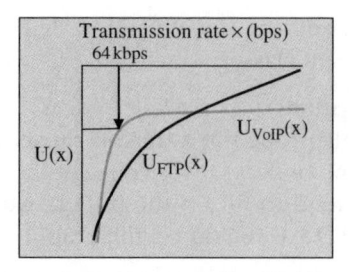

Figure 8.3. Utility functions.

Maximum utility fairness is the objective of choosing all of the source rates in the network, designated by the vector x, so that the sum of all of the utility values of each source is the maximum possible. This principle is summarized by the optimization

$$\max_x \sum_i U_i(x_i) \quad \text{st.} \sum_{i \in S(l)} x_i \leq C_l \quad l = 1, \ldots, L$$

This optimization is, of course, subject to the feasibility constraint that the set of flows at a link l, $S(l)$, does not exceed the link capacity, $C(l)$, for all L links in the network. Also, to avoid starving any applications (depriving them of network resources), the utility function $U_i(x_i)$ is usually a convex function.

- *Max–min fairness.* A network is max–min fair if the source flow rates are feasible and no flow can be increased while maintaining feasibility without decreasing a smaller or equal flow. A more general form of max–min fairness is weighted max–min fairness, which allows the flows to express a priority for bandwidth allocation.

8.2.5 CONGESTION CONTROL FEEDBACK SYSTEM

It has been shown that a similar feedback processes is found in basic supply and demand models, which allow buyers and sellers to negotiate price. Similarly, best effort networks allow applications to negotiate a solution that achieves either the maximum utility or max–min fairness.

In the case of networks, the rate of a TCP source is controlled by a congestion signal, or "price." that the network generates to regulate the demand for capacity. The source algorithm attempts to maximize the *net benefit*, B, of the consuming bandwidth at rate x_s, which is equal to the utility, $U(x_s)$, of having x_s units of resource less the cost of purchase:

$$B_s(x_s) = U_s(x_s) - x_s \cdot p \tag{8.1}$$

where p is the price of one unit of resource. So if the cost p increases, it may benefit the source to reduce its rate or, conversely, if the price decreases, the source would increase its rate. If a goal is to operate at the point of maximum benefit, by taking the inverse of the derivative of Equation (8.1) it is possible to find the demand function, which is the optimal transmission rate given the congestions signal p:

$$x_s = D_s(p) = U'^{-1}{}_s(p) \tag{8.2}$$

As explained earlier, in practice this price signal can be communicated to the source in a number of different ways, implicitly either as the packet loss rate or as the queuing delay, or explicitly in the packet as ECN or a multibit signal. Note also that, for most protocols, the utility function presents only a high-level model of its behavior, and the algorithm is, in fact, described in terms of particular events that occur upon the receipt of acknowledgment or packet loss. For example, with TCP/NewReno, if packet loss occurs, TCP halves its transmission window. For each ACK packet received, TCP grows its window. Despite these details, TCP's long-term rate is governed by the packet loss rate, and thus its response to this "price" signal can be modeled using the utility functions.

The link interfaces in routers and switches on the network generate this price signal to regulate the traffic arriving at the router/switch destined for a particular link. This signal has the dual function of making the source rates "feasible," that is, allowing them to control the traffic to ensure that it does not exceed the capacity, and driving the source rates toward the fairness objective. Each bottleneck link on the source to destination path of the connection contributes to the total signal received by the source. Figure 8.4 depicts this closed-loop CC system.

The links increase the price if the traffic bound for them exceeds their capacity, and decrease the price if the traffic is below their capacity. The prices can be generated passively, by the implicit behavior of the drop-tail queue, or actively, by an Active Queue Management (AQM) algorithm, which uses a formula to generate the price. A typical AQM algorithm which computes the price, $P(t)$, is the integrator function:

$$P(t+1) = P(t) + \alpha(X(t) - \mu C) \tag{8.3}$$

where $X(t)$ is the aggregate input traffic rate destined for the link of capacity C, μ controls the target utilization of the link, and α controls the responsiveness of the controller. Note that when $X(t) > \mu C$ the price $P(t)$ increases, and it decreases when $X(t) < \mu C$. The $P(t)$ is then used to determine the packet dropping rate or the packet ECN marking rate or it is communicated explicitly in the packet.

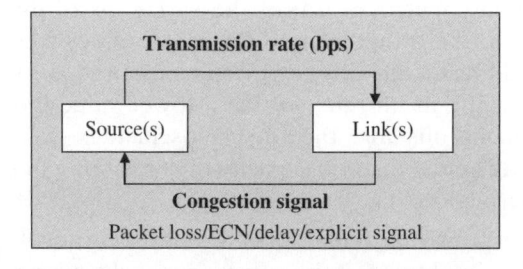

Figure 8.4. Logical congestion control feedback loop.

On today's Internet, the most common way of generating the feedback signal $P(t)$ is by packet loss, which occurs when the drop-tail queue at the head of the communication links overflows. When packet loss occurs, TCP Reno reduces its rate to better match it to the available capacity. Packet loss occurs whenever packets arrive at the link but the buffer is full, $B(t) = B_{max}$. The buffer occupancy is given by:

$$B(t+1) = B(t) + X(t) - C \quad B(t) \leq B_{max} \tag{8.4}$$

Clearly, this value grows when traffic $X(t)$ bound for the communications link exceeds the link's capacity, C. With the drop-tail queue, the loss signal, $P(t)$, is thus related to how heavily the link is loaded. The delay signal is directly proportional to $B(t)/C$.

Note the close relationship between the two ways of computing the signal $P(t)$, by the AQM (Equation 8.3) and by buffer occupancy (Equation 8.4). The key advantage of Equation (8.3) is that it introduces the α and μ parameters, which give control over the link utilization and the responsiveness of the controller. By controlling the utilization, the queuing delay through the link can be reduced.

8.2.6 CONGESTION CONTROL PERFORMANCE

The performance of the CC system that consists of the source TCP algorithm and the link congestion signal algorithm has many facets and these variables can impact the Quality of Service (QoS) that applications experience from the network. A variety of metrics that address different aspects of performance are used to evaluate a CC system. The key metrics are described in this list:

- *Fairness*. As there are different concepts of fairness, there is no one measure of fairness. The Jain index [3] is a popular fairness metric that measures how equally the sources share a single bottleneck link. A value of 1 indicates perfectly equal sharing, and smaller values indicate worse fairness. For a single-bottleneck network, both under max–min fairness and when maximizing total utility with sources having the same utility function, the sources should achieve equal rates.
- *Throughput*. Throughput is simply the data rate, typically in Mbps, delivered to the application. For a single source this should be close to the capacity of the link. When the BPDP is high, that is, when the link capacity or RTT or both are high, some protocols are unable to achieve good throughput.
- *Stability*. The stability metric measures the variations of the source rate and/or the queue length in the router around the mean values when everything else in the network is held fixed. Stability is typically measured as the standard deviation of the rate around the mean rate, so that a lower value indicates better performance. If a protocol is unstable, the rate can oscillate between exceeding the link capacity, and being below the link capacity thus resulting in poor delay jitter and throughput performance.
- *Responsiveness*. Responsiveness measures how fast a protocol reacts to a change in network operating conditions. If the source rates take a long time to converge to a new level, say after the capacity of the link changes, either the link may become

underutilized or the buffer may overflow. The responsiveness metric measures the time or the number of round trips to converge to the right rate.

- *Queuing delay.* As shown in Figure 8.2, once the congestion window is greater than the BPDP, the link is well utilized; however, if the congestion window is increased more, queuing delay builds up. Different TCP and AQM protocol combinations operate at different regions of the Figure 8.2 curve. Loss-based protocols when operated with the drop-tail queue typically operate in the A region, and delay-based protocols typically operated in the B region. It is desirable to avoid large queuing delays and operate in the B region to avoid large latencies in the network, for example in consideration of other users, who may be using interactive applications.
- *Loss recovery.* Packet loss can be a result both of overflowing buffers, which indicates network congestion, and also of transmission error, such as bit errors over a wireless channel. It is desirable that, when packet loss occurs due to transmission error, the source continues to transmit uninterrupted. However, when the loss is due to congestion, the source should slow down. Loss recovery is typically measured as the throughput that can be sustained under the condition of a certain random packet loss caused by transmission error. Loss-based protocols typically cannot distinguish between congestion and transmission error losses and slow down to both. As explicit signal and delay-based protocols do not rely on packet loss to detect congestion, they can to some extent ignore packet loss and maintain a better throughput over paths with packet loss.

Currently, network researchers are experimenting with different protocols that can be measured in using these metrics. In advanced of the discussion of the specific protocols, it is useful, to compare the variability among them in terms of these metrics.

Figures 8.5–8.8 show the fairness, throughput, stability, and responsiveness metrics for the Binary Increase Congestion control (BIC) [4], FAST [5], NewReno [6], High-Speed (HS) [7], and scalable [8] TCP protocols with the drop-tail queue. The figures show the Cumulative Distribution Function (CDF) of each metric. The CDF is obtained by compiling the results from numerous experiments performed on different networks. As an example of how to interpret the CDF, consider the fairness of the HSTCP protocol, which has a value of 0.8 for a CDF of 0.2. This means that 20% of the results had a fairness index of 0.8 or worse (lower).

Figure 8.9 shows the mean queue size under different protocols with a drop-tail buffer of various sizes. Notice that loss-based (BIC, H, HS, Reno, scalable) TCPs always fill the buffer regardless of its size. This results from their bandwidth-probing action, which increases the window until loss is detect. The delay measuring FAST TCP maintained a low queue size as it detected the increase of queuing delay.

Figure 8.10 shows the throughput achieved by different TCP protocols operating over a 10-Mbps 50-ms RTT link with a packet loss rate from 0 to 5%, simulating transmission error. The performance upper bound for a link of capacity C and loss rate l, $C(1-l)$, is also shown. As loss-based protocols reduce their rate due to loss, their performance decrease with increasing loss rate is predictable.

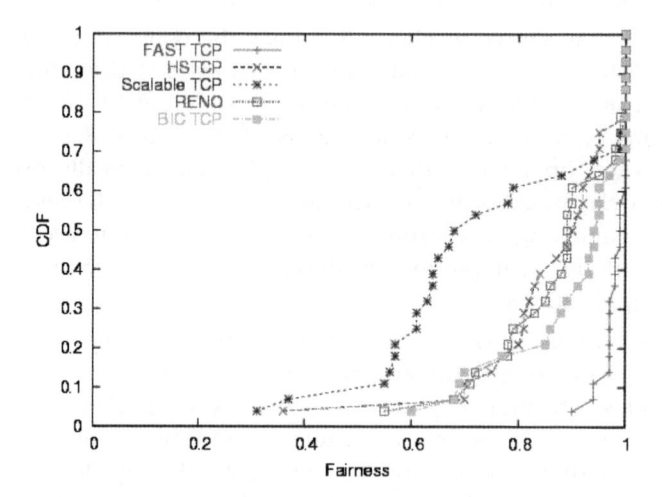

Figure 8.5. Fairness CDF.

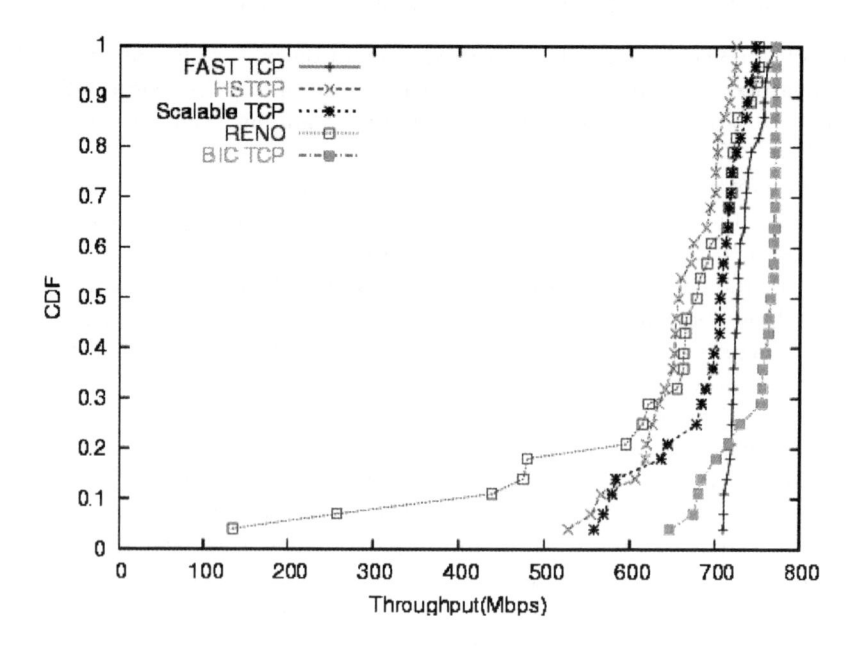

Figure 8.6. Throughput CDF.

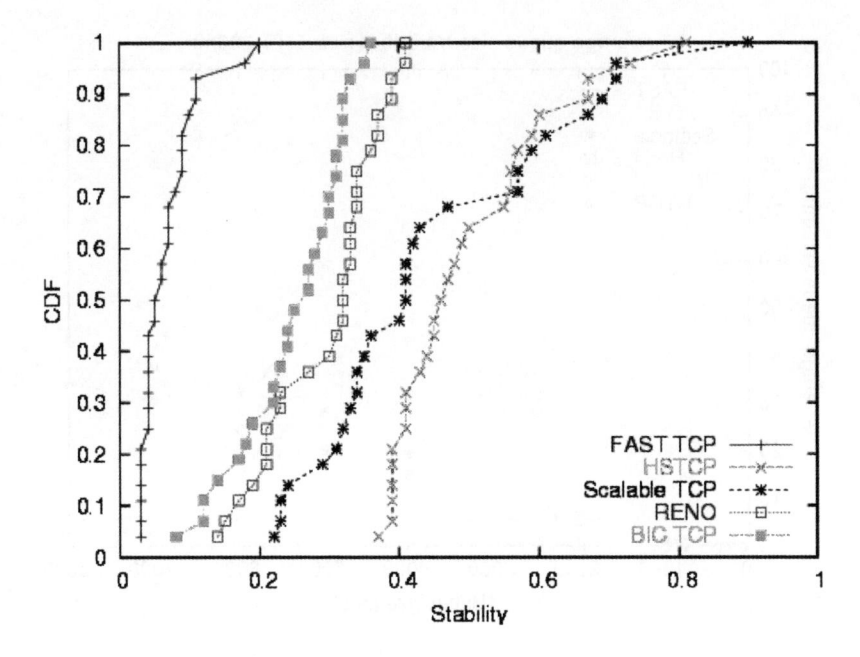

8.2 Background and Theoretical basis for Transport Layer Protocols

Figure 8.7. Stability CDF.

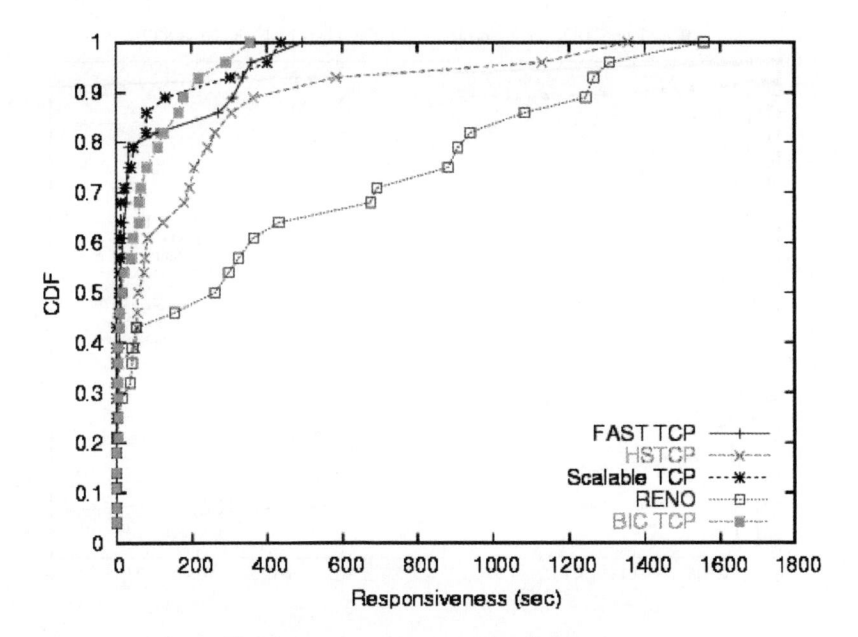

Figure 8.8. Responsiveness CDF.

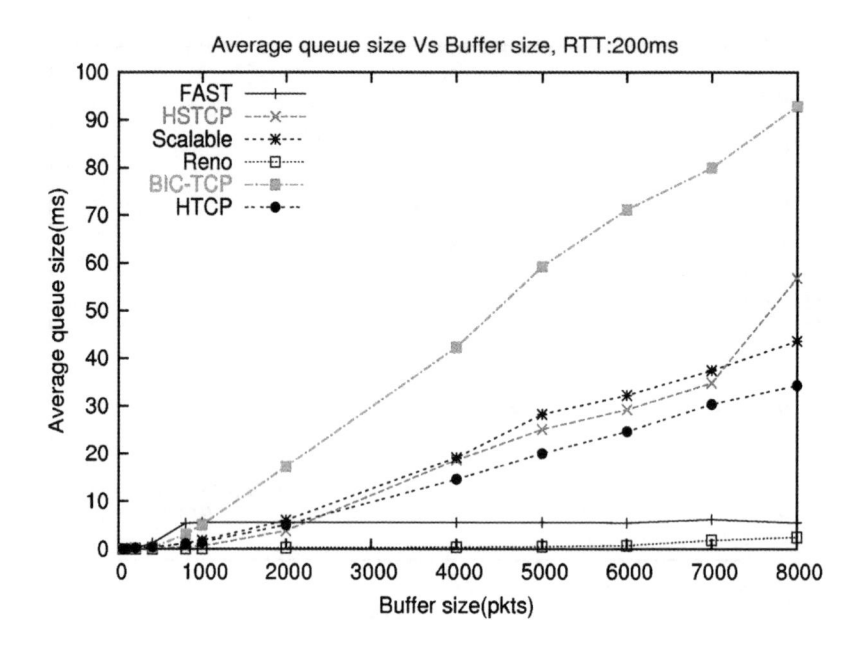

Figure 8.9. Queue size for different TCPs.

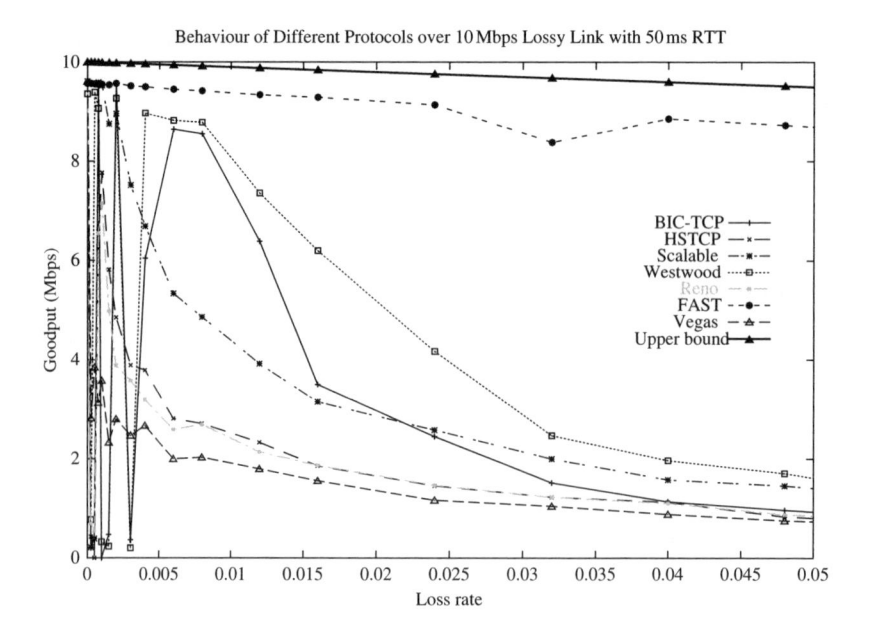

Figure 8.10. Loss response for different TCPs.

8.3 ENHANCED INTERNET TRANSPORT PROTOCOLS

As noted, a number of research projects have been established to investigate options for enhancing Internet transport protocol architecture through variant and alternative protocols. The following sections describe a selective sample set of these approaches, and provide short explanations of their rationale. Also described is the architecture for the classical TCP stack, which is useful for comparison.

8.3.1 TCP RENO/NEWRENO

TCP Reno's congestion control mechanism was introduced in 1988 [9] and later extended to NewReno in 1999 [6] by improving the packet loss recovery behavior. NewReno is the current standard TCP found in most operating systems. NewReno probes the capacity of the network path by increasing the window until packet loss is induced. Whenever an ACK packet is received, NewReno increases the window w by $1/w$, so that on average the window is increased by 1 every RTT. If loss occurs, the window is reduced by half:

$$\text{ACK} : w \leftarrow w + \frac{1}{w}$$
$$\text{Loss} : w \leftarrow \frac{w}{2}$$

This type of control algorithm is called Arithmetic Increase, Multiplicative Decrease (AIMD) and it produces a "sawtooth" window behavior, as shown in Figure 8.11.

Since the arrival of ACK packets and loss events is dependent only on the RTT and packet loss rate in the network, p, researchers [10] have described the average rate

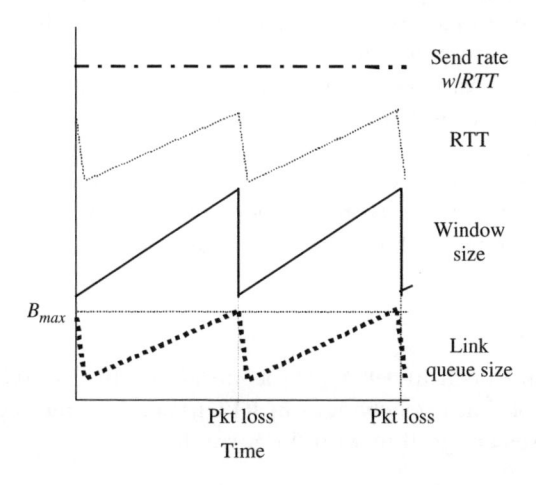

Figure 8.11. TCP Reno AIMD: throughput, RTT, windows size, queue size.

of the Reno by

$$x \leq \frac{1.5\sqrt{2/3} \cdot MSS}{RTT \cdot \sqrt{p}} \text{ (bps)}$$

where MSS is the packet size. Note that the rate depends on both the loss rate of the path and the RTT. The dependence on RTT means that sources with different RTTs sharing the same bottleneck link will achieve different rates, which can be unfair to sources with large RTTs.

The AIMD behavior actually describes only the "congestion avoidance" stage of Reno's operation. When a connection is started, Reno begins in the counterintuitively named "slow start" stage, when the window is rapidly increased. It is termed slow start because it does not immediately initiate transport at the total rate possible. In slow start, the window is increased by one for each ACK:

$$ACK : w \leftarrow w + 1$$

which results in exponential growth of the window. Reno exits slow start and enters congestion avoidance either when packet loss occurs or when the $w > ssthresh$, where $ssthresh$ is the slow start threshold. Whenever $w < ssthresh$, Reno re-enters slow start.

Although TCP Reno has been very successful in providing for Internet transport since the 1980s, its architecture does not efficiently meet the needs of many current applications, and can be inefficient when utilizing high-performance networks. For example, its window control algorithm faces efficiency problems when operating over modern high-speed networks. The sawtooth behavior can result in underutilization of links, especially in high-capacity networks with large RTTs (high bandwidth delay product). The sawtooth window decreases after a drastic loss and the recovery increase is too slow. Indeed, experiments over a 1-Gbps 180-ms path from Geneva to Sunnyvale have shown that NewReno utilizes only 27% of the available capacity. Newer congestion control algorithms for high-speed networks, such as BIC or FAST, described in later sections, address this issue by making the window adaptation smoother at high transmission rates.

As discussed earlier, using packet loss as a means of detecting congestion creates a problem for NewReno and other loss-based protocols when packet loss occurs due to channel error. Figure 8.10 shows that NewReno performs very poorly over lossy channels such as satellite links. Figure 8.11 illustrates how Reno's inherent reliance on inducing loss to probe the capacity of the channel results in the network operating at the point at which buffers are almost full.

8.3.2 TCP VEGAS

TCP Vegas was introduced in 1994 [11] as an alternative to TCP Reno. Vegas is a delay-based protocol that uses changes in RTT to sense congestion on the network path. Vegas measures congestion with the formula:

$$Diff = \frac{w}{baseRTT} - \frac{w}{RTT}$$

where baseRTT is the minimum RTT observed and baseRTT ≤ RTT and corresponds to the round-trip propagation delay of the path.

If there is a single source on the network path, the expected throughput is $w/\text{baseRTT}$. If w is too small to utilize the path, then there will be no packets in the buffers and RTT = baseRTT, so that $\text{Diff} = 0$. Vegas increased w by 1 each RTT until Diff is above the parameter α. In this case, the window w is larger than the BPDP and excess packets above the BPDP are queued in buffers along the path, resulting in the RTT being greater than the baseRTT, which gives $\text{Diff} > 0$. To avoid overflowing the buffers, Vegas decreases w by 1 if $\text{Diff} > \beta$. Thus, overall Vegas controls w so that $\alpha < \text{Diff} < \beta$.

If there are multiple sources sharing a path, then packets from other sources queued in the network buffers will increase the RTT, resulting in the actual throughput w/RTT being decreased. Since Diff is kept between α and β, the increase in RTT will cause w to be reduced, thus making capacity available for other sources to share.

By reducing the transmission rate when an increase in RTT is detected, Vegas avoids filling up the network buffers and operates in the A region of Figure 8.2. This results in lower queuing delays and shorter RTTs than loss-based protocols.

Since Vegas uses an estimate of the round-trip propagation delay, baseRTT, to control its rate, errors in baseRTT will result in unfairness among flows. Since baseRTT is measured by taking the minimum RTT sample, route changes or persistent congestion can result in an over- or underestimate of baseRTT. If baseRTT is correctly measured at 100 ms over one route and the route changes during the connection lifetime to a new value of 150 ms, then Vegas interprets this RTT increase as congestion, and slows down. While there are ways to mitigate this problem, it is an issue common to other delay-based protocols such as FAST TCP.

As shown by Figure 8.10, the current implementation of Vegas responds to packet loss similarly to NewReno. Since Vegas uses delay to detect congestion, there exists a potential for future versions to improve the performance in lossy environments by the implementation of a different loss response.

8.3.3 FAST TCP

FAST TCP is also a delay-based congestion control algorithm, first introduced in 2003 [5], that tries to provide flow-level properties such as stable equilibrium, well-defined fairness, high throughput, and link utilization. FAST TCP requires only sender side modification and does not require cooperation from the routers/receivers. The design of the window control algorithm ensures smooth and stable rates, which are key to efficient operation. FAST has been analytically proven, and has been experimentally shown, to remain stable and efficient provided the buffer sizes in the bottlenecks are sufficiently large.

Like the Vegas algorithm, the use of delay provides a multibit congestion signal, which, unlike the binary signal used by loss-based protocols, allows smooth rate control. FAST updates the congestion window according to:

$$w \leftarrow \frac{1}{2}\left(w + \frac{\text{baseRTT}}{\text{RTT}}w + \alpha\right)$$

where α controls fairness by controlling the number of packets the flow maintains in the queue of the bottleneck link on the path. If sources have equal α values, they will have equal rates if bottlenecked by the same link. Increasing α for one flow will give it a relatively higher bandwidth share.

Note that the algorithm decreases w if RTT is sufficiently larger than baseRTT and increases w when RTT is smaller than baseRTT. The long-term transmission rate of FAST can be described by:

$$x = \frac{\alpha}{q} \tag{8.5}$$

where q is the queuing delay, $q = $ RTT − baseRTT. Note that, unlike NewReno, the rate does not depend on RTT, which allows fair rate allocation for flows sharing the same bottleneck link. Note also from Equation (8.5) that the rate does not depend on the packet loss rate, which allows FAST to operate efficiently in environments in which packet loss occurs due to channel error. Indeed, the loss recovery behavior of FAST has been enhanced, and operation at close to the throughput upper bound $C(1-p)$ for a channel of capacity C and loss rate p is possible, as shown in Figure 8.10.

Like Vegas, FAST is prone to the baseRTT estimation problem. If baseRTT is taken simply as the minimum RTT observed, a route change may result in either unfairness or link underutilization. Also, another issue for FAST is tuning of the α parameter. If α is too small, the queuing delay created may be too small to be measurable. If it is too large, the buffers may overflow. It is possible to mitigate both the α tuning and baseRTT estimation issues with various techniques, but a definitive solution remains the subject of on-going research.

8.3.4 BIC TCP

The Binary Increase Congestion (BIC) control protocol, first introduced in 2004 [4], is a loss-based protocol that uses a binary search technique to provide efficient bandwidth utilization over high-speed networks. The protocol aims to scale across a wide range of bandwidths while remaining "TCP friendly," that is, not starving the AIMD TCP protocols such as NewReno by retaining similar fairness properties.

BIC's window control comprises a number of stages. The key states for BIC are the minimum, W_{min}, and maximum, W_{max}, windows. If a packet loss occurs, BIC will set W_{max} to the current window just before the loss. The idea is that W_{max} corresponds to the window size which caused the buffer to overflow and loss to occur, and the correct window size is smaller. Upon loss, the window is reduced to W_{min}, which is set to βW_{max}, where $\beta < 1$. If no loss occurs at the new minimum window, BIC jumps to the *target* window, which is half-way between W_{min} and W_{max}. This is called the "binary search" stage. If the distance between the *minimum* and the *target* window is larger than the fixed constant, S_{max}, BIC increments the window size by S_{max} each RTT to get to the target. Limiting the increase to a constant is analogous to the linear increase phase in Reno. Once BIC reaches the target, W_{min} is set to the current window, and the new target is again set to the midpoint between W_{min} and W_{max}.

Once the window is within S_{max} of W_{max}, BIC enters the "max probing" stage. Since packet loss did not occur at W_{max}, the correct W_{max} is not known, and W_{max} is set to a large constant while W_{min} is set to the current window. At this point, rather than increasing the window by S_{max}, the window is increased more gradually. The window increase starts at 1 and each RTT increases by 1 until the window increase is equal to S_{max}. At this point the algorithm returns to the "binary search" stage.

While BIC has been successful in experiments which have demonstrated that it can achieve high throughput in the tested scenarios, it is a relatively new protocol and the analysis of the protocol remains limited. For general networks with large number of sources and complicated topologies, its fairness, stability, and convergence properties are not yet known.

8.3.5 HIGH-SPEED TCP

High-Speed TCP (HSTCP) for large congestion windows, proposed in 2003 [12], addresses the problem that Reno has in achieving high throughput over high-BDP paths. As stated in ref. 7:

> On a steady-state environment, with a packet loss rate p, the current Standard TCP's average congestion window is roughly 1.2/sqrt(p) segments." This places a serious constraint on the congestion windows that can be achieved by TCP in realistic environments. For example, for a standard TCP connection with 1500-byte packets and a 100 ms round-trip time, achieving a steady-state throughput of 10 Gbps would require an average congestion window of 83,333 segments and a packet drop rate of, at most, one congestion event every 5,000,000,000 packets (or equivalently, at most one congestion event every 1&2/3; hours). This is widely acknowledged as an unrealistic constraint.

This constraint has been repeatedly observed when implementing data intensive Grid applications.

HSTCP modifies the Reno window adjustment so that large windows are possible even with higher loss probabilities by reducing the decrease after a loss and making the per-ACK increase more aggressive. Note that HSTCP modifies the TCP window response only at high window values so that it remains "TCP-friendly" when the window is smaller. This is achieved by modifying the Reno AIMD window update rule to:

$$\text{ACK}: w \leftarrow w + \frac{a(w)}{w}$$
$$\text{Loss}: w \leftarrow w(1 - b(w))$$

When $w \leq$ Low_window, $a(w) = 1$ and $b(w) = 1/2$, which makes HSTCP behave like Reno. Once $w >$ Low_window, $a(w)$ and $b(w)$ are computed using a function. For a path with 100 ms RTT, Table 8.1 shows the parameter values for different bottleneck bandwidths. Although HSTCP does improve the throughput performance of Reno over high-BDP paths, the aggressive window update law makes it unstable, as shown in Figure 8.7. The unstable behavior results in large delay jitter.

Table 8.1 Parameter values for different bottleneck bandwidths

Bandwidth	Average w (packets)	Increase $a(w)$	Decrease $b(w)$
1.5 Mbit/s	12.5	1	0.50
10 Mbit/s	83	1	0.50
100 Mbit/s	833	6	0.35
1 Gbit/s	8333	26	0.22
10 Gbit/s	83 333	70	0.10

8.3.6 SCALABLE TCP

Scalable TCP is a change to TCP Reno proposed in 2002 [8] to enhance the performance in high-speed WANs. Like HSTCP, scalable TCP makes the window increase more aggressive for large windows and the decrease after a loss smaller. The window update rule is:

ACK : $w \leftarrow w + 0.01$

Loss : $w \leftarrow 0.875w$

Like HSTCP, scalable TCP can fill a large BDP path but has issues with rate stability and fairness. Flows sharing a bottleneck may receive quite different rates, as shown in Figure 8.5.

8.3.7 H-TCP

H-TCP was proposed in 2004 [13] by the Hamilton Institute. Like HSTCP, H-TCP modifies the AIMD increase parameter so that

ACK : $w \leftarrow w + \dfrac{\alpha}{w}$

Loss : $w \leftarrow \beta \cdot w$

However, the α and β are computed differently to HSTCP. H-TCP has two modes, a low-speed mode with $\alpha = 1$, at which H-TCP behaves similarly to TCP Reno, and a high-speed mode at which α is set higher based on an equation detailed in ref. 13. The mode is determined by the packet loss frequency. If the loss frequency is high, the connection is in low-speed mode. The parameter β, where $\beta < 1$, is set to the ratio of the minimum to the maximum RTT observed. The intention of this is to ensure that the bottleneck link buffer is not emptied after a loss event, which can be an issue with TCP Reno, in which the window is halved after a loss.

8.3.8 TCP WESTWOOD

TCP Westwood (TCPW), which was first introduced by the Westwood-based Computer Science group at UCLA in 2000 [14], is directed at improving the performance of TCP over high-BDP paths and paths with packet loss due to transmission errors.

While TCPW does not modify the linear increase or multiplicative decrease parameters of Reno, it does change Reno by modifying the *ssthresh* parameter. The *ssthresh* parameter is set to a value that corresponds to the BPDP of the path:

$$ssthresh = \frac{RE \cdot baseRTT}{MSS}$$

where MSS is the segment size, RE is the path's rate estimate and baseRTT is the round-trip propagation delay estimate. The RE variable estimates the rate of data being delivered to the receiver by observing ACK packets. Recall that if the window is below *ssthresh*, slow start rapidly increases the window to above the *ssthresh*. This has the effect of ensuring that, after a loss, the window is rapidly restored to the capacity of the path. In this way, Westwood achieves better performance in high-BDP and lossy environments.

TCPW also avoids unnecessary window reductions if the loss seems to be caused by transmission error. To discriminate packet loss caused by congestion from loss caused by transmission error, TCPW monitors the RTT to detect possible buffer overflow. If RTT exceeds the $B_{\text{spike start}}$ threshold, the "spike" state is entered and all losses are treated as congestion losses. If the RTT drops below the $B_{\text{spike end}}$ threshold, then the "spike" state is exited and losses might be caused by channel error. The RTT thresholds are computed by

$$B\text{spike start} = \text{baseRTT} + \alpha(\text{max RTT} - \text{baseRTT})$$

$$B\text{spike end} = \text{baseRTT} + \beta(\text{max RTT} - \text{baseRTT})$$

where $\alpha = 0.4$ and $\beta = 0.05$ in TCPW. A loss is considered to be due to transmission error only if TCPW is not in the "spike" state and $RE \cdot baseRTT < re\_thresh \cdot w$, where *re_thresh* is a parameter that controls sensitivity. Figure 8.10 shows that, of the loss-based TCP protocols, Westwood indeed has the best loss recovery performance.

8.4 TRANSPORT PROTOCOLS BASED ON SPECIALIZED ROUTER PROCESSING

This section describes the MaxNet and XCP protocols, which are explicit signal protocols that require specialized router processing and additional fields in the packet format.

8.4.1 MAXNET

The MaxNet architecture, proposed in 2002 [15], takes advantage of router processing and additional fields in the packet header to achieve max–min fairness and improve many aspects of CC performance. It is a simple and efficient protocol, which, like other Internet protocols, is fully distributed, requiring no per-flow information at the link and no central controller. MaxNet achieves excellent fairness, stability, and convergence speed properties, which makes it an ideal transport protocol for high-performance networking.

```
┌─────────────────────┐
│ TCP/IP Packet       │
│  ┌───────────────┐  │
│  │ Price [32 bit]│  │
│  └───────────────┘  │
└─────────────────────┘
```

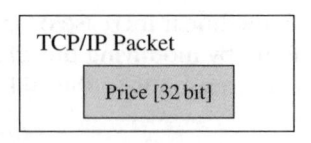

Figure 8.12. MaxNet packet header.

With MaxNet, only the most severely bottlenecked link on the end-to-end path generates the congestion signal that controls the source rate. This approach is unlike the previously described protocols, for which all of the bottlenecked links on the end-to-end path add to the congestion signal (by independent random packet marking or dropping at each link), which is termed "SumNet." To achieve this result, the packet format must include bits to communicate the complete congestion price (Figure 8.12). This information may be carried in a 32-bit field in a new IPv4 option, an IPv4 TCP option or in the IPv6 per-hop options field, or even in an "out-of-band" control packet.

Each link replaces the current congestion price in packet j, M_j, with the link's congestion price $P_1(t)$, if it is greater than the one in the packet. In this way, the maximum congestion price on the path is communicated to the destination, which relays the information back to the source in acknowledgment packets. The link price is determined by an AQM algorithm:

$$P_1(t+1) = P_1(t) + \alpha(Y_1(t) - \mu C_1(t))$$

where μ is the target link utilization and α controls the convergence rate and the price marked in packet j is $M_j = \max(M_j, P_1(t))$. The source controls its transmission rate by a demand function $D(.)$, which determines the transmission rate $x_s(t)$ given the currently sensed path price $M_s(t)$:

$$x_s(t) = w_s D(M_s(t))$$

where $D(.)$ is a monotonically increasing function and w_s is a weight used to control the source's relative share of bandwidth. Several properties about the behavior of MaxNet have been proven analytically:

- *Fairness*. It has been shown [15] that MaxNet achieves a weighted max–min fair rate allocation in steady state. If all of the source demand functions are the same, the allocation achieved is max–min fair, and if the function for source s is scaled by a factor of w_s, then w_s corresponds to the weighting factor in the resultant weighted max–min fair allocation.
- *Stability*. The stability analysis [16] shows that, at least for a linearized model with time delays, MaxNet is stable for all network topologies, with any number of sources and links of arbitrary link delays and capacities. These properties are analogous to the stability properties of TCP-FAST.
- *Responsiveness*. It has also been shown [17] that MaxNet is able to converge faster than the SumNet architecture, which includes TCP Reno.

8.4 Transport Protocols based on Specialized Router Processing 165

To demonstrate the behavior of MaxNet, the results of a preliminary implementation of the protocol are included here. Figure 8.13 shows the experimental testbed where flows from hosts A and B can connect across router 1 of 10 Mbps and router 2 of 18 Mbps capacity to the listening server and host C can connect over router 2. The round-trip propagation delay from hosts A and B to the listening server is 56 ms, and from host C it is 28 ms. Figure 8.14 shows the goodput achieved by MaxNet and Reno when hosts A, B, and C are switched on in the sequence, AC, ABC, and BC. Note that MaxNet achieves close to max–min fairness throughout the whole experiment (the max–min rate does not account for the target utilization μ being 96% and the packet header overhead). Note also that the RTT for MaxNet shown in Figure 8.15 is close to the propagation delay throughout the whole sequence. For TCP Reno the RTT is high as Reno fills up the router buffer capacity.

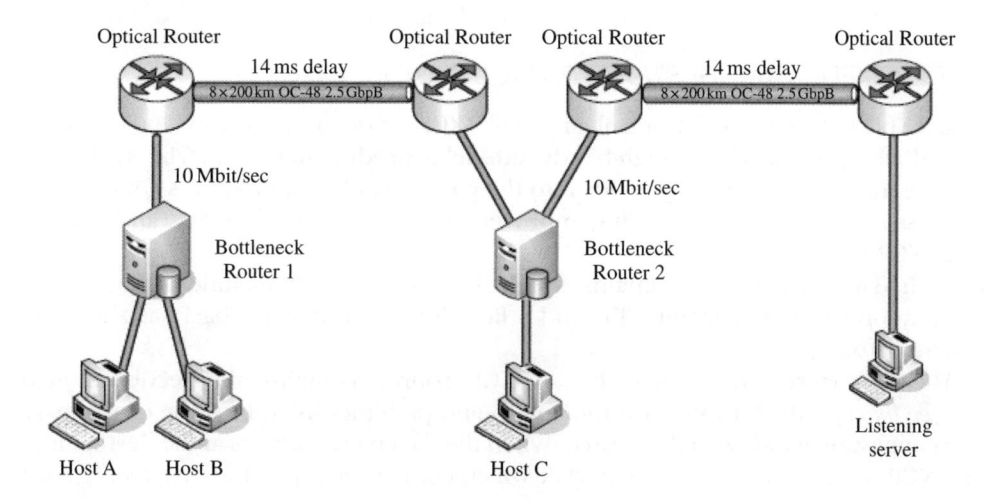

Figure 8.13. MaxNet experimental setup.

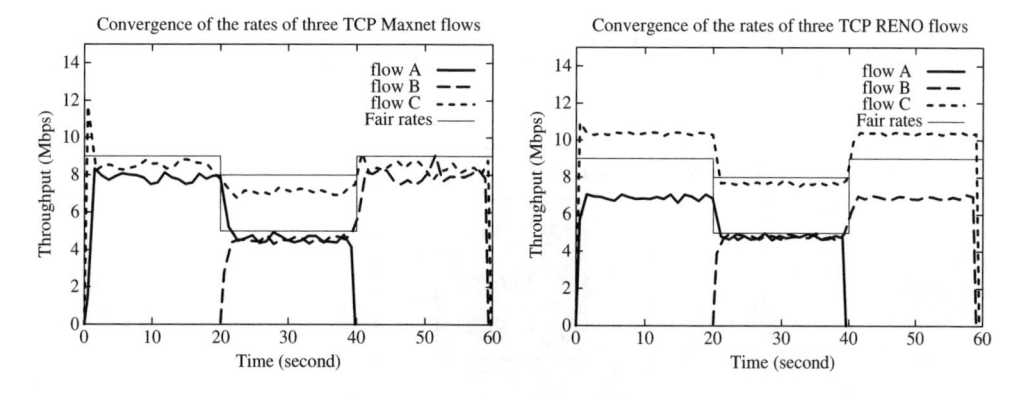

Figure 8.14. MaxNet (left) and Reno (right) TCP goodput and max–min fair rate.

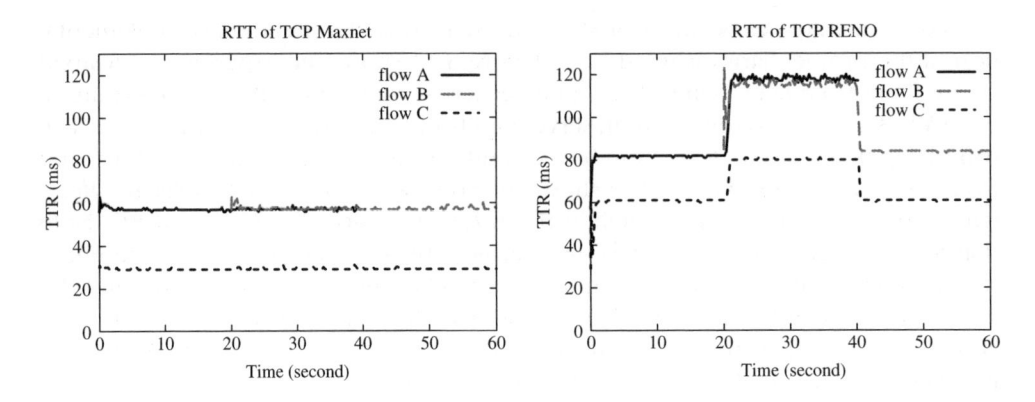

Figure 8.15. RTT for MaxNet (left) and Reno (right) TCP.

8.4.2 EXPLICIT CONGESTION CONTROL PROTOCOL (XCP)

The eXplicit Congestion Control Protocol (XCP), first proposed in 2001 [18], is aimed at improving CC on high-bandwidth-delay product networks. The XCP architecture introduces additional fields into the packet header and requires some router processing. XCP aims at providing improvements in fairness, efficiency, and stability over TCP Reno.

Each data packet sent contains the XCP header, which includes the source's congestion window, current RTT, and a field for the router feedback, as shown in Figure 8.16.

The kth packet transmitted by an XCP source contains the feedback field $H\_feedback_k$, which routers on the end-to-end path modify to increase or decrease the congestion window of the source. When the data packet arrives at the destination, the XCP receiver sends an ACK packet which contains a copy of $H\_feedback_k$ back to the source. For each ACK received, the source updates its window according to:

$$w \leftarrow \max(w + H\_feedback_k, s)$$

where s is the packet size. To compute $H\_feedback_k$, the router performs a series of operations which compute the lth router's feedback signal, $H\_feedback_l$. In the

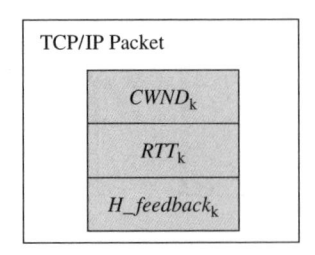

Figure 8.16. XCP packet header.

opposite way to MaxNet, the packet is remarked if the router's feedback is smaller than the packet's feedback:

$$H\_feedback_k \leftarrow \min(H\_feedback_k, H\_feedback_l)$$

The current version of XCP requires that each bottleneck router on the network path implements XCP for this CC system to work. The router computes the feedback signal based on the fields in the data packet using a process described in detail in ref. 19. Although the full process involves a number of steps, the main internal variable that controls the feedback increase or decrease is $\phi_l(t)$, which is computed by

$$\phi_l(t) = \alpha d(c_l - y_l(t) - \beta b_l(t))$$

where c_l is link l's capacity, $y_l(t)$ is the aggregate traffic rate for the link, d is the control interval, $b_l(t)$ is the buffer occupancy and α and β control stability [19] as well as fairness [20]. $\phi_l(t)$ is then used to compute $H\_feedback_l$. XCP has been simulated and analyzed, and some of its properties are:

- *Fairness.* The fairness properties of XCP were analyzed in ref. 21, and it was shown that XCP achieved max–min fairness for the case of a single-bottleneck network, but that for a general network XCP achieves rates below max–min fair rates. With the standard parameter settings suggested in ref. 19, link utilization is at least 80% at any link.
- *Stability.* The stability of XCP has also been analyzed in ref. 19, and for the case of a single bottleneck with sources of equal RTTs it was shown that XCP remains stable for any delay or capacity. For general heterogeneous delays stability is not known.
- *Responsiveness.* Simulation results in ref. 21 suggest faster convergence than TCP Reno.

Incremental deployment is suggested as taking one of two possible routes [19]. One way of achieving it is by using islands of XCP-enabled routers and having protocol proxies which translate the connections across these islands. Another way is for XCP to detect the presence of non-XCP enabled routers on the end-to-end path and revert back to TCP behavior if not all the routers are XCP enabled.

8.5 TCP AND UDP

This chapter has presented a number of the key topics related to the architecture of the TCP Reno protocol, primarily related to the congestion control algorithm, as well as potential algorithms that could serve as alternatives to traditional TCP. Early discussions of the congestion control issues [22] have led to increasingly more sophisticated analysis and explorations of potential responses. The next chapter presents other approaches, based on UDP, to these TCP Reno congestion control issues. These two approaches are not presented not as an evaluative comparison, but rather to further illustrate the basic architectural problems and potential alternatives for solutions.

ACKNOWLEDGMENTS

Bartek Wydrowski (primary), Sanjay Hegde, Martin Suchara (MaxNet section), Ryan Witt, Steven Low and Xiaoliang (David) Wei undertook proofreading.

REFERENCES

[1] E. He, P. Vicat-Blanc Primet, and M. Welzl (no date) 'A Survey of Transport Protocols other than "Standard" TCP,' www.ggf.org.
[2] K. Ramakrishnan, S. Floyd, and D. Black (2001) "The Addition of Explicit Congestion Notification (ECN) to IP," RFC 3168, September 2001.
[3] D.M. Chiu, and R. Jain (1989) "Analysis of the Increase and Decrease Algorithms for Congestion Avoidance in Computer Networks," *Computer Networks and ISDN Systems*, 17, 1–14.
[4] L. Xu, K. Harfoush, and I.Rhee (2004) "Binary Increase Congestion Control for Fast Long-Distance Networks", INFOCOM 2004.
[5] C. Jin, D.X. Wei, S.H. Low, G. Buhrmaster, J. Bunn, D.H. Choe, R.L.A. Cottrell, J.C. Doyle, H. Newman, F. Paganini, S. Ravot and S. Singh (2003) "FAST Kernel: Background Theory and Experimental Results", presented at the First International Workshop on Protocols for Fast Long-Distance Networks, February 3–4, 2003, CERN, Geneva, Switzerland.
[6] S. Floyd and T. Henderson (1999) "The NewReno Modification to TCP's Fast Recovery Algorithm", RFC 2582, April 1999.
[7] S. Floyd (2003) "HighSpeed TCP for Large Congestion Windows," RFC 3649, Experimental, December 2003.
[8] T. Kelly (2003) "Scalable TCP: Improving Performance in HighSpeed Wide Area Networks", First International Workshop on Protocols for Fast Long-Distance Networks, Geneva, February 2003.
[9] M. Allman, V. Paxson, and W. Stevens (1999) "TCP Congestion Control," RFC 2581, April 1999.
[10] S. Floyd and K. Fall (1997) "Router Mechanisms to Support End-to-End congestion control," LBL Technical Report, February 1997.
[11] L. Brakmo and L. Peterson (1995) "TCP Vegas: End to End Congestion Avoidance on a Global Internet", *IEEE Journal on Selected Areas in Communication*, 13, 1465–1480.
[12] S. Floyd (2002) "HighSpeed TCP for Large Congestion Windows and Quick-Start for TCP and IP," Yokohama IETF, tsvwg, July 18, 2002.
[13] R.N. Shorten and D.J. Leith (2004) "H-TCP: TCP for High-Speed and Long-Distance Networks." Proceedings of PFLDnet, Argonne, 2004.
[14] M. Gerla, M.Y. Sanadidi, R. Wang, A. Zanella, C. Casetti, and S. Mascolo (2001) "TCP Westwood: Congestion Window Control Using Bandwidth Estimation", In Proceedings of IEEE Globecom 2001, San Antonio, Texas, USA, November 25–29, Vol. 3, pp. 1698–1702.
[15] B. Wydrowski and M. Zukerman (2002) "MaxNet: A Congestion Control Architecture for Maxmin Fairness", *IEEE Communications Letters*, 6, 512–514.
[16] B.P. Wydrowski, L.L.H. Andrew, and I.M.Y. Mareels (2004) "MaxNet: Faster Flow Control Convergence," in *Networking 2004*, Springer Lecture Notes in Computer Science 3042, 588–599, Greece, 2004.
[17] B. Wydrowski, L.L.H. Andrew, and M. Zukerman (2003) "MaxNet: A Congestion Control Architecture for Scalable Networks," *IEEE Communications Letters*, 7, 511–513.
[18] D. Katabi and M. Handley (2001) "Using Precise Feedback for Controlling Congestion in the Internet", MIT-LCS Technical Report 820.

[19] D. Katabi, M. Handley, and C. Rohrs (2002) "Congestion control for high bandwidth-delay product networks," *Proceedings of the 2002 Conference on Applications, Technologies, Architectures, and Protocols For Computer Communications* (Pittsburgh, PA, USA, August 19–23, 2002). SIGCOMM '02. ACM Press, New York, pp. 89–102.

[20] S. Low, Lachlan L. Andrew, and B. Wydrowski (2005) "Understanding XCP: equilibrium and fairness ", IEEE Infocom, Miami, FL, March 2005.

[21] D. Katabi (2003) "Decoupling Congestion Control from the Bandwidth Allocation Policy and its Application to High Bandwidth-Delay Product Networks," PhD Thesis, MIT.

[22] V. Jacobson (1988) "Congestion Avoidance and Control", Proceedings of SIGCOMM '88, Stanford, CA, August 1988.

Chapter 9

Grid Networks and UDP Services, Protocols, and Technologies

Jason Leigh, Eric He, and Robert Grossman

9.1 INTRODUCTION

The previous chapter describes several issues related to the basic algorithms used by classical TCP Reno architecture, primarily those that involve congestion control. That chapter also presents initiatives that are exploring transport methods that may be able to serve as alternatives to TCP Reno. However, these new algorithms are not the only options for addressing these issues.

This chapter describes other responses, based on the User Datagram Protocol (UDP). As noted in the previous chapter, these approaches are not being presented as an evaluative comparison, but as a means of illustrating the basic issues related to Internet transport, and different approaches that can be used to address those issues.

9.2 TRANSPORT PROTOCOLS BASED ON THE USER DATAGRAM PROTOCOL (UDP)

As described in the previous chapter, TCP performance depends upon the product of the transfer rate and the round-trip delay [1], which can lead to inefficient link

---

Grid Networks: Enabling Grids with Advanced Communication Technology Franco Travostino, Joe Mambretti, Gigi Karmous-Edwards   © 2006 John Wiley & Sons, Ltd

utilization when this value is very high – as in the case of bulk data transfers (more than 1 GB) over high-latency, high-bandwidth, low-loss paths.

For a standard TCP connection with 1500-byte packets and a 100-ms round-trip time, achieving a steady-state throughput of 10 Gbps would require an average congestion window of 83,333 segments and a packet drop rate of at most one congestion event every 5 billion packets (or, equivalently, at most one congestion event every 1&2/3; hours) [2]. This situation primarily results from its congestion avoidance algorithm, which is based on the "Additive Increase, Multiplicative Decrease" (AIMD) principle. A TCP connection reduces its bandwidth use by half immediately a loss is detected (multiplicative decrease), and it would take 1&2/3; hours to use all the available bandwidth again in this case – and that would be true only if no more loss is detected in the meantime.

Certainly, over long-distance networks, the aggressive overfetching of data can be used as a means to lower the overall latency of a system by having the endpoints cache the data just in time for the application to use it [3]. Yet that approach also does not satisfy many transport requirements.

Consequently, a number of research projects are investigating mechanisms related to the UDP (RFC 768) [4].

9.2.1 UDP TRANSPORT UTILITY

UDP provides a datagram-oriented unreliable service by adding the following elements to the basic IP service: ports to identify individual applications that share an IP address, and a checksum to detect and discard erroneous packets [5]. UDP has proved to be useful for transporting large amount of data, for which the loss of occasional individual packets may not be important. However, because UDP includes no congestion control, its usage has to be carefully selected, especially when used on the commodity Internet, to prevent degrading the performance of TCP senders and, perhaps, appearing as a denial-of-service attack.

In the context of data-intensive Grid computing, UDP has become a popular protocol because of its inherent capabilities for large-scale data transport. For example, an emerging Grid model is one that connects multiple distributed clusters of computers with dedicated (and dynamically allocated) lightpaths to mimic a wide-area system bus. Within such an infrastructure, transport protocols based on UDP can be more attractive than TCP [6]. As more distributed Grid infrastructure becomes based on lightpaths, supporting essentially private network services consisting of 1–10 s of gigabits/s of bandwidth, it is advantageous for applications to be able to make full use of the available network resources.

UDP-based protocols exist that have adopted, augmented, or replaced portions of TCP (such as slow start and congestion control) to increase flexibility. Also, traditional UDP has been an unreliable transport mechanism. However, these new variations provide for reliability. Conceptually, UDP-based protocols work by sending data via UDP and reporting any missing packets to the senders so that the packets can be retransmitted.

The rate of transmission is determined by the particular requirements of the application rather than following TCP's AIMD mechanism. The first introduction of this concept dates back to 1985 with the introduction of NetBLT [7]. However, it is only recently, with the availability of high-bandwidth WANs, that this approach has been re-examined.

Three early contributions to this effort included Reliable Blast UDP (RBUDP) [8,9], the UDP-based data transfer protocol (UDT) [10], and Tsunami [11]. These contributions are described in the following sections.

For all of these protocols, implementations have primarily been at the application level rather than at the kernel level. This approach makes it possible for application developers to deploy usable systems without having to ensure that the same kernel patches have been applied at all locations that might run the application. Furthermore, situating the protocol at the application level allows opening up the API to a wider range of controls for applications – there is no longer the burden of having to provide the control within the constraints of the standard socket API – for which there is currently no declared standard.

9.2.2 RELIABLE BLAST UDP (RBUDP)

Reliable Blast [8,9] has two goals. The first is to network resource utilization, e.g., keeping the network pipe as full as possible during bulk data transfer. The second goal is to avoid TCP's per-packet interaction so that acknowledgments are not sent per window of transmitted data, but instead are aggregated and delivered at the end of a transmission phase. In the protocol's first data transmission phase, RBUDP sends the entire payload at a user-specified sending rate using UDP datagrams. Since UDP is an unreliable protocol, some datagrams may become lost as a result of congestion or an inability of the receiving host to read the packets rapidly enough. The receiver, therefore, must keep a tally of the packets that are received in order to determine which packets must be retransmitted. At the end of the bulk data transmission phase, the sender sends a DONE signal via TCP so that the receiver knows that no more UDP packets will arrive. The receiver responds by sending an acknowledgment consisting of a bitmap tally of the received packets. The sender responds by resending the missing packets, and the process repeats itself until no more packets need to be retransmitted.

Earlier experiments resulted in the recognition that one of the most significant bottlenecks in any high-speed transport protocol resided in a receiver's inability to keep up with the sender. Typically, when a packet is received by an application, it must be moved to a temporary buffer and examined before it is stored in the final destination. This extra memory copy becomes a significant bottleneck at high data rates. RBUDP solves this in two ways. First, it minimizes the number of memory copies. This is achieved by making the assumption that most incoming packets are likely to be correctly ordered and that there should be few losses (at least initially). RBUDP, therefore, uses the socket API to read the packet's data directly into application memory. Then, it examines the header for the packet and determines whether the data was placed in the correct location – and moves it only if it was not.

9.2.2.1 RBUDP, windowless flow control, and predictive performance

The second mechanism RBUDP uses to maintain a well-balanced send and receive rate is the use of a windowless flow control mechanism. This method uses packet arrival rates to determine the sending rate. Packet arrival rates at the application level determine the rate at which an application can respond to incoming packets. This serves as a good way to estimate how much bandwidth is truly needed by the application. To prevent this rate from exceeding available bandwidth capacity, packet loss rates are also monitored and used to attenuate the transmission rate.

One of the main contributions of this work was the development of a model that allows an application to predict RBUDP performance over a given network [9]. This is given by:

$$\frac{B_{best}}{B_{send}} = \frac{1}{1 + \frac{RTT^* B_{send}}{S_{total}}}$$

where B_{best} = theoretical best rate, B_{send} = chosen send rate, S_{total} = total data size to send (i.e., payload), and RTT=round-trip time.

This ratio shows that, in order to maximize throughput, an application should strive to minimize $(RTT \cdot B_{send})/S_{total}$ by maximizing the size of the payload to be delivered. For example, given that RTT for Chicago to Amsterdam is 100 ms, and B_{send} is 600 Mbps, if one wishes to achieve a throughput of 90% of the sending rate, then the payload, S_{total} needs to be at least 67.5 MB.

9.2.3 THE UDP-BASED DATA TRANSFER PROTOCOL (UDT)

The SABUL (simple available bandwidth utilization library)/UDT protocols are designed to supported data-intensive applications over wide-area high-performance networks, especially those with high-bandwidth-delay products [12,13]. These types of applications tend to have several high-volume flows, as well as many smaller standard TCP-based flows. The latter are used to pass control information for the data-intensive application, for example using Web Services.

Both SABUL and its successor, UDT, are application-layer libraries in the sense that a standard user can install them at the application layer. In contrast, the installations of new TCP stacks require modifications to the kernel, which in turn require that the user has administrative privileges. In addition, UDT does not require any network tuning. Instead, UDT uses bandwidth estimation techniques to discover the available bandwidth [10].

9.2.3.1 SABUL/UDT goals

The SABUL/UDT protocols are designed to balance several competing goals:

- *Simple to deploy.* SABUL/UDT are designed to be deployable at the application level and do not require network tuning or the explicit setting of rate information by the application.

9.2 Transport Protocols based on the User Datagram Protocol (UDP)

- *Speed and efficiency.* SABUL/UDT are designed to provide efficient, fast, and reliable transport over wide-area high-performance networks with high-bandwidth-delay products. In particular, SABUL/UDT are designed to quickly discover and utilize available bandwidth.
- *Intra-protocol fairness.* SABUL/UDT are designed to share the available bandwidth between multiple high-volume data flows (there may be up to dozens of such flows).
- *Inter-protocol friendliness.* SABUL/UDT are designed to be friendly to any TCP flows using the same link. SABUL/UDT flows need not only to share links with other applications, but also to control information for many data-intensive applications employ TCP-based Web Services.

Figure 9.1 shows UDT's intra-protocol fairness and inter-protocol friendliness. *October 22–25, 2000, Seoul, Korea*, pp. 8–15.

As an aggregate, all the TCP flows used about 155 Mbps of bandwidth. This illustrates the inter-protocol friendliness of UDT. There are also two UDT flows sharing

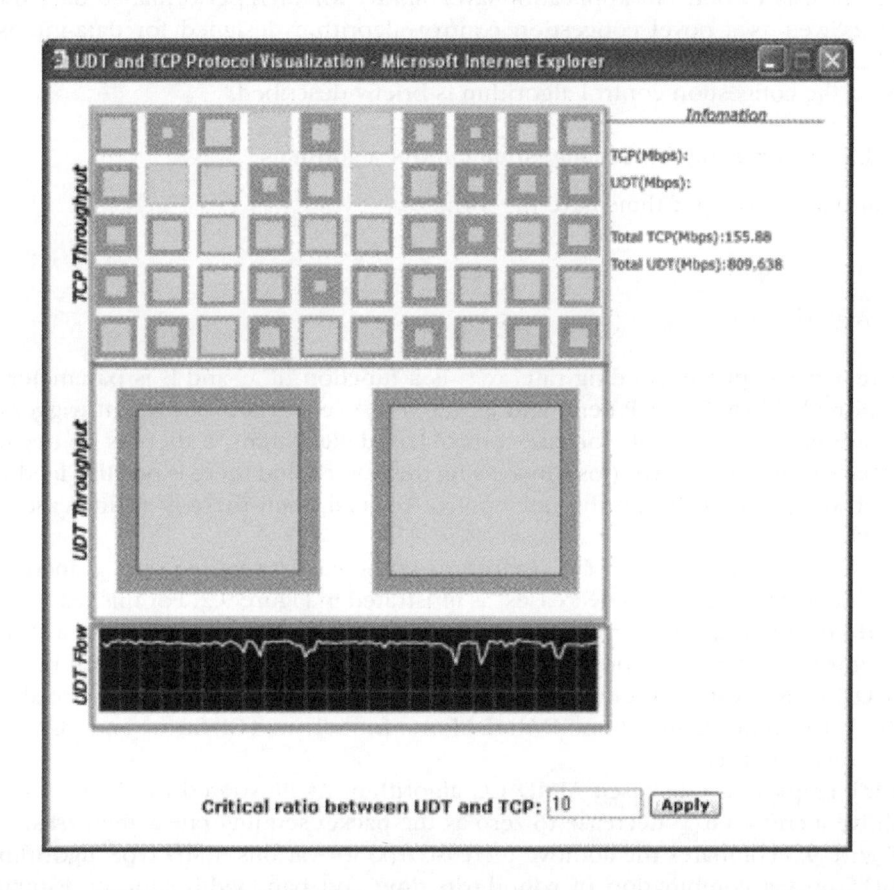

Figure 9.1. UDT intra-protocol fairness/friendliness. Each box represents a network flow. The size of the boxes represent the sending rate. This visualization helps to represent holistically the behavior of UDT in conjunction with other competing flows.

the link, each of which discovers the remaining available bandwidth automatically and shares it equally, as illustrated on the bottom half of the screen. This illustrates the intra-protocol fairness of UDT. Together, these two flows share about 809 Mbps of bandwidth. In total, over 95% of the available bandwidth is used over this high-bandwidth-delay product link.

For simplicity, SABUL used TCP as the control mechanism. For improved efficiency, UDT is entirely implemented using UDP.

The SABUL/UDT libraries were influential in part because, beginning in 2000, they were freely available as open source libraries and proved to be quite effective for developing and deploying data-intensive applications. Beginning in 2002 with version 2.0, SABUL was available via Source Forge. A year later, in 2003, the first version of UDT was released on Source Forge. The current version of UDT is 2.1, which was released in 2005.

9.2.3.2 UDT components

UDT consists of both an application-layer library for high-performance data transport as well as a novel congestion control algorithm designed for data-intensive applications over wide-area high-performance Grids.

Next the congestion control algorithm is briefly described.

9.2.3.3 Decreasing AIMD congestion control algorithms

Recall that AIMD algorithms have the form:

$$x \leftarrow x + \alpha(x)$$
$$x \leftarrow (1 - \beta)x$$

where x is the packet sending rate, $\alpha(x)$ is a function of x, and β is parameter. As mentioned above, for TCP Reno and its variants, every acknowledgment triggers an increase in x of $\alpha(x)$, while for rate control-based algorithms, if there is no negative feedback from the receiver (loss, increasing delay, etc.) and there is positive feedback (acknowledgments) during the rate control interval, then there is an increase in x of $\alpha(x)$.

UDT employs a rate-based CC algorithm. As the packet sending rate, x, increases, the additive increment $\alpha(x)$ decreases, as illustrated in Figure 9.2. For this reason, it is called a decreasing AIMD or DAIMD CC algorithm. In fact, a requirement of a DAIMD CC algorithm is that $\alpha(x)$ approaches zero as x increases. The particular formula used by UDT for $\alpha(x)$ can be found in ref. 14. In ref. 14, DAIMD CC algorithms are shown to be fair to other high-volume DAIMD flows, friendly to TCP flows, and efficient if properly configured.

UDT employs a rate-based AIMD CC algorithm. As illustrated in Figure 9.2, the additive increases $\alpha(x)$ decrease to zero as the packet sending rate x increases.

Figure 9.3 compares the additive increase $\alpha(x)$ for various AIMD type algorithms.

UDT uses a combination of round-trip time and bandwidth capacity estimates [10]. The sender sends periodic acknowledgment messages inn which are embedded a sequence number. This number is received by the receiver and sent back to

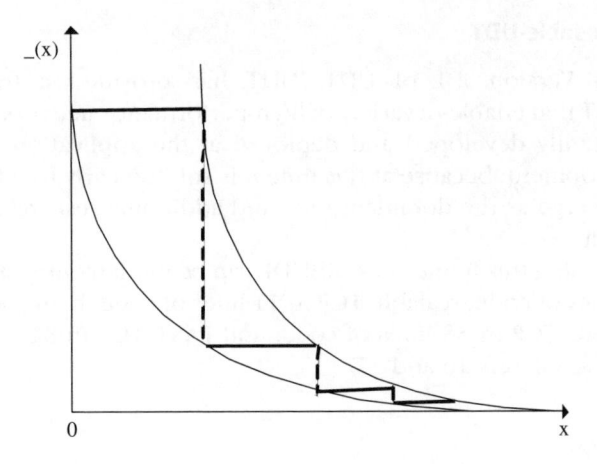

9.2 Transport Protocols based on the User Datagram Protocol (UDP)

Figure 9.2. Decreasing AIMD.

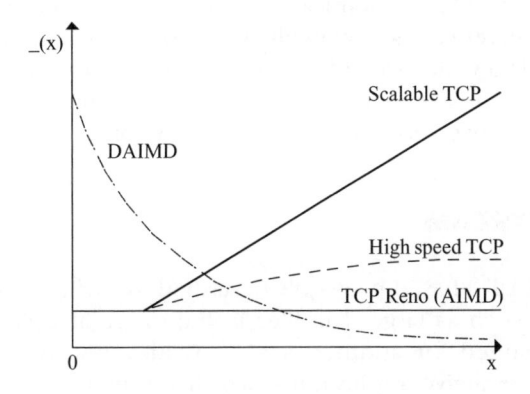

Figure 9.3. AIMD additive increases.

the sender to calculate round-trip time. As with RBUDP, UDT also sends negative acknowledgment messages to inform the sender of missing packets. Bandwidth estimates are achieved by calculating the median of the interval times between packet pairs (consisting of the application's payload) that are sent after every 16 packets. The bandwidth estimate is then used to control the sender's transmission rate. As UDT is constantly sensing the available bandwidth on the link, it is able to use the available bandwidth fairly in conjunction with multiple TCP transfers. UDT will use any available capacity that cannot be taken advantage of by a TCP flow.

To summarize, UDT employs a rate-based AIMD congestion control algorithm, whose additive increments $\alpha(x)$ decrease as the packet sending rate x increases. To improve performance, UDT also uses a window-based flow control mechanism. Finally, UDT provides a mechanism for automatically tuning the required parameters. For details, see ref. 15.

9.2.3.4 Composable-UDT

Beginning with Version 2.1 of UDT, UDT has provided a framework called composable-UDT that enables a variety of high-performance network transport protocols to be efficiently developed and deployed at the application level. This is an important development, because at this time it is still too early to tell which network protocol will emerge as the dominant one, and additional research and experimentation is still required.

For example, using this framework, RBUDP can be implemented at the application layer in three lines of code; scalable TCP in 11 lines of code; high-speed TCP in eight lines of code; BIC TCP in 38 lines of code; and FAST TCP in 31 lines of code. For additional details, see refs 16 and 17.

9.2.4 TSUNAMI

Whereas RBUDP and UDT provide both disk-to-disk and disk-to-memory transfer capabilities as an API, Tsunami's [12] primary goal is to provide a file delivery tool. Tsunami also provides authenticated data transfers whereby, upon connection establishment, the server sends a small block of random data to the client. The client XORs this random data with a shared secret, calculates MD5 checksum, and transmits the result to server. The server performs the same operation and verifies that the results are identical. Once verified, the file transmission can begin.

9.3 LAMBDASTREAM

While the TCP- and UDP-based protocols described in earlier sections work well for bulk data transfers such as large data files in the order of gigabytes and terabytes, they are not well suited for another class of Grid applications, namely real-time collaborative and interactive applications, which routinely need to stream comparatively small data objects but at much higher frequencies. For example, to stream high-definition microscopy images from a remote camera, high-definition video, or ultra-high-resolution images from a remote rendering cluster requires the transmission of image frames that are each of the order of a few megabytes but which are updated at 30 frames/second for an hour or more at a time. The total amount of data that can be transmitted in an hour can easily reach a terabyte.

These real-time applications often have tight constraints on latency and jitter because new updates need to be sent in response to user interaction. The TCP- and UDP-based protocols described perform very poorly in these applications because their error correction mechanisms introduce large amounts of latency and jitter. In the past, lossy transmission schemes have been acceptable because the applications have revolved around low-resolution video-conferencing. As Grid applications have begun to make greater use of distributed visualization resources, lossy transmission schemes have been found to be unacceptable because errors in transmission visually distort the content being transmitted. When used in scientific or medical applications, image distortions could lead to incorrect conclusions about the presented information.

LambdaStream [18] is a protocol intended to address this problem. In LambdaStream, lost packets are acknowledged immediately to reduce latency and jitter over long-distance networks. Link capacity estimation is used to find an optimal transmission rate [19]. When data loss occurs, packet interval times and packet sequence numbers are used to estimate whether the loss is due to the network infrastructure or due to the receiver. This technique borrows from the field of mobile and ad hoc networks, where a greater ability to predict the cause of packet loss can result in lower energy expenditures [20].

Often receivers incur incipient data loss when the CPU switches context to work on another task. These situations are characterized by data loss with no initial change in inter-packet arrival times. By detecting these situations, LambdaStream understands that they are temporary and will not decrease its transmission rate too aggressively. Conversely, when there is data loss due to congestion in the core network, inter-packet arrival times will increase and the transmission rate will be more aggressively reduced.

9.4 GRID APPLICATIONS AND TRANSPORT PROTOCOLS

The previous sections introduced a number of different transport protocols and indicated how they vary in terms of performance and implementation complexity. The following sections discuss the link between the application and the transport protocol.

9.4.1 BERKLEY SOCKETS

The universal API for communicating over the network available on most operating systems is Berkley sockets. The socket library provides a number of function calls that allow applications to establish connections using TCP or UDP or to gain access to constructing their own IP packet.

The *socket()* function initiates the communications interface, or "socket." The call allows the application to select between three types of sockets, *SOCK_STREAM* (TCP), *SOCK_DGRAM* (UDP), and *SOCK_RAW* (raw). In the existing implementations, when TCP is chosen, the default TCP NewReno protocol is used.

The *connect()* function actively connects the host to another host on the network. For the TCP protocol, this causes the host to send a SYN and perform a three-way handshake with the listening receiver. The *listen()* function allows the host to accept connections from other hosts, which actively open the connection with the *connect()* function. The function waits for the arrival of a SYN packet and the completion of the three-way handshake.

Once the connection is established, the *send()* and *recv()* functions are used to send and receive messages between the hosts.

9.4.2 FUTURE APIs

At present, there are no standard ways for an application to choose the type of TCP protocol used in the communication socket. Many of the TCP protocols discussed

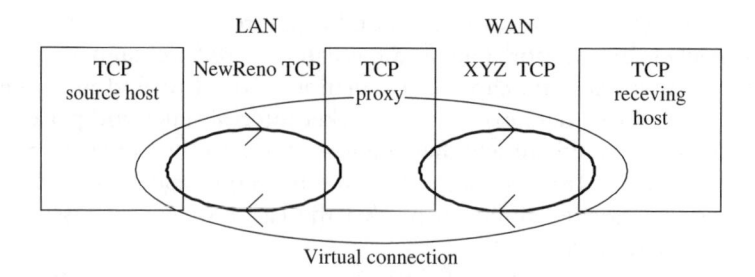

Figure 9.4. TCP proxy service.

in previous sections exist only as prototypes. They are available as patches to the operating system kernel. They either totally replace the inbuilt TCP protocol, or they co-exist with the default protocol in the operating system and they can be enabled by setting of system parameters such as /proc files under Unix.

Ideally, the socket interface will in the future extend to allow the selection of the underlying TCP behavior. One approach would be to extend the socket interface to allow the application to select a utility function. Recall that the utility function describes the application's relative bandwidth needs. A bandwidth-sensitive application could then simply select to have a steep utility function, which would give it a more aggressive use of the network's capacity.

Another approach would be to extend the socket interface to allow a selection of different TCP algorithms. This approach would allow the application to select between protocols such as BIC, FAST, NewReno, etc., depending on the desired performance characteristics.

9.4.3 TCP PROXIES

Another approach altogether is through the use of TCP proxies, as shown in Figure 9.4. While it may be difficult to modify the operating system of the host running the Grid application, a TCP proxy may be used to translate between the TCP implementation on the host and the desired TCP algorithm to be used over the network. This approach has been used particularly for communication over high-speed WAN and satellite networks, where TCP NewReno performance can be poor. An application would connect to the WAN through the proxy. The proxy terminates the TCP connection and establishes its own connection to the receiver using the desired TCP protocol. This process is transparent to the application on the originating host.

9.5 THE QUANTA TOOLKIT

An example of a method for more closely integrating Grid applications and network services can be seen in the design of Quanta. Quanta is an extensive cross-platform C++ toolkit consisting of Application Programming Interfaces (APIs) for accelerating data distribution in network-intensive collaborative applications [19]. Domains that have benefited from Quanta include computational science and engineering, earth sciences, and bioinformatic sciences [6, 22–24].

Quanta emerged from nearly a decade's experience in connecting immersive CAVE systems [25] to each other and to supercomputers – a concept called Tele-Immersion [26]. CAVEs are $10 \times 10 \times 10$ foot virtual reality rooms that create realistic 3D computer imagery through the use of rear-projected stereoscopic computer graphics. Quanta's predecessor, CAVERNsoft [26,27], has been widely used by the CAVE community to develop advanced tele-immersive applications. CAVERNsoft was unique in that it marked the first time a wide-area shared memory abstraction was used to synchronize data within a virtual reality application. CAVERNsoft was also one of the earliest examples of what is known today as peer-to-peer computing.

Quanta inherits from CAVERNsoft all of the data-sharing services that have been found to be essential for developing network-intensive applications. Capabilities include UDP and TCP network reflectors (needed to support collaborative applications), persistent shared memory (to save the state of a collaboration), message passing and remote procedure calls, high-performance remote file I/O (able to handle both 32- and 64-bit files), a variety of transmission protocols including forward error-corrected UDP (to provide low-latency, low-jitter streaming), parallel TCP, Reliable Blast UDP (RBUDP), and soon LambdaStream (a streaming high-throughput reliable protocol).

The key to Quanta's success has been in providing advanced networking techniques in a way that is accessible to application developers who are not networking experts. Tangible tools provide a systematic means to deploy advanced capabilities to the audiences that most desperately need them – but who do not have the expertise to interpret the complex network research literature.

9.5.1 TUNING AND OPTIMIZATION ISSUES

A combination of the dramatic decrease in network bandwidth costs and the rapid growth in the amount of data that scientific computing applications need to process has led considerable interest in using ultra-high-speed networks as wide-area backplanes for connecting multiple distributed cluster computers to form larger virtual computer systems. Beyond the middleware that is needed to enable the formation of these virtual computers and provision and manage their backplanes, there is a crucial need for an intelligent and automated way to optimally tune the connected systems. This requirement is especially critical for real-time interactive Grid applications, such as collaborative high-resolution visual data exploration.

9.5.2 COMMUNICATION SERVICES OPTIMIZATION

These applications typically require linking multiple large data repositories, data mining clusters, graphics rendering clusters, and remote displays. It will be impossible for future application developers to hand-optimize every resource configuration where their software might have to run. The next phase of the Quanta Project focuses on the *network* optimization problems. Specifically, it will address how to orchestrate concerted network streams from network-intensive cluster computing applications, which possess a wide variety of transmission characteristics, and which need to operate under a variety of network conditions.

This issue is a challenging problem, because for a Grid computing application each computing node of a cluster may have multiple network interface adapters, and the application may need to send and receive data over multiple network flows to and from multiple destinations simultaneously. Each of these flows may behave differently – some are bursty, some are constant flows, some transmit/receive data at a very high rate, some transceive data at a low rate but require low latency and low jitter, etc. This problem becomes even more complex when the nodes are distributed over a wide area, and must all share limited bandwidth resources. However, even in instances where the network may be overprovisioned to the point that there is no contention on the core networks, there is still the issue of differently varying WAN latencies that must be accounted for.

Furthermore, contention needs to be resolved at the end-hosts because each node must handle multiple network flows under the constraints of the available resources, such as network interface capacity (how much a network interface adapter can transceive), CPU (how much of the CPU is already loaded working on the computations for the application), memory (how much network buffer space is available), and system bus bandwidth (how much is the application using the system bus to simultaneous access other peripherals such as disks). At the present time few, if any, programming tools exist that will allow application developers to resolve these problems when building network-centered cluster computing applications.

9.6 GRIDS AND INTERNET TRANSPORT

Chapter 8 and this chapter have focused on layer 4 of the OSI model as it is implemented in Internet transport architecture and protocols. These chapters provide an overview of the basic Internet data delivery service protocols, issues related to those protocols that suggest changes are required, and emerging directions in research and development related to new architecture and methods for addressing those requirements.

These issues are important to Grid communities because many Grid applications today are demonstrating the need for an approach to Internet transport that can provide an alternative to the classical protocols. Although these protocols have been highly successful to date, and certainly will continue to serve many communities, alternative transport methods are required for multiple next-generation applications.

ACKNOWLEDGMENTS

Jason Leigh and Eric He are responsible for Sections 9.1 through 9.2.2.1 and Sections 9.3.3 through 9.7. Robert Grossman wrote Sections 9.3.2 through 9.3.2.4.

REFERENCES

[1] V. Jacobson, B. Braden, and B. Dorman (1992) "TCP Extensions for High Performance," RFC 1323, May 1992.

[2] S. Floyd (2003) "HighSpeed TCP for Large Congestion Windows", RFC 3649, Experimental, December 2003.

[3] C. Zhang, C., J. Leigh, J., T. DeFanti, M. Mazzucco, and R. Grossman (2003) "TeraScope: Distributed Visual Data Mining of Terascale Data Sets Over Photonic Networks," *Journal of Future Generation Computer Systems*, 19, 935–944.

[4] J. Postel, User Datagram Protocol, RFC 768, August 1980.

[5] M. Goutelle, Y. Gu, E. He, S. Hegde, R. Kettimuthu, J. Leigh, P. Primet, M. Weizi, C. Xiong, and M. Yousaf (2005) "A Survey of Transport Protocols other than Standard TCP," Grid Working Document, Data Transport Research Group, Global Grid Forum.

[6] L. Smarr, A. Chien, T. DeFanti, J. Leigh, and P. Papadopoulos (2003) "The OptIPuter,", special issue on "Blueprint for the Future of High-performance Networking," *Communications of the ACM*, 46, 58–67.

[7] D. Clark, L. Lambert, and C. Zhang (1988) "NETBLT: A High Throughput Transport Protocol ACM," *Proceedings of ACM SIGCOMM '88*, pp. 353–359

[8] J. Leigh, O. Yu, D. Schonfeld, R. Ansari, E. He, A. Naya, N. Krishnaprasad, K. Park, Y. Cho, L. Hu R. Fang, A. Verlo, L. Winkler, and T. DeFanti (2001) "Adaptive Networking for Tele-Immersion," Proceedings of Immersive Projection Technology/Eurographics Virtual Environments Workshop (IPT/EGVE), May 16–18, Stuttgart, Germany, 2001.

[9] E. He, J. Leigh, O. Yu, and T. DeFanti (2002) "Reliable Blast UDP: Predictable High Performance Bulk Data Transfer," Proceedings of IEEE Cluster Computing.

[10] Y. Gu, X. Hong, M. Mazzucco, and R. Grossman (2002) "Rate Based Congestion Control over High Bandwidth/Delay Links," White Paper, Laboratory for Advanced Computing, University of Illinois at Chicago.

[11] M. Meiss, Tsunami: "A High-Speed Rate-Controlled Protocol for File Transfer," http://steinbeck.ucs.indiana.edu/~mmeiss/papers/tsunami.pdf.

[12] R.. Grossman, Y. Gu, D. Hanley, X. Hong, D. Lillethun, J. Levera, J. Mambretti, M. Mazzucco, and J. Weinberger (2003) "Experimental Studies Using Photonic Data Services at iGrid 2002," *Journal of Future Computer Systems*, 19, 945–955.

[13] R. Grossman, Y. Gu, X. Hong, A. Antony, J. Blom, F. Dijkstra, and C. de Laat (2005) "Teraflows over Gigabit WANs with UDT," *Journal of Future Computer Systems*, 21, 501–513.

[14] Y. Gu, X. Hong and R. Grossman (2004) "An Analysis of AIMD Algorithms with Decreasing Increases," *Proceedings of GridNets 2004*. IEEE Press.

[15] Y. Gu, X. Hong, and R. Grossman (2004) "Experiences in Design and Implementation of a High Performance Transport Protocol", Proceedings of the ACM/IEEE SC2004 Conference on High Performance.

[16] Y. Gu and R. Grossman (2005) "Optimizing UDP-Based Protocol Implementations," Proceedings of the Third International Workshop on Protocols for Fast Long-Distance Networks, PFLDnet 2005.

[17] Y. Gu and R. Grossman, "Supporting Configurable Congestion Control in Data Transport Services," ACM/IEEE SC 2005 Conference (SC '05).

[18] C. Xiong, J. Leigh, E. He, V. Vishwanath, T. Murata, L. Renambot, and T. Defanti (2005) LambdaStream – a Data Transport Protocol for streaming Network-Intensive Applications over Photonic Networks, Proceedings of PFLDnet 2005.

[19] I. Stoica, S. Shenker, and H. Zhang (1998) "Core-Stateless Fair Queuing: Achieving Approximately Fair Bandwidth Allocations in High Speed Networks," Proceedings of ACM SIGCOMM '98, Vancouver, Canada, September 1998.

[20] C. Xiong and T. Murata (2004) "Energy-efficient method to improve TCP performance for Mobile and Ad hoc Networks," Proceedings of the International Conference on Computing, Communications, and Control Technologies (CCCT 2004), Austin, TX.
[21] E. He, J. Alimohideen, J. Eliason, N. Krishnaprasad, J. Leigh, O. Yu, and T. DeFanti, "Quanta: A Toolkit for High Performance Data Delivery over Photonic Networks," *Journal of Future Generation Computer Systems*, 19, 919–934.
[22] C. Zhang, J. Leigh, T. DeFanti, M. Mazzucco, and R. Grossman, "TeraScope: Distributed Visual Data Mining of Terascale Data Sets Over Photonic Networks," *Journal of Future Generation Computer Systems*, 19, 935–944.
[23] R. Singh, J. Leigh, and T. DeFanti (2003) "TeraVision: A High Resolution Graphics Streaming Device for Amplified Collaboration Environments," *Journal of Future Generation Computer Systems*, 19, 957–972.
[24] N. Krishnaprasad, V. Vishwanath, S. Venkataraman, A. Rao, L. Renambot, J. Leigh, and A. Johnson (2004) "JuxtaView – A Tool for Interactive Visualization of Large Imagery on Scalable Tiled Displays", Cluster 2004, San Diego, California, September 20–23, 2004.
[25] C. Cruz-Neira, D. Sandin, T. DeFanti, R. Kenyon, and J. Hart (1992) "The CAVE Automatic Virtual Environment," *Communications of the ACM*, 35, 64–72.
[26] J. Leigh, A. Johnson, and T. DeFanti (1996) "CAVERN: A Distributed Architecture for Supporting Scalable Persistence and Interoperability in Collaborative Virtual Environments," *Journal of Virtual Reality Research, Development and Applications*, 2, 217–237.
[27] K. Park, Y. Cho, N. Krishnaprasad, C. Scharver, M. Lewis, J. Leigh, and A. Johnson (2000) "CAVERNsoft G2: A Toolkit for High Performance Tele-Immersive Collaboration," *Proceedings of the ACM Symposium on Virtual Reality Software and Technology 2000*,

Chapter 10

Grid Networks and Layer 3 Services

Joe Mambretti and Franco Travostino

10.1 INTRODUCTION

The previous two chapters describe Internet transport protocols and show how various improved layer 4 techniques can be used to enhance Grid network services. This chapter presents several Internet layer 3 topics in a Grid context. Basic Internet layer 3 protocols are used to provide the core functionality of the Internet. They support services that reliably and efficiently route data through the network.

Key issues in this chapter involve topics related to integrating the basic flexibility attributes of Internet layer 3 protocols with the dynamic nature of Grid services requirements. The Internet was designed to be self-adjusting to changing conditions. To some degree, the original Internet assumed unstable infrastructure conditions, and, therefore, its protocols were designed to respond to them dynamically. The features that provide this flexibility are particularly useful for communication services within Grid environments.

10.2 THE INTERNET AND THE END-TO-END PRINCIPLE

Chapter 3 describes the end-to-end principle, which indicates that distributed systems, including networks, should have simple, powerful cores and intelligent edges. The Internet implements this principle in part through its fundamental protocols, TCP/IP. These protocols ensure that a highly distributed set of infrastructure resources can be used as a stable, robust network, which supports the delivery of

---

Grid Networks: Enabling Grids with Advanced Communication Technology   Franco Travostino, Joe Mambretti,
Gigi Karmous-Edwards   © 2006 John Wiley & Sons, Ltd

multiple data services over wide areas. As Chapter 8 explains, TCP is a basic Internet transport protocol (layer 4), which ensures a measure of reliable functionality. IP is a layer 3 protocol and provides for the Internet's basic connectionless data delivery service, without any firm guarantees for that service.

10.3 THE INTERNET AND LAYER 3 SERVICES

The development of the Grid has benefited substantially from the architecture of Internet layer 3 communication services [1]. Grid architecture and Internet architecture are highly complementary, and many of the basic design principles for both are very similar. For example, both provide high levels of abstraction from underlying infrastructure; both are decentralized; both were designed for highly dynamic conditions; both provide for simple, but powerful, cores, and for intelligent functionality placed within edge processes. Both have frequently been depicted as having an "hourglass design," indicating that the key functions are at the edges of the infrastructure and the middle is designed to expedite those functions.

The Internet is often described as an overlay network. Almost all of its current implementations consist of a packet-routing infrastructure, supported at lower levels with a foundation of layer 2 and layer 1 services, processes, and technologies. This architecture follows the mode of a classic overlay model, in part because currently there is almost no interaction between layer 3 and the other supporting layers. From the Internet layer 3 perspective, the other layers are usually treated as statically provisioned bit pipes.

Several major forces are changing this common model. One is the current general data communication trend toward the creation of multilevel services definitions and another is multilayer integration, which is enabling capabilities for signaling across layers to provide for enhanced functionality. This new multilayer services architecture is being developed in parallel with the creation of enhanced capabilities for layer 2 and layer 1 services, described in Chapters 11 and 12. These trends complement Grid architectural requirements, especially the requirement for flexibility. However, the following discussion focuses primarily on layer 3 Internet protocols.

10.3.1 IP CONCEPTS

One of the Internet's strengths is that it has been designed so that it can be used as a single network – one continuous global facility. However, with regard to its evolving architectural directions, it is important to understand that the Internet is not a single network but remains a network of a very large number of networks. The apparent seamless continuity of the Internet obscures the many different localized implementations of Internet network topologies and configurations, for example those segmented into the many Autonomous System (AS) domains, a part of the network managed under a single administration and set of policies.

This attribute also obscures the individual mechanisms that provide the Internet's basic functions. As earlier chapters have indicated, to investigate the potential for creating differentiated Grid services on the Internet, it is necessary to examine the basic processes that underlie such functions, i.e., to find mechanisms for directly addressing those processes, and to determine whether or not those processes can be used for specialized purposes related to Grid services.

IP provides a connectionless data delivery service, i.e., a process for sending data from a source to a destination, without concerns about the mechanisms by which that data will be handled between those two points or about the multiple networks that may be connecting those two points.

As explained in Chapter 8, TCP is used to verify reliable data delivery from senders to receivers. IP is used to ensure that packets are transmitted (forwarded) among the nodes of a network using best effort, not guaranteed, processes. The current widely implemented version of IP is IP version 4, or IPv4. (The future version is IPv6, discussed in a later section of this chapter.) Essentially, IP provides for addressing, marking datagrams (a unit of Internet information), and disassembling and assembling of datagrams.

10.3.2 IP COMPONENTS

It is not the responsibility of IP to ensure delivery; that is the role of other protocols, e.g., transport protocols such as TCP. IP simply transmits data, without assurances of delivery. A datagram, or IP data unit, is a series of bits comprising a header and the data content that is to be transported. The header field contains various pieces of information, including the source address, the destination address, protocol identifiers, data unit id, parameters related to service characteristics, time-to-live stamp, length, and other parameters.

IPv4 provides for a 32-bit address space, which is used to identify endpoints on the network; for example, these addresses define sources and destinations for transmitted data. The IP header field can be used for many functions, including, as described in Chapter 6, for marking packets in order to influence specific types of traffic behavior.

10.3.3 DIFFERENTIATED SERVICES

The IETF DiffServ standard provides a means to describe and implement a defined level of service quality, which is highly scalable [2,3]. The DiffServ architecture describes a mechanism for traffic classification, through individual packet marking, that can be linked to various forwarding behaviors (Per-Hop Behaviors, PHBs). These PHBs implement specific classes of services and service priorities. In addition, DiffServ describes mechanisms for continuous monitoring and change of those services, for example by using standard policing and traffic conditioning processes.

Differentiated Services (DS)-enabled routers support packets that have a DS field, which is an octet IP header, that can be used to identify flows within specific categories. The first six bits (capable of marking 64 different classes) – the Differentiated Services Code Point (DSCP) [4] – specifies the PHB [5], such as Expedited Forwarding (EF) [6,7]. The DSCP replaces the earlier IPv4 TOS (type of service) octet, which has been used only rarely. DiffServ, on the other hand, has often been used to provide for QoS-based services, both in production and experimentally, particularly with an EF implementation.

10.4 GRID EXPERIMENTATION WITH DIFFSERV-BASED QUALITY OF SERVICE

Many of the earliest experiments that explored providing Grid applications with adjustable DS were based on DiffServ QoS, using dedicated DS-enabled routers within a Grid environment. Often these routers were considered Grid first-class entities, in that they were fully controlled by Grid service processes. Specific goals of early experiments included developing mechanisms to serve Grid applications more precisely and to allocate available network resources more efficiently through application traffic flow identification by marketing packets within edge systems.

This approach generally involved four basic components: Grid-aware applications, implemented in the context of Grid services and application toolkits; the tool kits, which provide specialized application-related services; Grid services; and, integrated with those services, processes for determining QoS using DiffServ. Many early implementations used Grid services, often with GARA, to establish a process that could interrogate routers to discover available network resources, to schedule and reserve those resources, to allocate them, and to monitor their use.

The combination of Grid services and DiffServ techniques provides capabilities for governing many basic network process elements, including those related to policy-based service determination, priority setting, highly granulated (individual packet) behavior control (through DSCP marking), application classification, flow characteristic specification, service level specification, policy governance for services, resource requests (including those for router resources), dedicated allocation, use monitoring, and fault detection and recovery.

At network ingress points, a typical experimental implementation used DiffServ techniques, particularly EF, through router policy processes and mechanisms for packet classification and marking. For example, routers provided processes for rate limiting, setting IP precedence through DSCP, and traffic conditioning and policing through techniques such as packet drops. In such experiments, egress traffic was usually conditioned using standard traffic conditioning techniques such as variable, weighted queuing techniques. Other experiments incorporated these techniques in conjunction with Real-Time Transport Protocol (RTP), Real-Time Control Protocol (RTCP) to control RTP data transfers, and network management including instrumentation.

These early experiments demonstrated that combining Grid services with QoS tools, such as those based on DiffServ EF, can provide Grid applications with significant control over network behavior [8,9]. These initiatives also showed that this control can be implemented not only at network edge point, but also within edge hosts, that is the DiffServ packet marking could be accomplished within any DS-enabled edge device, not just within routers, providing the option of allowing DS capabilities to support any application. This technique can provide for a hierarchy of coordinated service governance capabilities within the network, e.g., specific flows within edge environments, within access paths, and aggregate flows within the network core.

10.5 INTERNET ROUTING FUNCTIONS

Transmitting data through the Internet is accomplished through routing, which determines the paths that support traffic flow among nodes. It could be argued that an instantiation of a network consists primarily of the design of its aggregate routing topology. Such designs can be extremely complex, especially in larger networks, because the overall design comprises a very large number of individual nodes, and each of these nodes can have a different configuration.

Routers, which constitute a basic device resource for the Internet, provide the functionality needed for delivering data packets. Two basic mechanisms within routers are forwarding and routing. Forwarding is the mechanism that sends the datagram to the next node. Routers use forwarding tables to determine this. Routing is a function that determines where to send datagrams based on their addresses.

To provide for the routing function, routers build and maintain routing tables using routing protocols. These protocols provide information on reachability, adjacency, and optimization of paths by examining a range of variables related to network conditions and configurations. The value of these parameters is provided to route calculation algorithms, and the resulting calculations are stored in routing tables.

Also important for Grid environments is the fact that almost all routers are implemented to support dynamic routing, which provides for continuous automatic updating of routing tables and for recalculations of optimal paths as traffic conditions and network topologies change, and as fault conditions are detected. Routers use this information to create appropriately formatted packets and transmit them to the next network node. These dynamic processes complement Grid requirements for adjustable communication service parameters.

Routers maintain the information in their routing tables updated by exchanging information with other, nearby, routers. This information exchange is based the function of several routing protocols, as is discussed in the following sections.

10.5.1 ROUTING PROTOCOLS

A network domain with a common administration and common policies is an Autonomous System (AS). Within an AS, the routing protocol that is used for routing table creation is the Interior Gateway Protocol (IGP), which is commonly used by organizations, as opposed to service providers. IGP protocols for the exchange of routing information include the Routing Information Protocol (RIP) [10]. RIP is a distance-vector protocol used when adjacent routers within an AS exchange a complete set of routing table information. To exchange that information with all routers within an AS domain, a link state IGP is used. Open Shortest Path First (OSPF) is a link state protocol [11], and Intermediate System to Intermediate System (IS-IS) is another [12].

The Border Gateway Protocol (BGP) is an Exterior Gateway Protocol (EGP) [13]. A version of BGP that was created for Unix computers is gated. BGP provides for the secure exchange of AS routing table information among multiple AS domains. BGP can also be used within an AS. If BGP is used to exchange information within an AS, it is labeled Interior BGP (IBGP), and when the information is exchanged among two or more AS domains, it is labeled Exterior BGP (EBGP).

10.5.2 COMMUNICATING ROUTING TABLE INFORMATION

These protocols all use the same basic communication mechanisms for exchanging information. For example, the BGP protocol works by first establishing a secure TCP connection between two AS domains, using the Open message. The two domains then exchange messages that open and confirm the connection variables.

If the connection is an initial one, the complete BGP routing table will be sent. Subsequent connections will send only the changed information. This information, which includes withdrawn routes and preferred paths, is sent using an update message. Periodically, "KeepAlive" messages are transmitted about twice a minute to maintain the connection. If the message is not received, the connection is terminated.

BGP also provides for exchanging information about problems through the Notifications message. If an error condition is detected, a message is sent indicating that a problem has occurred, and the connection is terminated.

10.5.3 ROUTE ADVERTISEMENT AND ROUTE STORAGE

Pairs of BGP-enabled routers advertise routes with Update messages. These messages contain a Network Layer Reachability Information (NLRI) field. This field holds the IP addresses, as well as an attributes field, containing path information.

Routing Information Bases (RIBs) contain the routing information by type according to use. RIBs contain the routes as Adj-RIBs-In (routes from other BGP routers), the Loc-RIB (routes used locally based on the Adj-RIBs-In data), and the Adj-RIBs-Out (routes that are advertised). The Loc-RIB next hop information is also sent to the local BGP device's forwarding information base.

10.5.4 ROUTING POLICIES

The interexchange of routing information is undertaken by router functions that manage the creation and implementation of routing policies. Such policies assist in creating and implementing routing topologies, which essentially define the network for particular segments. These policy creation mechanisms have an expansive set of options, which are used for traffic QoS assurances, engineering, redundancy, reliability, resource utilization, and related functions.

These policies are usually set manually and implemented as static policy definitions. However, through router APIs, they can be also adjusted through external processes, such as those integrated with Grid services. As an earlier section noted, it is possible to directly integrate router policy mechanisms with Grid service processes to better match specific application requirements to defined network behaviors.

10.5.5 ROUTING TOPOLOGIES

Given the complexity of the Internet, it would not be possible to construct an image of its overall routing topology in a way that would be meaningful. Even attempting to construct images of modestly sized segments can be challenging. However, it is possible to mine some information about network routing topologies by gathering, analyzing, and modeling available data, for example from routing tables, to create images of particular segments. Most of the prior research and development of tools and techniques in this area has focused on using these methods to examine functions related to general Internet services, for example to provide a higher quality of service or to make better use of resources.

However, these methods may also be of use within some Grid environments for addressing certain types of layer 3 communications requirements, for example those for dynamic provisioning services with precisely defined attributes. Within some Grid environments, it may possible to access directly and interact with the routers that support its network. Processes can be created that can gather and interrogate router topology information and that can make provisioning determinations based on that analysis. The basic data used would be those that constitute routing metrics.

10.5.6 ROUTING METRICS

In any network of modest size, multiple potential paths exist for traffic transit. Routing table information is used to provide information for determining which of the many path options should be used for sending data. A standard method used to determine the path is a simple one, a calculation of the shortest path with minimal latency, a function that is expressed in the distance vector protocol mentioned earlier, OSPF. However, other methods exist for calculating not necessarily the shortest path but the best path, as calculated by using router metrics.

The path is determined by the values related to a set of parameters, which constitute what is termed "routing metrics." It is notable that, for the primary Internet routing protocols, a route is defined as a "unit of information that pairs a destination with the attributes of a path to that destination" [13]. These attributes are variables that can be adjusted with control functions, and routing metrics can be used as basic data for decision-making.

Often these metrics are described as the "costs" that are associated with the path. The individual parameters include such variables as the path length (often described as the number of hops between the source and destination), the available bandwidth on a path, the delay (a measure of how long it would take for a packet to travel along the path), traffic volume, reliability, and others. Variables can include parameters that can be arbitrarily set by system administrators, who may select a preference for one path over another through setting values such as weights related to the paths. The weights can be used to provide relative values among alternate paths.

The number of hops is a basic parameter which simply indicates, for example, the number of segments that a datagram would have to pass through on a path; usually fewer is considered better. Of course, there are always other considerations, such as the total available bandwidth on the path at a given time, especially in comparison with the amount of traffic being transmitted on it. A large link is not the best path if it is congested. Delay is measured in the time required to go from source to destination. Reliability is usually a measure of the bit error rate on a particular path.

Research has been conducted in the area of maximizing router metrics. The number and types of variables used for router metrics suggest that standard optimization techniques could be used to provide for path determinations [14]. These complex techniques are difficult to apply to large-scale networks but may be useful for router interrogation and dynamic path weighting in modestly sized Grids which have access to router table information.

10.6 LAYER 3 ADDRESSING AND NETWORK ADDRESS TRANSLATORS (NATS)

Another important topic for the Grid community is addressing. As noted, IPv4 uses a 32-bit address space to identify endpoints. Although this address space has been sufficient for the Internet to date, it is segmented and structured such that its limitations for future growth have been recognized. The explosive growth of the Internet – its users now exceed 1 billion – is creating severe challenges for generating sufficient addresses. The number of Internet users is far exceeded by the number of total edge devices.

Two solutions that have been devised to respond to the growth in need for address space are Network Address Translators (NATs) and Network Address Port Translators (NAPT), which are becoming increasingly popular. These functions rewrite addresses (optionally, layer 4 port numbers also) in the packet header according to state information carried in the NA(P)T. This technique allows additional devices to be connected without using additional addresses.

Although this technique may conserve addresses, it can be problematic for Internet functionality. For example, some protocols that depend on layer 3 endpoint information within the content field of an IP packet may not work through intervening NA(P)Ts. The Grid middleware processes may be disrupted by intervening NA(P)Ts – the Globus toolkit has known problems with NA(P)T negotiation. Registering a "callback" service is a problem because the registration protocol may not transverse the complete path. The process forwards the registrant's endpoint information as data that is sent unchanged through NA(P)Ts, while the packet header endpoint information is actively mediated by the NA(P)Ts.

In general, Grids have participants joining and leaving virtual organizations within a large, possibly planet-scale, footprint. It is practically impossible to rule out the use of NA(P)Ts within such a large area. As has been demonstrated with widespread VoIP solutions, it is possible to build protocols that detect and work-around NA(P)Ts, although with the burden of a more complex, often cumbersome protocol.

The widespread adoption of IPv6 should eliminate such issues, because it guarantees that an endpoint address is always unique within a virtual organization.

10.7 IP VERSION 6

A primary motivation for the development of IPv6 was the recognition that that IPv4 did not provide for sufficient address space for the many billions of devices required for the rapidly growing Internet. IPv4 provides for only a 3- bit address space, while IPv6 provides for a 128-bit address space [15]. Essentially, this space can provide for 2 to 128 addresses.

IPv6 supports all common Internet services (unicast, multicast, anycast), and provides header fields that can be used to create new capabilities.

IPv6 also accommodates a 24-bit flow label field, which can be used to specifically identify individual flows in order to provide them with special treatments. These treatments can be related to forwarding, levels of quality of services, security, and optimization of flow management. IPv6 also provides for a priority field, which can be also used for special flow processing.

10.8 SUBSECOND IGP RECOVERY

Network reliability and survivability is an important consideration. Grid activities, as well as any other long-lived session across the network, are vulnerable to episodes of degraded behavior resulting from partial network failures.

Currently, recovery can be a lengthy process. For a distance vector protocol such as OSPF [11], for instance, the default value for the retransmission timer is set to 5 seconds, whereas the "router-dead" interval is set to 40 seconds. When these timers are triggered, storms of Link-State Advertisements (LSAs) have been known to break out in large topologies before the network converges to the use of new routing paths. Similar situations have been observed with IS-IS implementations [12].

Researchers have made the case that the goal of subsecond IGP recovery is both desirable and attainable [16]. Subsecond recovery would provide layer 3 restoration with values similar to layer 2 protection techniques such as SONET. The three points of intervention are:

(1) the mechanisms and time granularity with which a router monitors the liveness of the peer;
(2) the propagation and handling of LSAs;
(3) the execution of algorithms such as Dijkstra's shortest path first, with the goal being to make the algorithm's execution time proportional to log N, N being the number of nodes in the topology.

More recently, researchers have experimented with new methods for ensuring IP resilience, for example by manipulating techniques for transmitting failure information among nodes [17]. Some of these techniques are based on manipulating basic routing metrics described in an earlier section [18].

10.9 INTERNET SECURITY USING INTERNET PROTOCOL SECURITY

Grid activities are always likely to demand continuous confidentiality protection over long periods of time, during which many gigabytes or even terabytes are exchanged. In general, the IPsec protocol suite is well suited to meeting this type of requirements [19]. Because Grids are typically support high-volume, high-performance traffic, the use of IPsec sequence extensions [20] can be helpful. It provides a 64-bit sequence number, which is less likely to be exhausted during a session.

Some experimental results indicate that the insecurity of Cipher Block Chaining (CBC) – a popular technique to extend ciphers beyond block size – increases as $O(s^2/2^n)$, where n is the block size in bits, and s is the number of blocks encrypted [20]. Intuitively, this insecurity results from the dependency of the ciphertext of one block upon both the plain text for that block and the ciphertext of the preceding block. Should two blocks yield the same ciphertext, and the plaintexts of the next block are also identical, then the ciphertexts of the next block will be identical. This situation creates a vulnerability to certain types of malicious attacks.

As a matter of common practice, a rekeying event should occur any time B bytes have been sent through an encrypted security association whose crypto-transform uses a CBC mode of operation (e.g., 3DES) [21]. This event sets B to $(n/8)*2^{[n/2]}$, wherein n is the block size in bits. A security association that uses 3DES ($n = 64$) at 1 Gbps requires a keying event every 274.9 seconds. At 10 Gbps, it requires a keying event every 27.5 seconds. As speeds increase, designers will need to take into account the challenge of executing a rekeying event at shorter intervals. It should be noted that a keying event typically requires a very large integer exponentiation, which is a very demanding challenge when compared with ordinary message crypto-processing.

10.10 IP MULTICAST

The IP multicast extensions [22] were introduced to relieve the network from forwarding as many copies of a set of data as there are receivers. Furthermore, its receiver-driven style of operation is meant to relieve the sender (or publisher) from tracking the subscribers to the data. In general, the promise to mitigate traffic volumes and complexity is likely to appeal to Grid communities, especially when multicast aligns with a "push" style of data diffusion in the Grid.

In practice, however, few networks have IP multicast enabled on the scale that would be significant to Grids. The difficulties in policy and security enforcement, in scaling reliable services above IP multicast [23], in cross-provider support, and providers' additional operating costs in running extra control plane software have greatly hindered a widespread roll-out of IP multicast services. An informational document by the GGF surveys the IP multicast landscape in greater detail [24].

On a confined footprint, IP multicast has proved to be an important enabler of group communication for the increased reliability of systems [25,26]. As Grids mature and the expectations of their being dependable grow, it is expected that IP multicast will continue to play an important role at the inter-component level (and replicas thereof).

Chapter 3 has substantiated the case for profound architectural flexibility in the Grid infrastructure (whether it is general or network specific). More than ever, Grid designers are empowered to specialize into realizations of multicast semantics at the layer 7 (e.g., application-level multicast) and/or layer 1 (e.g., optical multicast).

10.11 INTERNET LAYER 3 SERVICES

This chapter presents only a small number of topics related to Internet layer 3 protocols and services. One point emphasized in this chapter is that these protocols were designed specifically to support services in dynamic environments. Consequently, they have inherent features and functions that can be integrated with Grid processes in order to customize specific types of network services.

ACKNOWLEDGMENTS

Joe Mambretti developed, Sections 10.1–10.5 and Franco Travostino for Sections 10.6–10.1.

REFERENCES

[1] B. Carpenter (1996) "Architectural Principles of the Internet," RFC 1958, June 1996.
[2] S. Blake, D. Black, M. Carlson, E. Davies, Z. Wang, and W. Weiss (1998) "An Architecture for Differentiated Services," RFC 2475, December 1998.
[3] K. Nichols, V. Jacobson, and L. Zhang (1999) "A Two-bit Differentiated Services Architecture for the Internet, RFC 2638, July 1999.
[4] K. Nichols, S. Blake, F. Baker, and D. Black (1998) "Definition of the Differentiated Services Field (DS Field) in the IPv4 and IPv6 Headers, RFC 2474, December 1998.
[5] D. Black, S. Brim, B. Carpenter, and F. Le Faucheur (2001) "Per Hop Behavior Identification Codes," RFC 3140, June 2001.
[6] B. Davie, A. Charny, J.C.R. Bennet, K. Benson, J.Y. Le Boudec, W. Courtney, S. Davari, V. Firoiu, and D. Stiliadis (2002) "An Expedited Forwarding PHB (Per-Hop Behavior)," RFC 3246, March 2002.
[7] A. Charny, J. Bennet, K. Benson, J. Boudec, A. Chiu, W. Courtney, S. Davari, V. Firoiu, C. Kalmanek, and K. Ramakrishnan (2002) "Supplemental Information for the New Definition of the EF PHB (Expedited Forwarding Per-Hop Behavior,)" RFC 3247, March 2002.
[8] T. DeFanti and M. Brown, "EMERGE: ESnet/MREN Regional Science grid Experimental NGI Testbed – Final Report," DOE Office of Scientific and Technical Information.
[9] V. Sander (2003) Design and Evaluation of a Bandwidth Broker that Provides Network Quality of Service for grid Applications, NIC Series Vol. 16, John von Neumann-Institut für Computing (NIC).
[10] G. Malkin (1998) "RIP Version 2," RFC 2453, November 1998.
[11] J. Moy (1998) "OSPF Version 2," RFC 2328, April 1998.
[12] D. Oran (1990) "OSI IS-IS Intra-domain Routing Protocol," RFC 1142, February 1990.
[13] Y. Rekhter and T. Li (1995) "A Border Gateway Protocol 4 (BGP-4)," RFC 1771, March 1995.
[14] M. Gouda and M. Schneider (2003) "Maximizing Router Metrics," *IEEE/ACM Transactions on Networking*, 11, 663–675.
[15] S. Deering and R. Hinden (1998) "Internet Protocol, Version 6 (IPv6) Specification," RFC 2460," December 1998.
[16] C. Alaettinoglu, V. Jacobson, and H. Yu (2000) "Toward Millisecond IGP Convergence", NANOG 20, October 22–24, 2000, Washington, DC.
[17] S. Rai, B. Mukherjee, and O. Deshpande (2005) "IP Resilience within an Autonomous System: Current Approaches, Challenges, and Future Directions," *IEEE Communications*, 43(10), 142–149.
[18] B. Fortz and M. Thorup (2002) "Optimizing OSPF/IS-IS Weights in a Changing World," *IEEE JSAC*, 20, 756–767.
[19] S. Kent and K. Seo (2005) "Security Architecture for the Internet Protocol," RFC 4301, December 2005.

[20] S. Kent (2005) "IP Encapsulating Security Payload (ESP)," RFC 4303,"December 2005.
[21] M. Bellare, A. Desai, E. Jokippi, and P. Rogaway (1997) "A Concrete Treatment of Symmetric Encryption: Analysis of the DES Modes of Operation", http://www-cse.ucsd.edu/users/mihir/papers/sym-enc.html.
[22] S.E. Deering and D.R. Cheriton (1990) "Multicast Routing in Datagram Internetworks and Extended LANs", *ACM Transactions on Computer Systems*, 8(2), 85–110.
[23] Reliable Multicast Transport, Internet Engineering Task Force (2004) http://www.ietf.org/html.charters/rmt-charter.html.
[24] V. Sander, W. Allcock, P. CongDuc, I. Monga, P. Padala, M. Tana, F. Travostino, J. Crowcroft, M. Gaynor, D. Hoang, P. Primet, and M. Welzl (2004) "Networking Issues for grid Infrastructure," Grid Forum Document, No. 37, November 2004.
[25] K. Birman (2005) Reliable Distributed Systems: Technologies, *Web Services, and Applications*. Springer Verlag.
[26] F. Travostino, L. Feeney, P. Bernadat, and F. Reynolds (1998) "Building Middleware for Real-Time Dependable Distributed Services," the First IEEE International Symposium on Object-Oriented Real-Time Distributed Computing, p. 162.

Chapter 11

Layer 2 Technologies and Grid Networks

John Strand, Angela Chiu, David Martin, and Franco Travostino

11.1 INTRODUCTION

A consistent theme of this book is architectural flexibility of the infrastructure. Grid environments are designed and implemented to provide a suite of directly usable services and to present options for customization. Grids allow infrastructure resources, including network services, to be configured and reconfigured so that they can meet the exact requirements of applications and higher layer services.

This chapter focuses on how Grids can benefit from flexible, reconfigurable layer 2 network services. The Grid community has been exploring options for supplementing traditional network services with those at layer 2 that can provide exact service qualities, while also providing flexibility through dynamic allocations. One capability inherent in most layer 2 service implementations is switching, a capability for establishing a nonpermanent connection between two or more points. Traditionally, layer 2 switching capabilities have been provisioned with an understanding that such connections would be semipermanent. However, Grids are motivating the development of dynamic switching capabilities for layer 2 services.

11.2 LAYER 2 TECHNOLOGIES AND GRID REQUIREMENTS

Grids have typically used layer 2 services as a statically provisioned resource. More recently, Grid environments are being created that utilize dynamic layer 2 resources

---

Grid Networks: Enabling Grids with Advanced Communication Technology Franco Travostino, Joe Mambretti, Gigi Karmous-Edwards © 2006 John Wiley & Sons, Ltd

to meet requirements that cannot be easily met by layer 3–4 services. Primarily, this trend is motivated by application requirements. However, it is also being motivated by the technology advances in layer 2 environments that are allowing for enhanced ad-hoc layer 2 service provisioning and reconfiguration.

This chapter presents a number of layer 2 technologies that are important to the Grid community. Layer 2 of the OSI model is defined as the layer at which data is prepared for transmission required by a specific physical protocol. The data is "packaged" so that it can be sent and received across an infrastructure reliably. Functions include those for link control, such as general flow control, error correction, and for Media Access Control (MAC), which governs placement of data on specific physical infrastructure.

This last point should be understood in the context of the current network trend, discussed in earlier chapter, toward network virtualization. There are many different implementations of layer 2 services, including virtual services. For example, it is possible to create from higher level processes emulations of network services that resemble physical links with regard to functionality, but actually provide those capabilities at higher layers. Virtual Private Networks (VPNs) created using the TCP/IP protocols have been popular for many years. Although many of these virtual services make use of higher layers, they simulate a layer 2 network and are thus included in this chapter.

The IETF is developing several architectural approaches that provide for virtualization functions for layer 2-type services, including "virtual line services." One of the most popular protocols developed by the IETF is Multiprotocol Label Switching (MPLS) [1], which is discussed in the next section. It is notable that MPLS and related technologies have been described as layer 2.5 technologies because they do not strictly adhere to the classic definition set forth by the OSI model.

11.3 MULTIPROTOCOL LABEL SWITCHING (MPLS)

As explained in Chapter 8, in a typical IP network, traffic is sent from source to destination through a series of routers. Each router relies on the information provided by IP routing protocols (e.g., Open Shortest Path First (OSPF) or Border Gateway Protocol (BGP)), or static routing, to make an independent forwarding decision at each hop within the network. The forwarding decision is based solely on the destination IP address, and this decision is stored in a *routing table*. When a packet arrives at a router, the router examines the destination address of the incoming packets, looks this address up in the routing table, and then sends the packet to the specified next-hop router. If incoming traffic to the router is heavy, packets are stored until they can be processed. If outbound links are busy, packets are queued until they can be sent. This simple process was the technique used in earliest IP routers and, as other chapters have noted, is still in widespread use today. All packets are treated equally and each packet is analyzed independently.

However, from the earliest days of the Internet, network engineers noticed that many Internet applications tend to generate streams of packets with identical source and destination information. These point-to-point streams are called "flows." Having a router make the same routing decision over and over for the packets in a flow

11.3 Multiprotocol Label Switching (MPLS)

is an inefficient use of resources. Besides this inefficiency, the simple and robust destination-based hop-by-hop routing has the following limitations:

- It is insensitive to traffic conditions in the network and is therefore inefficient in the way it utilizes network capacity.
- Destination-based routing does not allow packets belonging to different classes of services to be routed and protected separately to meet their QoS needs.
- Destination-based routing is vulnerable to security attacks by anyone who learns the destinations of other users. The common approach today of filtering incoming packets for disallowed destinations is prone to provisioning errors and costly to implement.

To address these limitations, the IETF formed the MPLS Working Group to combine various approaches from the industry into one common standard [1] that can run over any media (e.g., peer-to-peer, frame relay, Asynchronous Transfer Mode (ATM)). With MPLS, each packet is associated with a label that identifies its Forwarding Equivalence Class (FEC). FEC can be defined by a common egress router and optionally by other characteristics (e.g., a common VPN or class of service). The label is encoded in the connection identifier when available (e.g., ATM Virtual Path Identifier (VPI)/Virtual Channel Identifier (VCI) and Frame Relay Data-Link Connection Identifier (FR DLCI)), otherwise a "shim" is inserted between IP header and link layer header. Based on the label of an incoming packet, a router determines the packet's next hop as well as the label to use on outgoing interface and performs the label swapping. Label Switched Paths (LSPs) can be established based on the IP address of a router that is downstream on the hop-by-hop route. On the other hand, LSPs can also be established using explicit routing determined offline or by a separate routing protocol. Explicit routing can take the bandwidth availability and traffic conditions into account. MPLS has many potential advantages compared with the traditional hop-by-hop destination-based routing. They are as follows:

- MPLS increases routers' forwarding capacity by switching based on short labels. Although this was one of the main motivations of introducing MPLS, this advantage is diminishing due to many new ASIC designs for fast and scalable prefix-matching algorithms.
- MPLS provides a way to perform Traffic Engineering (TE) [2] by overlaying logical connectivity on top of the physical topology to distribute loads more efficiently and in a more scalable way than the traditional IP over ATM overlay model, which was used for traffic engineering for large IP backbones during the early years.
- MPLS is connection oriented, and thus is capable of providing faster and more reliable protection and restoration than what current Interior Gateway Protocol (IGP) (e.g., OSPF and IS-IS) rerouting can provide [3].
- MPLS can map traffic with different class of service requirements onto different LSPs, which are routed and protected independently [4].
- MPLS provides added security if packets sent on outgoing interfaces are limited to those from particular connections.
- MPLS labels may be stacked to implement tunnels, which is useful in constructing VPNs [5].

The last two in particular enable service providers to offer IP services with enhanced security over a shared network infrastructure, as described below.

11.3.1 MPLS AND SHARED NETWORK INFRASTRUCTURE

As other chapters have explained, many applications are not well served by a technique that treats all packets equally as in standard "best effort" Internet services. Also, many communities, including Grid communities, want private networks on which they can determine many key service parameters. One response to this requirement is to build separate infrastructures for each community, but building and maintaining private physical networks for multiple communities is cost prohibitive. The real goal is to design, implement, and maintain a physical network that can be shared, but which is able to treat traffic differently based on individual communities. MPLS has become an ideal tool for meeting this goal. Each community of users sees a VPN, while the network service provider sees a shared network.

Using MPLS, traffic from a specific community can be tagged and treated uniquely throughout the network in accordance with that community's requirements. Each router uses the MPLS labels to make custom routing decisions. For example, a router may send traffic over a moderately loaded path or give preferential treatment together with some key QoS mechanisms, e.g., buffer management and queue scheduling, to those communities that have paid for a premium service.

11.3.2 MPLS AND VIRTUAL PRIVATE NETWORKS

Given the cost and complexity of deploying and managing private networks, more and more enterprise customers are implementing IP services to connect multiple sites over a service provider's shared infrastructure with the requirement that the service provides the same access or security policies as a private network. This requirement motivated the creation of VPNs. Two implementation models have gained widespread use:

- the overlay model, in which the service provider provides emulated leased lines to the customer;
- the peer-to-peer model, in which the service provider and the customer exchange layer 3 routing information and the provider relays the data between the customer sites on the optimum path between the sites and without customer's involvement.

The overlay model is easy to understand and implement since the service provider simply offers a set of Virtual Connections (VCs) connecting customers sites either by using some layer 2 WAN technologies such as frame relay and ATM, or with some IP-over-IP tunneling such as Generic Route Encapsulation (GRE) tunneling and IP security (IPsec) encryption. However, it has a few major drawbacks:

- It is well suited for enterprise customers with a few central sites and many remote sites. However, it becomes exceedingly hard to manage in a more meshed configuration due to scalability limitation.

- Proper provisioning of the VC capacities requires detailed knowledge of site-to-site traffic profiles, which are usually not readily available.

The peer-to-peer VPN model was introduced to alleviate the above drawbacks. A main solution in this space is BGP/MPLS IP VPNs standardized by the IETF [5]. It utilizes MPLS's connection-oriented nature and its ability to stack labels, together with Multiprotocol BGP (MP-BGP). MP-BGP offers rich routing policies and the capability to exchange VPN routing information between routers that are not directly connected, such that core provider routers need not to be VPN aware. Note that this network-based VPN solution does not preclude customers' or the provider's desire to use IPsec for added security through encryption.

11.3.3 GRID NETWORK SERVICES AND MPLS

As noted, Grid environments place unique stresses on network services because of the dynamic nature of Grid systems. The Grid demands that services including data flows be dynamically configured and quickly reconfigured. However, many network implementations are based on statically configured systems with known patterns of behavior. For example, in a standard MPLS implementation, a label switched path is established manually unless Label Distribution Protocol (LDP) [6] is used to set up hop-by-hop LSPs based on shortest path routing. Grid networks demand capabilities for temporary capacity requirements and dynamic topologies and routing. In addition, network services must have QoS guarantees and security assurances.

One approach would be to establish MPLS paths for all possible configurations of a Grid. This method is being assisted by equipment vendors, who are improving the performance of routers with large numbers of MPLS paths both by increasing memory and processing power and by new methods for the intelligent management of unused paths. The approach assumes that all Grid nodes are known in advance. Although, fortunately, this is often the case, it does not provide the techniques for rapid reconfiguration.

Nothing in the MPLS standard precludes rapid reconfiguration, but, at this time, the tools to accomplish this function are still being researched. Some of the basic tools required are emerging from initiatives established by Grid middleware architecture.

11.4 ETHERNET ARCHITECTURE AND SERVICES

Since its invention in 1973 at Xerox Corporation in Palo Alto, California [7], Ethernet has become the most popular Local Area Network (LAN) technology, connecting hundreds of millions of computers, printers, servers, telephone switches, pieces of laboratory equipment, etc., worldwide. Most of today's network data traffic is generated from Ethernet interfaces. In fact, it is not exaggerating to say that nearly all the IP traffic consists of Ethernet frames. The inventor of Ethernet, Dr. Robert Metcalfe, credited the success to several main factors, including packet switching, distributed nature, speed-up in data rate, and a business model that includes commitment to standardization, fierce competition, preserving install base, etc.

The IEEE 802 project has developed a highly defined standard specification for Ethernet in accordance with layer 2, and to some degree layer 1. These specifications describe the required attributes and measures for service quality, based on service availability, out-of-order frames, frame duplication, frame loss, priority, and other parameters. An important part of the specification is the IEEE 802.1D standard, which defines attributes of MAC bridges, such as traffic classification. One measure of success for this standard is its ubiquitous deployment. Another is that switches based on this standard are being implemented on chips, including 10-Gbps rate switches.

Owing to fierce competitions based on an open standard, the Ethernet community has adopted the price and performance model of the PC industry. For each new generation, the speed of Ethernet has been improved 10-fold with a targeted price increase of only three to four times.

As with the TCP/IP architecture, a key strength of Ethernet's architecture has been its simplicity and distributed nature. There is no central controller required for Ethernet. The auto-negotiation feature allows seamless interconnection of Ethernet interfaces at different rates and eliminates human errors.

Over the last 30 years, the Ethernet medium changed from the original shared coaxial bus in the first generation to dedicated point-to-point fiber-optic links. The medium access has also changed from half-duplex to full-duplex. This change removed the distance limitation imposed by the MAC protocol and enabled Ethernet packets to travel extended distances in their native format. The changes in bandwidth and MAC opened the path for Ethernet to penetrate from LANs into backbones and WANs.

Despite all the changes, the format of Ethernet frames has been kept invariant in all the Ethernet formulations. An end-to-end Ethernet solution eliminates much of the format conversion inefficiencies when different technologies such as SONET/SDH and ATM are used for backbone transport. Unlike SONET/SDH and ATM technologies, which tend to be vendor dependent, Ethernet products from various vendors are highly compatible. In order to support time-sensitive traffic such as voice and video as well as many grades of data traffic, Ethernet now requires additional attributes to continue its development, especially in the areas of QoS, reliability, and management. Many industry groups, including the IEEE, the ITU, the Metro Ethernet Forum (MEF), and the IETF are working diligently on relevant standards to address challenges in each of these areas. Over the next decade, Ethernet will continue to augment or even replace traditional WAN technologies.

The merits of using large-scale metro and regional layer 2 networks to suit Grid applications have been demonstrated in research initiatives on several testbeds, described in the appendix. These experiments and demonstrations have shown that dynamic layer 2 path provisioning over Dense Wavelength-Division Multiplexing (DWDM) channels can bring capabilities to distributed infrastructure that cannot be easily provided through traditional routed network services.

11.4.1 ETHERNET ARCHITECTURE FEATURES AND CHALLENGES

In recent years, many network operators have started migrating their transport network away from SONET/SDH-based infrastructure to a design optimized around

11.4 Ethernet Architecture and Services

IP and Ethernet. The goal is to create a "flatter" network with overall reduced costs of equipment and maintenance, permitting more rapid, flexible service introduction, provisioning, and control. This trend has moved from LAN to MAN, and yet need to happen in the WAN. Key deficiencies of Ethernet that are restricting its use in the wide area include its use of broadcast signaling for resource discovery, its flat address space, its lack of fast protection and restoration capability as well as sophisticated management tools for remote operations and fault detection and resolution, the limitations of its spanning tree capabilities, its minimal security attributes, and its limited capabilities for virtualization. All of these areas are important to implementing flexible Ethernet services within Grid environments and are all being addressed through IEEE standardization initiatives.

One important activity is the IEEE 802.1ah [8], which is defining an architecture and bridge protocols that will allow for the interconnection of multiple bridge networks (defined in 802.1ad [9]) to ensure VLAN scalability and associated management capabilities, e.g., through SNMP. The 802.1ad effort addresses issues such as layer 2 control protocol tunneling, VLAN stacking ("UNI QinQ"), and spanning tree segmentation. The 802.1d and 802.1w define the standards and protocols for the MAC spanning tree specification. Spanning tree prevents loops and assists in fault isolation. The 802.1s ("virtual bridge local area networks") extends the standard to include support for multiple spanning trees for metro areas.

Unlike SONET/SDH which provides fast protection and restoration, existing Ethernet technologies still lack such recovery capabilities from failures. The Resilient Packet Ring (RPR), a ring-based protocol, which was standardized by the IEEE 802.17 work group [10], has been designed particularly to address this need for Ethernet transport. See a brief overview of RPR in Section 11.7. RPR's ability to guarantee SONET-level protection and restoration time, and its efficient statistical multiplexing capability, together with its support for multiple classes of services with strict performance guarantees, make it an ideal traffic management layer for Ethernet-based services.

Unlike LAN/enterprise networks, in which Ethernet dominates, carrier networks are known for sophisticated and effective management capabilities, and set a high standard for reliability, availability, fast failover, and recovery to ensure that Service Level Agreements (SLAs) made by providers can be met. One of the greatest challenges facing service providers as they deploy Ethernet-based solutions in WAN environments lies in achieving the same level of Operations, Administration, and Maintenance (OAM) support that users are accustomed to receiving with traditional carrier networks. Specifically, SLAs attached to specific services need to be met without regard to the underlying technologies used to provide them. Ethernet OAM is one of the capabilities required to meet SLAs. It includes link performance monitoring, fault detection and fault signaling, and loopback testing. Many standard bodies and forum are addressing this challenge:

- ITU-T Y.1730 – Ethernet-based networks and services – provides the Ethernet OAM objectives, requirements, and a quick view of what types of functions need to be implemented, as well as some of the underlying reasons for implementing the function.

- IEEE 802.3ah – Ethernet link OAM – and IEEE 802.1ag – Ethernet maintenance domain connectivity fault management – together specify how OAM is to be implemented within the Ethernet network to achieve the objectives of Ethernet OAM and interoperability.
- MEF Fault and Management.
- IETF L2 VPN OAM.

11.4.2 ETHERNET AS A SERVICE

The IEEE initiatives are being complemented by the Metro Ethernet Forum [11], which is attempting to move Ethernet beyond a technical specification for basic transport. Specifically, the MEF develops Ethernet service standards, in the sense of a delivered service from a service provider. This forum is creating standard definitions of service classes so that Ethernet can be implemented not simply as a type of physical connection with an associated bandwidth, but as an actual service with well-defined and directly measurable service parameters, including traffic parameters, performance parameters, and class of service, etc.

The MEF has defined two categories of general services: a point-to-point service, Ethernet line service (E-line); and multipoint-to-multipoint Ethernet LAN service (E-LAN, also termed transparent LAN service, TLS). Within these categories, other more specific services and service attributes are defined. For example, the E-line service can be implemented as Ethernet Wire Service (EWS) on a shared switched infrastructure using VLANs to segment traffic. Another E-line service is the Ethernet Private-Line (EPL) service, which has associated service guarantees.

The current Ethernet service trend is moving from point-to-point to a multipoint E-LAN service, which can interconnect large numbers of sites with less complexity than meshed or hub and spoke connections implemented using point-to-point networking technologies. Furthermore, an E-LAN service can be used to create a broad range of services such as private LAN and virtual private LAN services such as VPLS described in Section 11.5. Both E-line and E-LAN services can be utilized to provide scalable and cost-effective transport for Grid services

11.4.3 10 GBPS ETHERNET AND BEYOND

Ever since the newest Ethernet standard, 10-Gbps Ethernet (10GE), was standardized by IEEE P802.3ae 10 Gbps Ethernet Task Force in 2003, it has been adopted worldwide by users and operators that run networks in data centers, financial institutions, universities, government agencies, airports, etc., and on the way to MANs and WANs. This issue results in a question: "Where does it go next?"

There are two potential next steps: 40-Gbps Ethernet (40GE) and 100-Gbps Ethernet (100GE). 40GE can be viewed as a logical next step if one wants to leverage SONET/SDH OC-768/STM-64 PHYs, as well as Optical Internetworking Forum (OIF) efforts under way to standardize OC-768 interfaces. It also has an advantage in dealing with transmission impairments compared with 100GE. However, the current cost per port of state-of-the-art OC-768 technology is much higher than the $4 \times 10\text{GE}$ pricing.

It may not have a significant enough performance improvement to warrant the cost of development/adoption. If 100GE has the same development cost as 40GE, people may opt for the higher bandwidth, which will scale better for future applications. There is no clear winner at this point in either standard process or the marketplace. Whatever the next step will be, it will face the challenge of riding on the same performance – price curve as previous generations to enable wide adoption.

11.5 PSEUDO-WIRE EMULATION (PWE) AND VIRTUAL PRIVATE LAN SERVICES OVER MPLS (VPLS)

The IETF has established several efforts to standardize methods for layer 2 type services over layer 3 infrastructure. Layer 3 VPNs have been discussed in Section 11.3.2. The Pseudo-Wire Emulation edge-to-edge initiative (PWE3 [12]) provides encapsulation capabilities that allow for multiple types of protocols to be transported over routed infrastructure, using common standard control planes for multiple protocol stacks.

Another initiative is defining virtual private LAN services over MPLS (VPLS), also called Transparent LAN Services (TLS) [13]. This architecture allows for an emulated LAN segment to be partitioned for a network segment. A VPLS supports a layer 2 broadcast domain that can discover Ethernet MAC addresses and can forward using those addresses within that domain. This architecture describes both the data plane and the control plane functions for this service and describes a measure of availability capabilities (e.g., multihoming). VPLS encapsulation methods are detailed in refs 12 and 14.

11.6 LAYERS 2/1 DATA PLANE INTEGRATION

The connectivity and performance of layer 2 WANs are constrained by the underlying layer 1 technology, topology, and operational capabilities. For high-bandwidth applications, the ubiquitous technology is Time-Division Multiplexed (TDM) based. TDM time-interleaves multiple (constant bit rate) signals to produce a composite multiplexed signal. The framing structures and protocols used to do this have evolved over the last 40 years. This evolution has culminated in the Synchronous Optical Network (SONET [15]) and Synchronous Digital Hierarchy (SDH [16]) standards.[1] Either stand-alone or as the interface to optical networks, they are ubiquitous in today's layer 1 networks.

[1] SONET dominates in the USA, SDH in Europe and much of the rest of the world. For our purposes, they are essentially equivalent.

The basic SONET entity is the Synchronous Transport Signal-1 (STS-1). STS-1 operates at 51.84 Mbps, of which 49.5 Mbps is usable payload (called "payload envelope") and the rest overhead. The STS-1 frame structure is byte oriented, with a total of 810 bytes, 36 of which (4.4%) are used for overhead purposes. The frame rate is 8000/s (125 µs/frame). Normally all SONET signals are bidirectional and run at the same rate in each direction. An STS-N signal ($N = 3, 12, 48, 192, 768$) is formed by byte-interleaving n STS-1s together. This leads to n separate payloads, which is not desirable for data; hence an "STS-Nc" with a single (×Nin size) payload has been defined.[2]

[2] For a detailed description of SONET and SDH, see ref. 17.

SONET/SDH is the ubiquitous interface to optical networks in today's WAN networks, both public and private. There are several reasons for this:

- All wireline carriers conform to the standard, both internally and as the basis for defining private line services.
- As a consequence, there is a vast amount of conforming hardware (from chip sets to network elements and turnkey networks) available, with prices that have been driven down by competition and by learning curve effects.
- Fault detection, management and other Operations, Administration, and Maintenance (OAM) functionalities are mature.

For some Grid applications (e.g., very large file transfers), SONET/SDH has a number of attractive features: low overhead; straightforward achievement of data transfer rates up to multiple gigabits per second; very low bit error rates; deterministic throughput rates; no out-of-sequence data; and mature, reliable fault handling and OAM. Packet-based networking alternatives currently available are built on top of SONET/SDH and hence find it difficult to compete on either costs or performance with an *adequately utilized* SONET/SDH alternative.

Unfortunately for Grid networks, the deployed SONET/SDH networks historically have only provided "pipes": Static connections are reconfigurable only on a timescale of weeks or longer, and (if a network operator is involved) via a provisioning request and negotiation with the operator. Since the durations of even the largest file transfer operations are much less than this, adequate utilization levels have only been achievable by use of a more dynamic overlay (layer 2/layer 3) network. Furthermore, the bandwidth granularity is quite coarse (basically, 55 Mbps, 155 Mbps, 622 Mbps, 2488 Mbps, and 9953 Mbps including overhead) and hence not efficient for important transport technologies such as 1Gbps Ethernet.

11.6.1 SONET AND TDM EXTENSIONS FOR ETHERNET-OVER-SONET (EOS)

ANSI, ITU-T, and other bodies such as the IETF have significantly extended and improved the basic SONET/SDH protocols to deal with these problems. These extensions have been made both to the data plane and to the control plane.

These extensions are now widely implemented by all major equipment vendors and thus are available to "roll your own" network operators such as large Grid networks. Much of the equipment uses commercially available chip sets, which could also be incorporated into custom-designed equipment for unique applications.

Some leading commercial operators have widely deployed equipment incorporating these extensions. However, the motivation for these deployments has largely been internal cost reduction and efficiency and the capabilities have not been visible to overlay network operators such as Grid networks. However, the hardware and control structures to support flexible, dynamically reconfigurable optical networks are largely in place, and could support customer-controlled applications if desired. The middleware to allow applications to take advantage of this capability is in an early development stage (Chapter 7).

11.6.2 VIRTUAL CONCATENATION

Virtual Concatenation (VCAT) [16] breaks an integral payload of up to 40 Gbps into multiple discrete 50-Mbps or 155-Mbps component payloads. These can then be separately routed through a SONET/SDH network and recombined at the endpoint of the transmission. VCAT functionality is required only at the two SONET endpoints (path terminations, in SONET parlance), and the intervening SONET network(s) need not be VCAT aware. This allows the integral payload to be larger than the maximum payload supported by the intervening networks. Instead of the five broadband capacity options available in traditional SONET, 256 are available with VCAT, ranging in increments based on the base payload size (e.g., N^*150 Mbps, $N \leq 256$ if 150-Mbps payloads are used).

11.6.3 LINK CAPACITY ADJUSTMENT SCHEME

Link Capacity Adjustment Scheme (LCAS) [18] builds on VCAT. It allows the dynamic hitless increase or decrease in the capacity of a VCAT connection in 50- or 155-Mbps increments, pending the availability of the intervening links. It can also gracefully decrease capacity autonomously if there is a partial failure and restore the capacity when the fault is repaired. Like VCAT, LCAS capabilities are required only at the endpoints of the SONET connections; the intervening networks need not be VCAT aware.

11.6.4 GENERIC FRAMING PROCEDURE

Generic Framing Procedure (GFP) [19] is a generic mechanism to carry any packet over a fixed-rate channel. It provides a flexible encapsulation framework and has the potential to replace a plethora of proprietary framing procedures for carrying data on SONET/SDH with a simpler, more robust, and often more efficient framing procedure. It provides frame delineation, multiplexing (both frame and client), and client data mapping.

11.7 RESILIENT PACKET RINGS (RPR)

Resilient Packet Ring (RPR) is a recent layer 2 development defined in ref. 10 for LANs and MANs primarily. It specifies a new layer 2 MAC including a "reconciliation sublayer" to allow use of existing IEEE 802 Physical Layer (PHY) specifications for SONET and Ethernet. In addition to new SONET-like resilience features, which are

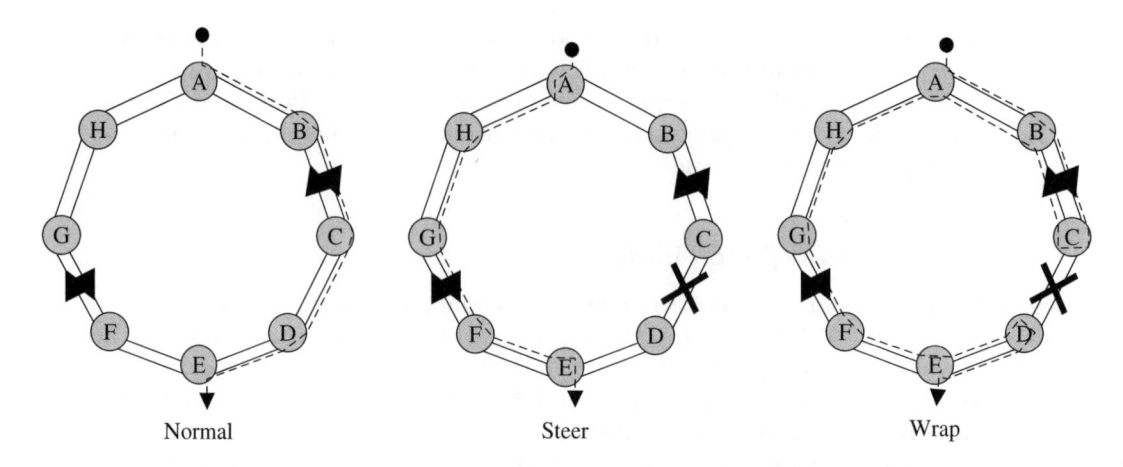

Figure 11.1. Resilience options in resilient packet ring.

discussed below, RPR offers improved bandwidth utilization, bandwidth management, and multiple service classes (see ref. 20 for a comprehensive overview).

The physical network underlying RPR is assumed to be dual unidirectional counter-rotating "ringlets". RPR resilience options are illustrated in Figure 11.1. Figure 11.1(a) shows an eight-node RPR ring with a virtual connection from A to E in a clockwise direction. Figure 11.1(b) and 11.1(c) show the two RPR restoration options for this VC in the event of a failure between nodes C and D:

- "*Steering*" (Figure 11.1b). In this case, upon detection of the failure, signaling messages back to A causes the VC to be routed counter-clockwise to E.
- "*Wrapping*" (Figure 11.1c). In this case, C immediately routes the VC in the reverse direction all the way around the ring to D, where the initial path is resumed. Thus the total route is A-B-C-B-A-H-G-F-E-D-E.

Steering is clearly more efficient. Wrapping, however, can be initiated more rapidly and so will have less packet loss. RPR requires steering; wrapping is optional. When it is available, it might be used only during the transient until steering can be invoked.

11.8 USER–NETWORK INTERFACES

A User–Network Interface (UNI) is defined by the ITU [21] to be "An interface that is used for the interconnection of customer equipment with a network element of the transport network." The "dial tone" interface to the traditional telephone network is a simple example. In more complex settings, the UNI allows the customer to specify the connection(s) including bandwidth and QoS desired within service parameters advertised by the network. The network also provides connection state information back to the customer. Reachability information may also be provided, particularly for nodes associated in a VPN. Commercial networks and other networks servicing customers between whom there is not necessarily a "trust relationship" normally do

11.8 User–Network Interfaces

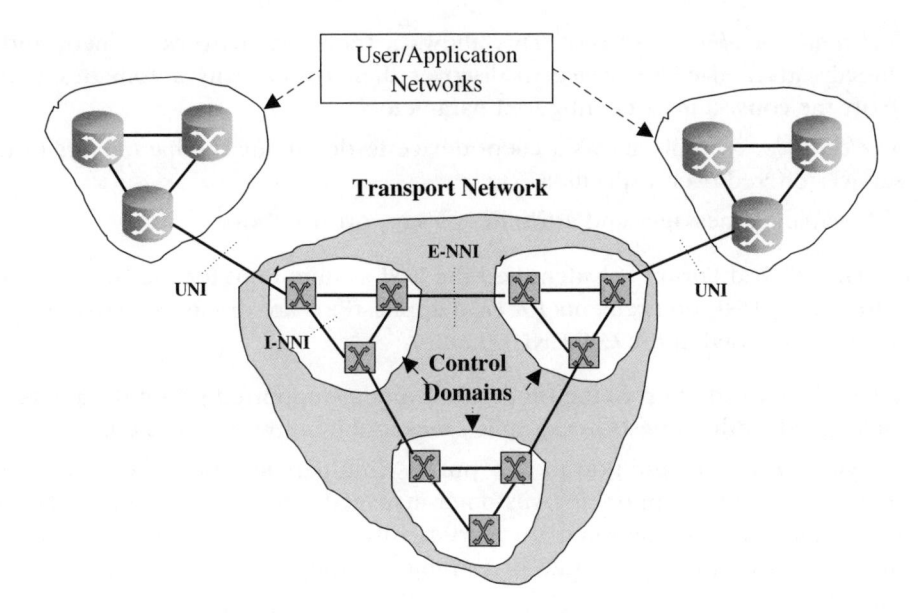

Figure 11.2. A network topology featuring a UNI interface. I-NNI (E-NNI) stands for "internal (external) network–network interface." Functionally UNIs and NNIs are quite similar, but the specific information flows differ.

not advertise global network state information across the UNI: in Internet terms, information sharing is more like BGP than OSPF in this respect.

UNIs are highly desirable because they decouple the user/application and the transport network and allow each to evolve and be managed independently. This is consistent with the requirements for site autonomy and the Services-Oriented Architecture (SOA) vision introduced in Chapter 3.

UNIs have been defined for many protocols, including ATM [22], Ethernet [21], and optical networks (e.g., ref. 23). A complete UNI for a transport service would specify data plane, control plane, and management plane interfaces. However, the term is most frequently applied to the control plane interface, the focus of this section. The relationship of the UNI and other key interfaces is illustrated in Figure 11.2.

The Optical Interworking Forum's UNI 1.0 Implementation Agreement [23] is supported by a number of vendors and carriers. It defines:

- *Signaling reference configurations.* Two sets of configurations covering direct (between the boxes) and indirect (one or both boxes represented by proxies) service invocation are defined.
- *Services offered.* Initially various types of framing, transparency, and concatenation of SONET and SDH signals up to OC-768 were supported. Various SONET/SDH protection and restoration schemes are also supported, as is a limited ability to control routing.
- *Addressing schemes supported.* These are IPv4, IPv6, and Network Service Access Point (NSAP) (ITU X.213).

- *Automatic neighbor discovery.* This allows a transport network element and a directly attached client device to discover their connectivity automatically and verify the consistency of configured parameters.
- *Service discovery.* This allows a client device to determine the parameters of the services offered over a specific UNI.
- UNI signaling messages and attributes to support the above.

The Global Grid Forum has identified the UNI as important for the Grid community. In ref. 24, UNI requirements for Grid applications are discussed. Two requirements not addressed in the OIF UNI 1.0 are:

- *Scheduling.* In addition to the on-demand model supported presently, a need to make bandwidth requests for specific times in the future is identified.
- *Private networks.* In addition to the "public" (multiapplication) services targeted by telecom UNIs, support for customer-managed networks or other networks where there is trust between the parties involved is desired. In this instance there could be more exchange of routing (topology state) information across the UNI.

Both these topics are under study by groups like the OIF and ITU. Work in these bodies on layer 1 VPNs is particularly relevant (see Section 12.2.3.1). One popular layer 1VPN model would have a subnetwork of a carrier network dedicated to a specific application. Sharing topology state information about the subnetwork across the UNI is then much less problematic, as is application scheduling of future connections within the subnetwork.

11.9 OPTICAL INTERWORKING FORUM INTEROPERABILITY DEMONSTRATION

Effective integration of optical networking into Grid computing requires standardization of both horizontal (inter-network) and vertical (inter-layer) interworking. The OIF contributes to this effort by developing "inter-operability agreements" to facilitate the use of ITU and IETF standards. This primarily involves making specific choices among the myriad alternatives consistent with the standards.

A demonstration of some of the capabilities described here was provided by the OIF-sponsored Interop demonstration at Supercomm 2005 [25], which focused on dynamic Ethernet services utilizing underlying intelligent optical networking. The multilayer architectures utilized are illustrated in Figure 11.3.

End-to-end Ethernet connections were dynamically established by piecing together Ethernet segments and a SONET/SDH (VCAT) connection. On the edges of the network, signaling used the OIF UNI specification; this was transported transparently between the edge network UNIs.

In addition to Ethernet private line service, Ethernet virtual private line and virtual private LAN services were demonstrated.

11.10 Infiniband

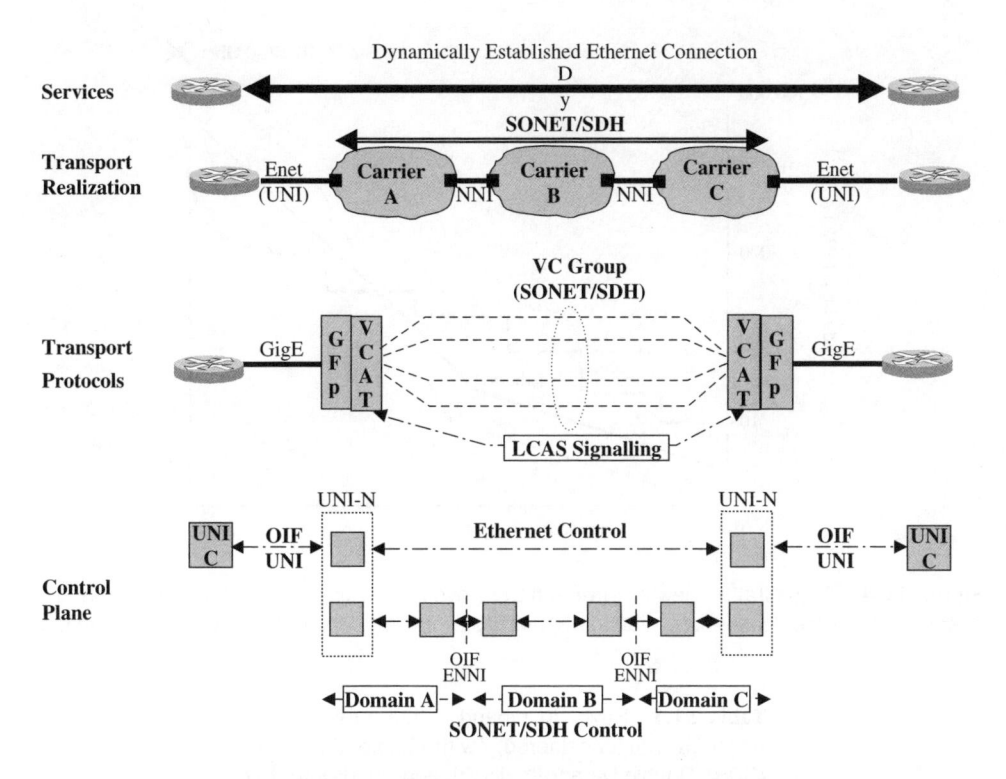

Figure 11.3. OIF SuperComm '2005 Interoperability Demo (adapted from ref. 25).

11.10 INFINIBAND

This chapter closes with a brief account of the Infiniband technology and its appeal to Grid networks. It is well understood that Infiniband is a comprehensive technology suite which spans several layers of the OSI stack, rather than just layer 2. Unlike the layer 2 technologies just reviewed, Infiniband operates in the realm of high-performance fabric interconnects for servers.

In the late 1990s, the IT industry experienced a momentous inflection point. The breakthroughs in optical communication catapulted the throughput of WAN links well ahead of the I/O rates experienced in and out of commodity end-systems (Figure 11.4). This inflection point is widely regarded as one of the key contributors to the seminal views of disaggregated IT functions and Grid computing that are gaining in popularity in this decade.

With the launch of the Infiniband Trade Association [26] in 1999, several companies joined forces to rise above the limitations of traditional shared, parallel I/O buses and restore I/O to a more fitting growth curve. Furthermore, the OpenIB Alliance [27] has played a pivotal role in forming and driving an open source community around the software aspects of the Infiniband technology, resulting in broader support, interoperability, and turnkey toolkits, lowering the initial costs of ownership and migration.

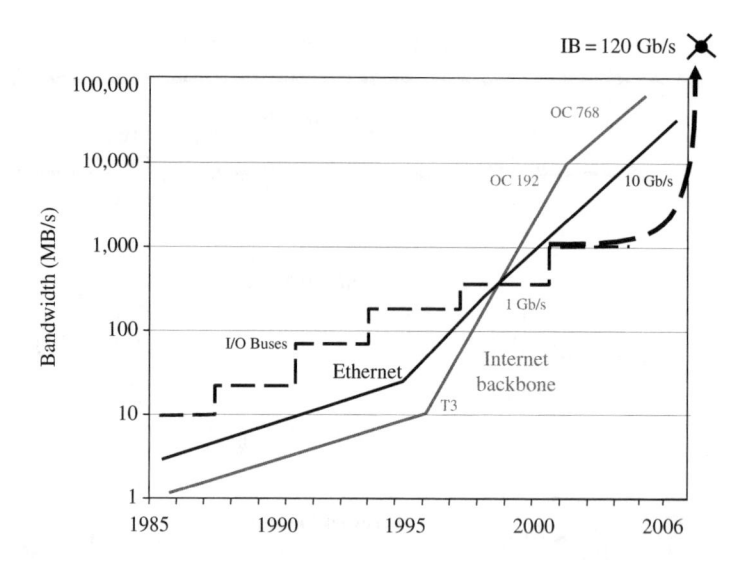

Figure 11.4. A qualitative view of three different evolution curves (Ethernet, optical, and I/O) over more than two decades. Infiniband (IB) holds promise of giving I/O bandwidth back the lead.

Table 11.1 Raw Infiniband data rates (line encoding not considered), with Single Data Rate (SDR), Double Data Rate (DDR), and Quad Date Rate (QDR), and one-way, four-way, or 12-way aggregation

	SDR	DDR	QDR
1×	2.5 Gbit/s	5 Gbit/s	10 Gbit/s
4×	10 Gbit/s	20 Gbit/s	40 Gbit/s
12×	30 Gbit/s	60 Gbit/s	120 Gbit/s

Infiniband is a serial, switched fabric that interconnects servers, storage, and switches. It utilizes either copper wires or fiber-optics. Infiniband excels in low switching latency and high bandwidth. It owes low latency to cut-through switching, which makes switching time independent of frame size and at least one order of magnitude faster than store-and-forward switching. Table 11.1 shows the range of standardized rates.

In a way, Infiniband revives the concept of channelized I/O that mainframes made popular. An Infiniband-capable server is equipped with one or more Host Channel Adapters (HCAs,) which implement an I/O engine and a point of attachment to the switched fabric. Similarly, storage devices and various I/O adaptors (e.g., for SCSI, Ethernet, fiber channel, graphics, etc.) connect to the switched fabric by way of Target Channel Adapters (TCAs). Both HCAs and TCAs promote a hardware-intensive implementation of functions similar to the ones traditionally associated with protocol stacks such as TCP/IP.

11.10 Infiniband

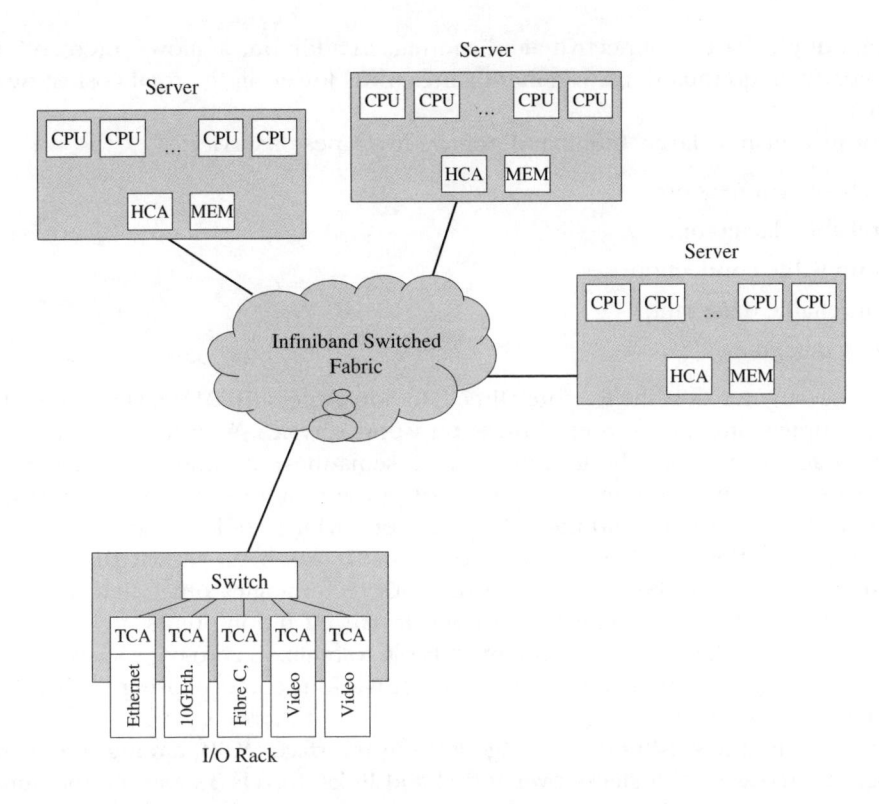

Figure 11.5. Effective sharing of resources (e.g., an I/O rack) among servers within the data center.

HCAs, TCAs, and switches form the Infiniband network fabric. From one edge of the Infiniband fabric, a server can seamlessly reach onto an Ethernet NIC or a graphic device connected at the other edge of the fabric, thus enabling dynamic allocation and sharing of assets within a data center (Figure 11.5). The Infiniband specification defines the management layers through which it is possible to manage dynamic allocation schema. Building upon standardized layers and abstractions, it is possible for Grid communities to virtualize the agile fabric and represent it as a Grid resource by way of Web Services and a stateful model such as the Web Services Resource Framework (WSRF) [28].

In Infiniband, the link layer protocol can yield strong guarantees of reliability and frame ordering. For this, it uses a credit-based style of flow control. It also uses the concept of a Virtual Lane (VL) to support multiple logical channels within an individual physical link. VLs mitigate the well-known syndrome of head-of-line blocking, wherein traffic is needlessly held back behind a frame whose transit or destination are momentarily precluded. With regard to routing a frame through a set of Infiniband switches, researchers [29] have explored adaptive routing formulations that increase throughput, resiliency, and utilization of the network.

For the network layer, Infiniband has chosen to use an IPv6 header. This choice is meant to enable straightforward composition of "fabrics of fabrics" by way of commodity IP-based connectivity and routing. In addition, it allows interception of the nearly ubiquitous IP management suites, thus lowering the total cost of ownership.

For its transport layer, Infiniband defines five types of service:

- reliable connection;
- reliable datagram;
- unreliable connection;
- unreliable datagram;
- raw datagram;

Of special interest is the Remote Direct Memory Access (RDMA) style of operation, which applies only to a subset of these transport services. With RDMA, a server can directly access memory (be it read or write semantics) at another remote system. It does so without burdening the processor at the remote system with interrupt(s) or with the execution of additional instructions which could negatively affect the system's execution pipeline or cache contents. It has been shown that the same RDMA semantics can be extended to network technologies other than Infiniband (e.g., ref. 30), albeit with different cost/performance fruition points.

With Infiniband, it becomes quite practicable to realize a zero-copy stack (e.g., ref. 31) for message-oriented communication, yet without compromising on hardware-level memory protection.

The communities using the Message Passing Interface (MPI), among others, have experimented with MPI stacks over Infiniband links. In refs 32 and 33, the authors characterize the complementary techniques of smearing and striping MPI traffic over Infiniband subchannels, which can be realized as paired physical links or logical lanes.

The multipronged edge in low-latency, high line rates, and native RDMA naturally extend Infiniband's appeal to the Grid communities. Since Grids operate over metro and wide-area networks, it is desirable that Infiniband be effectively distance extended beyond the raised floor of a computer room.

ACKNOWLEDGMENTS

David Martin is responsible for Sections 11.2–11.4, Angela Chiu for Sections 11.3–11.8, John Strand for Sections 11.6–11.9, and Franco Travostino for Section 11.10.

REFERENCES

[1] E. Rosen, A. Viswanathan, and R. Callon (2001) "Multiprotocol Label Switching Architecture," RFC 3031, January 2001.
[2] D. Awduche (1999) "MPLS and Traffic Engineering in IP Networks,"*IEEE Communication Magazine*, 37(12), 42–47.
[3] V. Sharma and F. Hellstrand (eds)(2003) "Framework for MPLS-based Recovery," Internet Engineering Task Force, RFC 3469, 2003.

References

[4] F. Le Faucheur and W. Lai (2003) "Requirements for support of Differentiated Services (Diff-Serv)-aware MPLS Traffic Engineering", RFC 3564, July 2003.
[5] E. Rosen, et al. (2004) "BGP/MPLS IP VPNs" (work in progress), draft-ietf-l3vpn-rfc2547bis-03.txt, October 2004.
[6] L. Andersson, P. Doolan, N. Feldman, A. Fredette and R. Thomas (2001) "LDP Specification," RFC 3036, January 2001.
[7] R.M. Metcalfe and D.R. Boggs (1976) "Ethernet: Distributed Packet Switching for Local Computer Networks," *Communications of the ACM*, 19, 395–404.
[8] IEEE, "Backbone Provider Bridges," IEEE 802.1ah, http://www.ieee802.org/1/pages/802.1ah.html.
[9] IEEE, "Provider Bridges," IEEE 802.1ad, http://www.ieee802.org/1/pages/802.1ad.html.
[10] IEEE (2004) "Resilient Packet Ring (RPR) Access Method and Physical Layer Specifications," IEEE 802.17.
[11] Metro Ethernet Forum, www.metroethernetforum.org.
[12] IETF Pseudo Wire Emulation Edge to Edge (pwe3) Working Group, draft, http://www.ietf.org/html.charters/pwe3-charter.html.
[13] M. Lasserre and V. Kompella (eds) (2005) "Virtual Private LAN Services over MPLS," draft, IETF draft-ietf-ppvpn-vpls-ldp-08.txt, November 2005.
[14] L. Martini, E. Rosen, N. El-Aawar, and G. Heron (2005) "Encapsulation Methods for Transport of Ethernet Frames Over IP/MPLS Networks," IETF draft-ietf-pwe3-Ethernet-encap-11.txt, November 2005.
[15] American National Standards Institute (2001) "Synchronous Optical Network (SONET) – Basic Description including Multiplex Structure, Rates, and Formats," T1.105-2001.
[16] ITU-T (2000) "Network Node Interface for the Synchronous Digital Hierarchy," Recommendation G.707, October 2000.
[17] G. Bernstein, B. Rajagopalan, and D. Saha (2004) *Optical Network Control: Architecture, Protocols, and Standards*, Addison-Wesley.
[18] ITU-T (2004) "Link Capacity Adjustment Scheme (LCAS) for Virtual Concatenated Signals," Recommendation G.7042/Y.1305, February 2004.
[19] ITU-T (2001) "Generic Framing Procedure (GFP)," Recommendation G.7041/Y.1303, October 2001.
[20] F. Davik, M. Yilmaz, S. Gjessing, and N. Uzun (2004) "EEE 802.17 Resilient Packet Ring Tutorial," *IEEE Communications Magazine*, 42(3), 112–118.
[21] ITU-T (2005) "Ethernet UNI and Ethernet NNI," Draft Recommendation G.8012.
[22] ATM Forum (2002) "ATM User-Network Interwork Interface (UNI)." Specification Version 4.1.
[23] Optical Interworking Forum (2004) "User Network Interface (UNI) 1.0 Signaling Specification – Common Part".
[24] D. Simeonidou, R. Nejabati, B. St. Arnaud, M. Beck, P. Clarke, D.B. Hoang, D. Hutchison, G. Karmous-Edwards, T. Lavian, J. Leigh, J. Mambretti, V. Sander, J. Strand, and F. Travostino (2004) "Optical Network Infrastructure for Grid," Grid Forum Document No. 36, August 2004.
[25] J.D. Jones, L. Ong, and M. Lazer (2005) "Interoperability Update: Dynamic Ethernet Services Via Intelligent Optical Networks", *IEEE Communications Magazine*, 43(11), pp. 4–47.
[26] The Infiniband Trade Alliance, http://www.infinibandta.org/home.
[27] The OpenIB Alliance, http://www.openib.org/.
[28] Y. Haviv (2005) "Provisioning & Managing Switching Infrastructure as a Grid Resource", GridWorld, October 2005, Boston, MA.

[29] J.C. Martinez, J. Flich, A. Robles, P. Lopez, and J. Duato (2003) "Supporting Fully Adaptive Routing in InfiniBand Networks", International Parallel and Distributed Processing Symposium (IPDPS '03).
[30] IETF Remote Direct Data Placement Working Group, http://www.ietf.org/html.charters/rddp-charter.html.
[31] D. Goldenberg, M. Kagan, R. Ravid, and M.S. Tsirkin (2005) "Transparently Achieving Superior Socket Performance Using Zero Copy Socket Direct Protocol over 20Gb/s InfiniBand Links", RAIT Workshop, IEEE International Conference on Cluster Computing (Cluster 2005), Boston, MA, USA, September 26, 2005.
[32] MPI over Infiniband Project, http://nowlab.cse.ohio-state.edu/projects/mpi-iba/.
[33] J. Liu, A. Vishnu, and D.K. Panda (2004) "Building Multirail InfiniBand Clusters: MPI-Level Design and Performance Evaluation," SuperComputing 2004 Conference (SC 04), November, 2004.

Chapter 12

Grid Networks and Layer 1 Services

Gigi Karmous-Edwards, Joe Mambretti, Dimitra Simeonidou, Admela Jukan, Tzvetelina Battestilli, Harry Perros, Yufeng Xin, and John Strand

12.1 INTRODUCTION

This chapter continues the discussion of expanding Grid networking capabilities through flexible and efficient utilization of network resources. Specifically, it focuses on the important topic of integrating Grid environments directly with layer 1 optical network services. Included in this chapter is a discussion on layer 1 services involving Optical-to-Optical (OOO) switching as well as Optical-to-Electrical-to Optical (OEO) switching of optical signals [1]. The layer 1 network consists of many types of resources, including Optical Cross-Connect (OXC), Time Division Multiplexing (TDM) switches, and Fiber Switches (FXCs) interconnected via fibers. The layer 1 service can provide dedicated end-to-end connectivity in the forms of time-slots, wavelength lightpath(s), or wavebands (multiple wavelengths), either static or dynamic.

This chapter is organized as follows. The next section, 12.2 provides a context to the consideration of layer 1 Grid network services. A general introduction to layer 1 Grid network service is given in Section 12.2.1. Network control and management issues are discussed in Section 12.3. Section 12.4 deals with the current technical challenges facing the layer 1 networks. The Grid network service with an all-optical network infrastructure is presented in Section 12.5 with a focus on Quality of Service

(QoS) control. Section 12.6 is dedicated to a new optical networking technology, Optical Burst Switching (OBS), with a focus on the integration with Grid.

12.2 RECENT ADVANCES IN OPTICAL NETWORKING TECHNOLOGY AND RESPONSES

There have been many recent advances in optical networking technologies at all levels: services capabilities, control planes, devices, elements, architectures, and techniques. For example, recent advancements in optical network technology have built upon the 20-year history of successful deployment of Wavelength Division Multiplexing (WDM) technology. Also, the IETF has standardized methods of providing IP-based optical control plane protocols, such as the Generalized MultiProtocol Label Switching (GMPLS) suite, which has allowed direct access and control to optical network (layer 1) resources. On the research front, simplified network architectures with much fewer layers, and novel architectures such as Optical Burst/Packet Switching (OBS/OPS), have emerged [3]. Meanwhile, other recent innovations can be seen in new devices such as high-performance, low-cost optical switches, described in Chapter 15, Reconfigurable Optical Add-Drop Multiplexers (ROADMs), tunable lasers and amplifiers, high-performance gatings, and many others.

During the last several years, considerable Grid research and development activity has focused on the question of how to take advantage of these new optical technologies and architectures for the purpose of meeting the demanding application requirements. New methods in this area have quickly moved from research laboratories to testbeds to prototype deployments and, in some cases, even to production networks.

> Optical networks can be viewed as essential building blocks for a connectivity infrastructure for service architecture including the Open Grid Service Architecture (OGSA), or as 'network resources' to be offered as services to the Grid like any other resources such as processing and storage devices [3].

Central to this development are the new challenges in the area of optical control planes, providing results that promise to extend the concept of Grid networking into optical networking, with unique features that could not be achieved otherwise.

The optical control plane, based on widely accepted IP-based signaling and routing protocols, can automate resource discovery and dynamic provisioning of optical connections, offering support for a broad range of differentiated, and highly dynamic reconfigurable services. Instead of the traditional static service delivery, with manual provisioning and long-term bandwidth allocation, service providers have begun aspiring to intelligent capacity utilization, automated delivery of value-added services and on-demand circuit provisioning over optimized network paths. Most recent development in optical network control plane methodologies is represented by the GMPLS suite of control plane protocols.

Drivers for optical control plane evolution and enhancements initially came from the service provider sector in the interest of reducing operational costs for configuring and provisioning the network by adding distributed intelligence to the network elements. However, in the data-intensive Grid community, another driving force also emerged: the requirement for end-users and applications to have access to and to control high-capacity, deterministic end-to-end connections (lightpaths).

A new generation of scientific applications is emerging that couples scientific instruments, distributed data archives, sensors, and computing resources, often referred to as "E-science." Grid middleware and applications couple distributed instruments, scientific data archives, and computing facilities with high-speed networks, enabling new modes of scientific collaboration and discovery. As an example, the Large Hadron Collider at CERN will support multiple particle detectors, constructed and managed by huge international teams of researchers who must hierarchically capture, distribute, and analyze petabytes of data world-wide [4]. Many large-scale Grid applications requiring direct access to layer services are being demonstrated at specialized international and national conferences [5].

Further on the horizon, today's E-Science applications are a view of what future enterprise applications will soon require. The banking and medical industries provide just a few examples of the emerging requirements for large scale enterprise applications that will be able to benefit from Grid services based on advanced L1 networking.

12.2.1 LAYER 1 GRID NETWORK SERVICES

To date, there have been many efforts to develop network infrastructures that support the Grid applications. A major recent innovation in layer 1 service provisioning has been the creation of capabilities that allow edge processes, including applications, to directly access low-level network resources, such as lightpaths in optical networks. With the advancement of optical networking technology, high-performance, Optical Circuit-Switched (OCS) networks featuring dynamic dedicated optical channels that can support high-end E-science applications are emerging [6–11]. On these networks, long-lived wavelength or subwavelength paths between distributed sites could be established – driven by the demands of applications.

Together with efforts in building Grid network infrastructure, an enormous amount of work has also been accomplished in establishing an optical control plane for Grid networks. Most of the definitions and capabilities regarding recent technology development can be found in the documents produced by international standards bodies, such as the IETF, the ITU, the GGF, and the OIF, as well as a few other organizations.

12.2.2 BENEFITS OF GRID LAYER 1 SERVICES

Layer 1 or physical layer in the OSI stack is responsible for defining how signals are actually transmitted through the physical network and, therefore, deals with the description of bit timing, bit encoding, and synchronization. The combination of IP control plane protocols with high-capacity WDM provides several benefits:

- low Bit Error Rate (BER);
- increasing capacity at a rapid rate;
- reduction in Operational Expenditure (OpEx);
- determinism, QoS;

- faster service deployment, on-demand provisioning;
- ENABLING two-layer networking IP/OPTICAL;
- reduction in manual intervention and human mistakes.

In the case of all-photonic capabilities:

- complete transparency;
- protocol independent;
- simpler switching – no electronics (OEO);
- cheaper devices – less complexity;
- transmission of analog signals.

Low BER. In comparison with other physical transmission media, optical fiber communication has a significantly lower BER, 10^{-9} to 10^{-15} BER, in contrast to a wired medium [12]. Also, the cost–benefit advantage of layer 1 switching compared with routers is significant [13].

High capacity. High capacity in optical fibers makes it possible to achieve transmission speeds greater than 1 terabyte per second [34]. Channel capacity of 10 Gbps is deployed and continues to increase adaptation. (In part, this use is being driven in the Grid community by the availability of low-cost 10-Gbps Network Interface Cards (NICs) for nodes in compute clusters.) Capacity of 40 Gbps is on the horizon, and experiments with 160 Git/s are under way. Deployed technology today offers 80 channels per fiber, while experimentation in the laboratory has successfully achieved 1000 channels per fiber [15]. This type of rapid technological advance builds a strong case that layer 1 services are a good match for the ever more data-intensive applications.

Reduced OpEx. Introducing distributed intelligent in the network elements themselves through IP-based optical control plane migrates functionality traditionally handled by operators and management applications. Reducing the amount of work an operator is required to do to manage and control the network results in a decrease in operational costs. Also, direct access enables the implementation of cost-efficient ad hoc provisioning.

Determinism, QoS. Layer 1 services are connection oriented (excluding OBS/OPS), which means that a dedicated path is set up prior to data transmission. Data transmission over a dedicated path means that all data arrives at destination in order. Nondeterministic delay, jitter, and bandwidth are usually a result of variable queues in routed networks, which is not the case in a dedicated layer 1 service. Layer 1 services can be used to ensure the precise level of quality required by Grid applications.

On-demand provisioning. Use of GMPLS and other control plane signaling protocols for lightpath establishment and deletion provides mechanisms to dynamically reconfigure the network in timeframes of seconds to subseconds.

The advantages of the layer 1 services described here may be useful for the Grid community as a whole, and already several experimental testbeds and research activities on a global basis are achieving success in integrating layer 1 services with Grid middleware to support data-intensive Grid applications. Although there are many benefits to using layer 1 services, there also exist many obstacles and challenges which need to be addressed before a wider deployment of these services, can be observed. Section 12.4 will describe several challenges to be explored by the Grid and network research communities.

12.2.3 THE ROLE OF NETWORK STANDARDS BODIES

As previously mentioned in Chapter 4, there are many standards organization efforts as well as less formal initiatives that are relevant to the Grid control plane. Many of the benefits described in the previous section are a result of the relevant standard bodies described in this section. Regarding standardized work, the IETF has developed the GMPLS protocol suite, which can be used for resource discovery, link provisioning, label switched path creation, deletion, property definition, traffic engineering, routing, channel signaling, and path protection and recovery [16].

The OIF has developed OIF UNI 1.0 followed by OIF UNI 2.0 UNI. UNI stands for user–network interface and is the signaling service interface between an edge client device and the transport network. UNI 1.0 specifies methods for neighbor and service discovery, and allows a client to request a connection, including bandwidth, signal type, and routing constraints. In addition to UNI, OIF has developed a service interface to connect different networks, which is the Network-to-Network Interface (NNI).

The ITU developed the overall architecture for optical networking, called Automatic Switched Optical Network (ASON) [17]. Currently, the OIF's Network-to-Network Interface (NNI) is a standard architecture for connecting different parts of a network of a single carrier. Other efforts have been established to extend the provisioning between carriers [18].

12.2.3.1 Optical (layer 1) virtual private networks

Another optical services topic important to the Grid community that is receiving considerable attention in the primary standards bodies is layer 1 Virtual Private Networks (VPNs). Layer 1 VPNs can combine many of the attractive features of a dedicated physical network (SONET/SDH or OTN) – customer control, security, flexibility in choosing higher layer protocols, extremely high capacities and throughputs – without incurring the cost and complexity of procuring and managing an extensive physical infrastructure. In a Grid context, they may be particularly useful for transient "virtual organizations" (described in earlier chapters) requiring high-bandwidth connections for a limited period to a geographically dispersed set of locations.

Standards for layer 1 VPNs are starting to emerge from standards organizations:

- The ITU has developed service requirements (ITU Y.1312) and architecture (draft Y.1313). They are focusing on multiple connection service (one customer, three or more customer interfaces) utilizing either customer-controlled soft permanent or switched connections.
- The Telemanagement Forum (TMF) standard TMF 814 covers some important control interfaces.

There are many different types of layer 1 VPNs possible (see ITU Y.1312). A rough realization of a layer 1 VPN at the wavelength level might be as follows:

- The Grid application contracts a specific numbers of wavelengths on designated links in a provider network, including OXC capacity.
- The detailed state of this capacity, including service-affecting outages, is advertised back to the application in real time by the network.
- The application (and only the application) can add and remove connections routed over the contracted capacity. Routing can be controlled if desired by the application. In effect, this would behave like a customer-owned network.

12.2.3.2 IETF and layer 1 VPNs

Related to the layer 1 VPN initiatives in the ITU-T are IETF efforts that are developing standards for layer 1 VPN. The IETF has a Layer 1 VPN Working Group, which is specifying mechanisms necessary for providing layer 1 VPN services (establishment of layer 1 connections between customer edge (CE) devices) over a GMPLS-enabled transport service provider network. (GMPLS is described in a later section.) With a layer 1 VPN the network connects clients' devices by private collections of layer 1 network resources (wavelengths, interfaces, ports) over which clients have certain level of control and management [19]. Clients can add and remove connections over their established layer 1 VPN.

The IETF is developing a "basic mode" specifying only path setup signaling between a customer and a network, and has plans for an "enhanced mode" providing limited exchange of information between the control planes of a provider and clients to assist with such functions as discovery of reachability information in remote sites, or parameters of the part of the provider's network dedicated to the client [20].

12.2.3.3 Related research initiatives

This capability is closely related to the work being pioneered by CANARIE (e.g., User-Controlled Lightpath Provisioning (UCLP) and Optical Border Gateway Protocol (OBGP; described in Chapter 5) for customer-owned networks. Regarding less standardized work, there is the UCLP of CANARIE, Inc., which allows clients to establish end-to-end lightpaths across different network boundaries – even boundaries between different carriers' networks from the management plane [21]. Another experimental architecture being used for dynamic lightpath provisioning for Grid applications is the Optical Dynamic Intelligent Networking (ODIN) architecture, described in Chapter 5. This method has been used in Grid environments to provide applications with a means of directly provisioning lightpaths dynamically over advanced optical networks.

12.2.3.4 Enumeration of Grid requirements and challenges for layer 1

Common to all solutions related to the layer 1 network control plane is the need to accommodate the dynamic aspects of Grid applications. These applications frequently have one or more of the following requirements:

(1) very active use of end-to-end optical networking resources;
(2) global transfers of large datasets (often in the order of several terabytes) across great distances;
(3) coordination of network resources with other vital Grid resources (such as CPU and storage);
(4) the capability of having advanced reservation of networking resources;
(5) deterministic end-to-end connections (low jitter, low latency);
(6) time scales of a few micro-seconds to long-lived wavelengths; and
(7) near-real-time feedback of network performance measurements to the applications and Grid middleware.

As recognized by the standards bodies, new challenges involved in the Grid-based optical control plane will need to address concepts associated with (i) application-initiated optical connections, (ii) interaction with higher layer protocols, (iii) interaction with Grid middleware, (iv) inter-domain operation and operation between different control planes, (v) integrating novel optical networking technologies and architectures such as high-rate transceivers, novel wavelength conversion, Optical Packet Switching (OPS) and Optical Burst Switching (OBS), and)vi) resource discovery and coordination of network, CPU, and storage resources.

12.2.3.5 The role of Global Lambda Integrated Facility (GLIF)

The Global Lambda Integrated Facility (GLIF) [22] represents a growing global community interested in the sharing of Grid resources, especially optical networking resources on a global scale. In essence, the goal of the GLIF community is to provide scientists with a global laboratory that resides within an agile high-capacity distributed facility. This distributed facility is capable of supporting multiple overlay networks based on lightpaths deployed on optical networks. This distributed facility has been implemented and is currently supporting large-scale science projects. The first components of its infrastructure were based on those supported by the Trans-Light research project [7].

This community mainly comprises the Grid/E-science research community as well as many National Research and Education Networks (NRENs) from around the world. Because most of the experimental Grid testbeds are part of one or more NRENs, this community represents the strongest driving force behind research and technological advancements for layer 1 and 2 Grid services.

The GLIF community is currently addressing several challenges, including integration of Grid middleware and optical control plane. A new working group, the Grid Middleware Integration and Control Plane Working Group, was established in September 2005 to address these issues and challenges. The mission of this group is: "To agree on the interfaces and protocols to automate and use the control planes of the contributed Lambda resources to help users on a global scale access optical resources on-demand or prescheduled" [23]. The network connectivity challenge is how to bring GLIF resources to the desktop of any scientist who wants to use the global resources.

The vision of the GLIF community is the formation of a resource-sharing global laboratory. An example of the value-add that the sharing of these global resources will have on future scientific discovery is the use of the Korea Basic Science Institute's (KBSI) High-Voltage Electron Microscope (HVEM) [24]. This instrument has the unique capability to analyze the 3D structure of new materials at an atomic scale and nano-scale by utilizing its atomic resolution high tilting feature. The 3D disintegration capability of this HVEM is one of the best in the world. The availability of this instrument to researchers around the world is achievable only through high-capacity optical networking and Grid computing middleware.

However, through the layer 1 Grid services (lambda grid), the research of many scientists around the world can benefit from the amazing capabilities of the KBSI microscope. A scientist can use the regular mail to send a specimen to Korea and then use the lambda e-science Grid capabilities to remotely examine the data from the microscope. The remote scientists are able to view the visualization data from the microscope and remotely steer the instrument. For this experiment to be successful requires a large-bandwidth end-to-end lightpath from the microscope to the remote scientist's machine. That is because the visualization data produced by the HVEM is very large and the best effort connectionless Internet of today cannot provision and guarantee a large, reliable, and secure data pipe. The Korean Institute of Science and Technology Information (KISTI) is already trying to use this HVEM for distributed, remote e-science through the GLORIAD project [25]. Creating and deleting end-to-end lightpaths on demand by applications, or by researchers in coordination with other resource availability, is essential for the realization of this Global E-science Laboratory (GEL).

12.3 BEHAVIORAL CONTROL OF LAYER 1 NETWORKS

Layer 1 service definition and description is provided in Section 12.2. In this section the management and control of the network infrastructure is described. Research challenges associated with the control and management and Grid integration are some of the most critical aspects for building the future Grid. Behavioral control of networks is defined as basically the roles and mechanisms of entities involved in operating and controlling the network. In today's Grid environment that role is shared among three complementing entities: (i) the management plane, (ii) the control plane, and (iii) Grid middleware. Today, there is no consensus on how to define these functional areas, where to place those functions, or how to separate them into different domains. It is in fact a key area of research within the Grid networking community as well as the future of the Internet in general to determine the complex and interrelated roles of each of these entities [3].

The success of the layer 1 Grid services, referred to by some as lambda Grid, is dependent on the seamless vertical integration between the applications, the Grid middleware, and the optical network. The Grid middleware acts as an intermediary between the application and the network. It provides APIs to the applications so that they can take advantage of the overall Grid capabilities, such as dynamic resource scheduling, monitoring, data management, security, information, adaptive services, etc. In addition, the Grid middleware has to be aware of the status of the network and successfully provision the necessary on-demand, guaranteed connections. To solve these challenges, the optical networking community, in conjunction with the Grid community, has to rethink the role of both the management plane and the optical control planes and their interaction with Grid middleware for future Grid computing.

12.3.1 MANAGEMENT PLANE

Traditional network management functionality is often referred to as FCAPS, an acronym for fault management, configuration management, performance management, and security management. Traditional optical networks utilize a centralized operator-controlled means for creating optical connections from one end to the other. Management plane mechanisms rely on client–server model, usually involving one or more management applications (structured hierarchically) communicating with each network element in its domain via a management protocol (i.e., SNMP, Tl1, XML, etc.). Information exchanged between the network elements and the management application is usually in the form of an information base (i.e., a Management Information Base (MIB), Resource Description Framework (RDF), etc.) or in the form of a Command Line Interface (CLI). The Grid research community and the networking research community are revisiting the role of the management plane for next-generation behavioral control of optical networks.

In today's carrier networks, configuration management involving requests for connections is routed to the operations center. Such requests must be initiated by an operator and implemented through a network management application. Often, this type of connection also requires an operator physically interchanging connections on a patch panel. This manual provisioning is a time-consuming and operationally expensive task and usually results in a rather static layer 1 connection. Although some of the management functionality has migrated to the more distributed control plane, the management plane will continue to play a vital role in the behavioral control on Grid network services. The management plane must also manage and control the "control plane" of a network. Ongoing research will continue to decipher which behavioral control functions should continue to exist in a centralized management plane and which ones will be more efficient as a distributed control plane.

Initiating connections linking end stations, instruments, and sensors, rather than network aggregators (edge nodes), on demand is a critical requirement for the Grid community. Such requirements cannot be met through traditional control and management of optical networks. Control plane signaling is more capable than the management plane of handling the dynamicity required for creating and deleting layer 1 connections. The next section will introduce concepts on the control plane and the research involved in meeting these and other Grid requirements.

12.3.2 CONTROL PLANE

Control plane is defined as follows:

> Infrastructure and distributed intelligence that controls the establishment and maintenance of connections in the network, including protocols and mechanisms

to disseminate related information, and algorithms for engineering an optimal path between end points.

Another way of defining the optical control plane is to describe it as the capability that provides the functionality traditionally relegated to the centralized network management plane but which is now provided using more distributed methods (this is consistent with the requirements of Grid computing as well). Many of these functions are low-level capabilities that require direct interaction with physical devices, and which require extremely fast responses in comparison with traditional manual operations.

Implementing specific control functions in the distributed control plane rather than in the centralized management plane should have the following benefits:

- a speedup of reaction time for most functions;
- increase in control scalability;
- a reduction in operational time and costs;
- more agility in the behavior of the optical network.

For the communications industry, the motivation behind control plane progress mainly lies in the reduction of operational cost and better service delivery for carriers, which differs from the motivation for the use of such protocols with Grid computing. In the Grid computing community, research on the optical control plane and its progress is mainly aimed at meeting end-user and application requirements – not necessarily service providers' requirements. Most networks today provision end-to-end connections using manual centralized management applications, in which the connection's end-points are usually an edge device on the network and are rather static in nature. With Grid computing, applications, sensors, and instruments may request on-demand end-to-end optical connections with endpoints at the application host (supercomputer, clusters, instruments, workstations, sensors). These connections will be more dynamic in nature than today's more static optical environment.

The role of the control plane continues to evolve as more and more intelligence is added to network elements and the edge devices. Extending the control plane features to the outer edges of networking results in complete paradigm shifts in how networking is viewed. Current control plane functions include:

- routing, both intra-domain and inter-domain;
- automatic topology and resource discover;
- path computation;
- signaling protocols between network switches for the establishment and maintenance, and tear-down of connections;
- automatic neighbor discovery;
- local resource management to keep track of available bandwidth (and buffer in connection-oriented packet switches) resources on the switch's interfaces.

12.3.2.1 GMPLS, a unified control plane

Traditionally, each network layer has had its own management and control planes. At this time, standards organizations are attempting to build consensus for a uniform

12.3 Behavioral Control of Layer 1 Networks

control plane based on the IP protocol that could be used for multiple network layers, particularly layers 1–3. This unified control plane would be responsible for switching of packets, flows, layer 2 paths, TDM channels, wavelengths, and fibers. One objective of this control plane would be to allow for dynamic provisioning. Currently, the GMPLS protocol suite [16] which is being developed by the IETF has gained significant momentum as a contender to become the basis of this uniform control plane.

GMPLS RSVP-TE (GMPLS reservation protocol with traffic engineering) is the generalized extension of MPLS-TE, a signaling protocol with a flexible framework that specifies separating forwarding information from IP header information, allowing for forwarding through label swapping and various routing options. Signaling is defined as standard communications protocols between network elements for the establishment and maintenance of connections. GMPLS can be used to directly control network elements at layers 1–3. It can also be the basis for services based on interlayer integration.

RSVP-TE provides for extensions to MPLS that include forwarding planes, such as traditional devices based on TDM (e.g., SONET ADMs) and newer devices based on wavelengths (optical lambdas) and spatial switches (e.g., inflow port or fiber to outflow port or fiber). Consequently, GMPLS allows for forwarding decisions to be based on time slots, wavelengths, or ports. In Grid networks, initiation of the RSVP-TE signal comes from the end-user or application making a request via Grid middleware.

The IETF has established another key initiative related to GMPLS, one that is creating a protocol suite as a generic interface based on RSVP-TE signaling that would support Label Switched Path (LSP) provisioning and recovery. The GMPLS UNI allows groupings of multiple LSPs with protection paths. This suite of protocols can be used as an alternative to the OIF UNI, to allow for enhanced provisioning, reliability, protection, and scalability for optical paths [26].

Chapter 7 describes the components of Grid middleware and interactions with the applications and the network resources and control in more details. A key idea is that, once a request is initiated for Grid resources, the Grid resource manager and resource allocation components coordinates and processes the request and integrates seamlessly, with the Grid control plane behaving as a RSVP-TE client. The RSVP-TE signaling then continues across the network on a hop-by-hop basis to create a unidirectional connection-oriented point-to-point reserved path observing the requested QoS, as seen in Figure 12.1 [2].

Most networking equipment providers support some of the GMPLS protocols; however, carriers have not shown wide support yet and deployment in that industry is still quite small. Most carriers continue to test GMPLS protocols in the laboratory

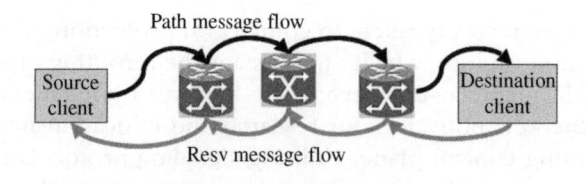

Figure 12.1. RSVP-TE message flow.

environment. RSVP-TE is currently being deployed on experimental research testbeds across the globe. Other signaling protocols are also being investigated for use in the Grid environment [2].

12.3.2.2 Routing

As described in Chapter 8, the Internet is basically a layer 3 network focused on packet routing. Research in the Grid community regarding routing functions and protocols relies heavily on IP protocols with Traffic Engineering (TE). TE routing consists of route computation of lightpaths based on multiple constraints. Examples of constraints used in Grid computing are as follows: bandwidth, latency, jitter, and link cost. Path computation is based on each node having access to resource information including topology, link, and node characteristics, and resource availability, within the routing domain in a distributed manner.

Grid researchers are revisiting constraint-based path computation with regards to (i) other resource coordination, such as CPUs and instruments, (ii) advanced scheduling, and (iii) physical layer impairments in all-photonic networks.

There are three models for routing domain interactions [27]: (i) the peering model, (ii) the overlay model, and (iii) the hybrid model. Each has its pros and its cons. Most Grid-based testbeds follow the peering model. Below are some of the key concepts for each model.

- *Overlay model.* Separate routing instances between domains.
- *Augmented model.* Separate routing instances, but some information is passed, i.e., IP destination address.
- *Peer model.* Single routing instance between client and network.

Inter-domain routing is still a key challenge for carriers adopting control plane signaling for provisioning. This is due to the carrier's environment of economic competitiveness. Information exchange between carrier domains is a sensitive issue for carriers: relevant information about the network architecture and topology may expose competitive information to a carrier's competitor. Some believe that this is why carriers are reluctant to deploy the GMPLS control plane. Another major reason for this widespread reluctance is the loss of operator control for provisioning (management plane versus control plane).

In contrast, NRENs are deploying control plane technologies on experimental testbeds. Research on inter-domain solutions within the Grid environment is currently being explored [5,21,28,29].

12.3.2.3 Recovery

Layer 1 control plane recovery refers to connection protection or restoration in the event of a link or node failure, which often means the rerouting of links after a failed connection. Establishing recovery mechanisms for layer 1 implemented at the control plane is one of the key motivators for research and education networks migrating toward implementing control plane. Utilizing signaling protocols to re-establish an end-to-end connection results in much quicker recovery time than can be achieved with the current management plane fault management system. This reduction in recovery time is significant in the e-science and Grid community, since a link failure over a high-capacity optical connection could result in an enormous amount of data loss. Using control plane recovery mechanisms, the application chooses the recovery mechanism most suited for its purpose. Below are some of the key differences between the two mechanisms. More information can be found in ref. 30.

Protection
- Recovery resources are preconfigured prior to failure.
- Protection could have the form of dedicated or shared recovery resources.
- Protection is less efficient use of network resources.
- Protection provides faster recovery times than restoration and is easier to implement.

Restoration
- Recovery resources are dynamically configured after a failure has occurred.
- Restoration makes efficient use of network resources.
- Restoration usually has slower recovery times than protection

It is widely believed that the NRENs will lead commercial industry in deployment of the GMPLS control plane. Recovery is thought of as the key driving force behind this adaptation of control plane, providing a considerable reduction in downtime for layer 1 services for their production networks. The Grid community is further exploring means of extending recovery mechanisms to provide more adaptive responses within the Grid context.

12.4 CURRENT RESEARCH CHALLENGES FOR LAYER 1 SERVICES

12.4.1 APPLICATION-INITIATED CONNECTIONS

An increasing number of Grid applications today require accessibility to the globally distributed Grid resources. Many of these applications represent the work of research teams world-wide collaborating through optical Grid infrastructure for advancing scientific discovery. Grid technologies using high-performance optical networks for data-intensive applications are essential to the collaborative research efforts. This use includes enabling those applications to provision their own lightpaths. The capability of large-scale Grid applications to provision their own lightpaths globally across multiple domains was first demonstrated at an international conference in 2002 [31,32].

A prime example of such research is the high-energy physics community experiment at CERN's Large Hadron Collider. Data collection is due to begin in 2007 – thousands of scientists in over 60 regions world-wide are forming a "collaboratory," sharing and analyzing petabytes of complex data. The large amounts of data to be exchanged across these different regions require the capacity available today only with optical networks, using 10-Gbit links between clusters. Similar examples can be found in astrophysics, molecular biology, and environmental sciences.

The typical scenario where applications initiate network connections involves an application that cannot be executed within the local cluster environment and needs access to a supercomputing facility. Furthermore, an application may require access to storage and visualization resources, and the connection to these sites may be reserved not in advance but during the application run-time, i.e., in near real time. For example, researchers in the astrophysics community require highly demanding applications simulating rare events in the universe, such as colliding black holes [33,34].

The numerical relativity group at CCT at Louisiana State University and astrophysicists at the Albert Einstein Institute of Potsdam, engaged in black hole research, are experimenting with several highly dynamic e-science applications using the Grid Application Toolkit (GAT). The simulations they run typically require well over 100 GB of memory, generate terabytes of data, and require days of computational time on thousands of processors. Such research is highly computational and data-intensive, with computing needs that cannot be satisfied in a typical cluster environment. Rather, this type of research could take advantage of several clusters, which are distributed and connected via high-capacity optical networks.

Considered here is a typical scenario as an example of new generation of composite Grid applications (Figure 12.2), inspired by the astrophysics examples. Through a portal, a researcher submits the request for a computation simulation somewhere on a global Grid (1), using an abstract, application-oriented API. The researcher's application will contact an underlying Grid information service, to determine where to launch the requested simulation, and, once the initial target is decided, the application migrates the code to the target and spawns the simulation; the newly spawned

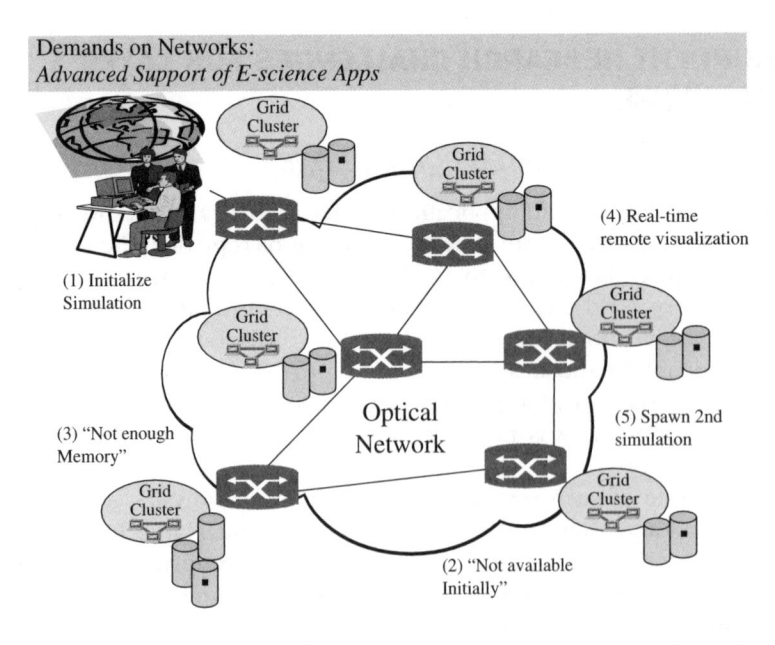

Figure 12.2. An example of six different application-initiated connections for one Grid application.

12.4 Current Research Challenges for Layer 1 Services

code then registers the new location with the Grid information service. In some cases, the requested computational resource may not be available (2). If that happens, the application further migrates the code until the request is accepted. New, more powerful resources could become available (e.g., the application becomes aware of newly available resources from the feedback loop with Grid resource management middleware), leading the simulation to migrate to a faster cluster. As the initial simulation runs, the application may perform these or one or more of the following scenarios.

In the scenario denoted with (3), huge amounts of simulation result data will require storage, either local to the simulation or at geographically dispersed locations. If remote storage is necessary, then the application itself creates an on-demand network connection and streams data to the site. In scenario (3), the application has tried to access the remote storage site but the storage resources were not available at the time of the request. When that happens, the application may either schedule the storage services or migrate the request to another available storage site.

As the amount of data collected or generated for simulations continues to grow, data mining techniques increasingly require the use of high-end visualization tools as an alternative to analysis methods based on statistical or artificial intelligence techniques. Visualization exploits the human perception to identify structures, patterns, and anomalies by visually presenting abstract data, letting the user explore the complex information and draw his or her own conclusions, as seen with the black hole data from simulations. State (4) shows, for example, the final stages of black hole simulations in the real-world astrophysics example, which ran on machines at the National Center for Supercomputing Applications (NCSA) and the National Energy Research Scientific Computing Center (NERSC).

As the initial simulation runs, the application might also perform near real-time analysis of the output data and detect important events, specific to the application. In some cases, this detection can trigger the spawning of a new finer grain simulation at a remote cluster, according to the availability on the global Grid (5). A slow part of the simulation runs asynchronously, so the application might spawn that part separately. Finally, an end-user could interact with the simulation and perform computational steering, i.e., interactive control over computational process during run-time (6).

As this simple example shows, there may be at least six different application-initiated connections:

- connection for the code migration;
- connection to the data retrieval/storage;
- connection for alternative computing sites;
- connection for parallel simulations;
- connection to visualization resources with real-time characteristics;
- connections between the researchers and all involved Grid resource sites for steering.

These and similar scenarios demonstrate that there are unique and challenging demands on the optical network connection provision, as dictated by applications.

12.4.2 INTERACTION WITH GRID MIDDLEWARE

Grid middleware can be defined as software and services that orchestrate separate resources across the Grid, allowing applications to seamlessly and securely share computers, data, networks, and instruments. Several research initiatives investigating the integration of the optical control plane with Grid middleware are underway. There are several key research challenges that must be addressed:

- exchange of information between middleware and the optical control plane;
- how often the status information is updated;
- coordination of layer 1 network resources and other Grid resources per request;
- inter-domain exchange of information;
- integrating Grid security for the network resources.

12.4.3 INTEGRATING NOVEL OPTICAL TECHNOLOGIES

In recent years, there have been several advances in optical technologies which may have a significant impact on how networks are designed and implemented today and in the near future, for example laboratory experiments with 1000 high-capacity channels per fiber and electronic dispersion compensation.

Integration of advanced optical prototypes into Grid network and computing research testbeds is rare in today's research environments. This is clearly one of the great obstacles of future network research as reported in ref. 35. Interdisciplinary research is the key to integration of advanced optical technologies into current state of the art as well as current Grid research on network architecture and protocols. For example, as advanced breakthroughs in the handling of optical physical layer impairments occur, it will become more likely that larger deployments of all-photonic islands will be seen.

Experimenting with such prototypes could lead to radical architectural advances in network design. Another consideration is that all-photonic switches are continuously reducing their reconfiguration times. Today Microelectromechanical System (MEMS)-based switches have reconfiguration times of several milliseconds. However, some silicon optical amplifiers reconfigure in the nanosecond timescale. Integrating this technology with Grid experimental testbeds may lead to more advances on a completely different type of network control plane, such as OBS networks.

Below is a short list of some key research areas for Grid computing on experimental testbeds:

- experiments with 1000 channels per fiber;
- experimentation with 160 Gbps per channel;
- All-optical switches with nanosecond reconfiguration times;
- control plane protocols, Service Oriented Architecture (SOA);
- dispersion compensation;
- fiber, optical impairments control;
- optical signal enhancement with electronic Forward Error Correction (FEC);

12.4 Current Research Challenges for Layer 1 Services

- cheap wavelength converters;
- optical packet switches;
- physical impairment detectors and compensators;
- optical 3R devices;
- tunable lasers and amplifiers;
- optical/photonic devices;
- optical monitoring for SLAs.

12.4.4 RESOURCE DISCOVERY AND COORDINATION

The resources in a typical Grid network are managed by a local resource manager ("local scheduler") and can be modeled by the type of resource (e.g., switch, link, storage, CPU), location (e.g., in the same network, outside), or ownership (e.g., inter-carrier, metro, access). The use of distributed Grid resources is typically coordinated by the global Grid manager ("meta-scheduler"). The negotiation process between the global and local Grid resource schedulers must reach an agreement in a manner that offers efficient use of all the resources and satisfies the application requirements. The bulk of this process is still manual, and control plane automation is an important challenge and a necessity if Grid networks are to operate in an efficient manner. Furthermore, the applications are becoming more and more composite, thus requiring an additional level of coordination. Therefore, the implementation of the resource discovery mechanisms and the coordination of resource allocation is of central importance in Grid resource management. It is illustrated in Figure 12.3.

The complexity associated with coordinated resource allocation within the optical control plane is depicted with respect to three basic dimensions: applications, Grid

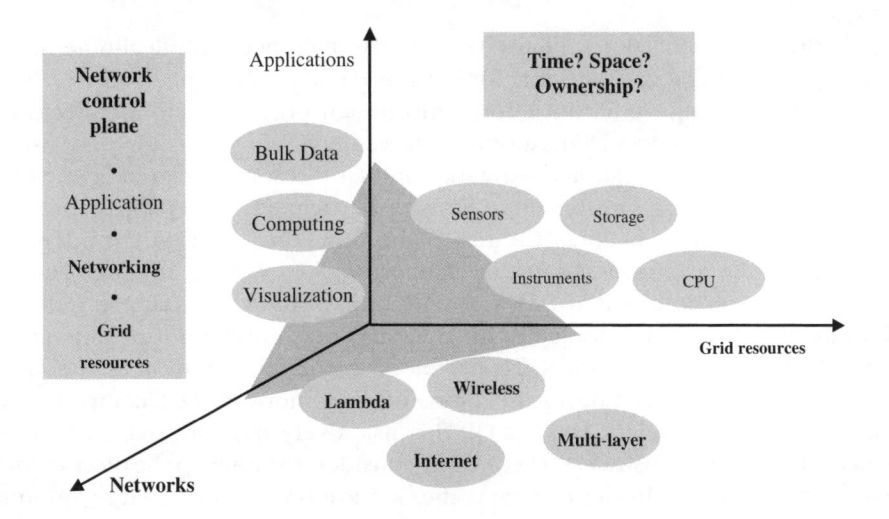

Figure 12.3. The design space for coordinated resource allocation in Grid environments.

resources, and networks. As shown, each dimension consists of multiple components that need discovery and coordination. Depending on the Grid network system in place, the combination of various resources and their significance in call setup varies. Consider the scenario in which a Grid application requires a connection with guaranteed bandwidth and least-cost computation cycles. In this case, connectivity within the network is established end-to-end from users to computation and storage resources with the condition of lowest (monetary) cost for their usage.

The operation mode is across the Grid resources axis: the location of computation resources is not as important as the cost of their use. At the same time, connections are required that guarantee a certain performance in terms of bandwidth, which requires the network resource coordination. In another scenario, the Grid application may require guaranteed bandwidth and scheduled access to remote visualization, which in the context of coordinated resource management illustrated in Figure 12.2 operates in the Grid resources networks plane, since the remote visualization is provisioned with guaranteed bandwidth on a specific location within the network. In addition, since the use of remote visualization resource is scheduled, the dimension of time must be considered too. In the previous two modes, the bandwidth was assumed to be always available at no cost. Conversely, in the scenario of the least-cost bandwidth/least-cost computation, the dimension of network optimization must be coordinated.

Advance reservation and scheduling of Grid resources pose a number of interesting research problems. In Figure 12.3, this is illustrated by the dimension *time*. If the bandwidth or computational resources are not instantaneously available, the resources have to be scheduled. The scheduling can be done for Grid resources (such as CPU time), for networking resources (such as available bandwidth), or for both simultaneously. One of the open architectural questions is how to design the coordinated scheduling of Grid and networking resources given a number of constraints.

Furthermore, the applications themselves can also be scheduled. Real-time interactive applications can be given priority for both Grid and networking resources. The GGF is currently putting significant efforts into design protocols and architectures for local and meta-schedulers [36]. Another interesting dimension is the dimension of *ownership*, whereby applications, networks, and Grid resources can be owned by different parties and their interrelations have to be defined and enabled through advanced control plane functions. Control plane functions can also consider the *locality (space)* of the resources as a further dimension. For example, in a high-energy physics community experiment at CERN, the location of the Large Hadron Collider as well as the distance to the storage of the data may be an important parameter.

The third important dimension of coordinated resource allocation is the Grid application. Today the applications are more often composite, i.e., composed of two and more interdependent tasks [21]. This has a very large impact on coordinated management. To illustrate this concept, consider a simple application model composed of three tasks. In the first task, the application requires a large amount of data (from a remote storage location) to be sent to a computing resource. After the computing has been accomplished (second task), the resulting data needs to be sent to the visualization site (third task).

Even this simple example poses a number of far-reaching research questions. How does the placement of computational nodes and network connectivity impact the performance of network and application? If the computing resources can be arbitrarily chosen within the network, what is the best algorithm to select the CPU and visualization sites? Is the network or the CPU congestion more important for scheduling consideration? These and other questions are critical architectural considerations, and quantifying some of these factors is essential in determining the interactions between the applications and networks, and the coordination and discovery of Grid resources.

12.5 ALL-PHOTONIC GRID NETWORK SERVICES

12.5.1 ALL-PHOTONIC GRID SERVICE

Having a Grid service that can provide an all-photonic end-to-end connection may provide capabilities that are of great interest to the Grid community. All-photonic network connection provides the following advantages: (i) transparent transport capability where only the two end-point transceivers need to understand the format, protocol, data rate, etc. of the data transmitted; (ii) low latency across the network (assuming that application-level latency and jitter requirements are handled at the edges) as a result of the lack of OEO transformation and buffering; (iii) simplified control and management plane,; and (iv) efficient performance predictability, QoS, and fault tolerance capability.

All-photonic network service can be either circuit switching based (wavelength routed network) or burst/packet switching based (OBS/OPS). The most fundamental network service that an all-photonic network can provide for Grid applications is the dynamic connection provisioning with QoS guarantees. In this section, the following three interrelated dimensions for all-optical connection provisioning are presented:

- *Switching granularity.* The bandwidth required by an application can be subwavelength, wavelength, or multiple wavelengths [37], and the connection can be long-term (circuit) or short-term (burst). Optical packet-switched network service may also be possible in the future.
- *Connection type.* The connection can be either unicast (lightpath) [38] or multicast (lighttree) [39] in the optical domain.
- *Quality of service.* Delay, data loss, jitter, fault tolerance. In all-photonic networks, quality requirement in the optical domain is important.

Many studies have been conducted on the optical switching technologies with different granularities, connection type, and QoS constraints [21]. In the following, the focus the discussion on the QoS issues of optical transport networks.

For large-scale optical networks covering large geographical areas, a unique feature is that the quality of the physical layer signal is critical to the QoS provisioning of optical connections. Although there are many benefits to keeping an end-to-end connection in the pure all-optical domain, OEO conversion is sometimes necessary because of the degradation of signals due to physical layer impairments. As signals travel longer distances without OEO regeneration, the accumulated effects on BER will increase. Therefore, optical layer quality monitoring and optical quality-based network service provisioning (routing and resource allocation) become more critical in an all-photonic network for connection SLA assurance and fault detection. It can be concluded that a Grid service providing an all-photonic connection should interact closely with a Grid service that provides optical physical layer monitoring information on a per-channel basis.

Before proceeding to the provisioning of optical connection with QoS requirements, first a brief introduction of some important Grid application scenarios that may benefit from all-photonic connection provisioning.

12.5.2 GRID SERVICE SCENARIOS FOR ALL-PHOTONIC END-TO-END CONNECTIONS

Today, the majority of data transfers within the Grid community involve large file transfer between sites using IP applications such as GridFTP. A basic all-photonic connection service that can be provided to Grid applications is the ultra-high-speed pipe for the transfer of a large amount of scientific data. For example, the current high-energy physics projects at CERN and the Stanford Linear Accelerator Center (SLAC) already generate petabytes of data. Apparently, the IP-based Internet would be extremely inefficient in this scenario. Furthermore, new latency-sensitive applications are starting to appear more frequently in the Grid community, e.g., remote visualization steering, real-time multicasting, real-time data analysis, and simulation steering. Collaborative projects analyzing the same dataset from remote instrumentation may be inclined to send raw digital data across the network via an all-photonic connection, so that processing of data can be done remotely from the data collection instrument. This will only require compatible transceivers, while the network will be completely unaware of the contents of the transmitted payload.

It can be concluded that the basic optical connections, either lightpath or lighttree with different bandwidth granularities and QoS requirements, are excellent service candidates for a broad range of Grid applications.

12.5.3 PHYSICAL LAYER QUALITY OF SERVICE FOR LAYER 1 SERVICES

Application QoS is usually concerned with end-to-end performance measurements, such as latency, jitter, BER, dynamic range (for analog signals), and bandwidth. However, for a high-bit-rate all-optical lightpath, the increased effect of optical layer impairment can severely limit the effective transmission distance. On the other hand, different application streams have different signal quality requirements, e.g., 10^{-4} BER for voice signal and 10^{-9} for real-time video.

The majority of applications as well as application developers are not aware of Optical QoS (OQoS) and the effects of the optical plane on the performance of the application. It is therefore necessary to provide a means for mapping application QoS requirements to the optical layer's QoS classifications.

Jitter, latency, and bandwidth of application data are dependent not on the optical plane's QoS but rather on the protocol layers above the optical plane (e.g., the transport layer). Optical bandwidth (OC-48, OC-192, etc.) in an optical network is controlled by the fixed bandwidth of the two end-nodes. The optical plane has no control over bandwidth, and has no access to measure it (to assure proper delivery). A distinction is made between optical bandwidth (OC-48, OC-192, etc.) and application bandwidth (related more to I/O capacity at the end-nodes). However, optical bandwidth does have an effect on the optical plane's QoS.

BER and dynamic range are very dependent on the optical plane's QoS; however, these parameters cannot be measured in the optical layer. Both BER and dynamic range are parameters evaluated within the electrical plane. BER is specified for digital signals and dynamic range is specified for analog signals. BER is the ratio of the number of bits in error over the number of bits sent (e.g., 10^{-12} bit errors per terabit of data transmitted). Dynamic range is the ratio of highest power expected signal to the lowest signal, which must be resolved. Both parameters are measurements of the QoS required for a particular signal transferred (i.e., end-to-end). A general approach to defining the OQoS is by considering the effects of various linear and non-linear impairments [40]. The representative parameters are Optical Signal to Noise Ratio (OSNR) and Optical jitter (Ojitter). This allows both analog and digital signals to be represented accurately as both amplitude (noise) and temporal (jitter) distortions can be accounted for independently.

OSNR is the strongest indicator of optical layer QoS. It is a measure of the ratio of signal power to noise power at the receiving end. The SNR of an end-to-end signal is a function of many optical physical layer impairments, all of which continue to degrade the quality of the signal as it propagates through the transparent network. It is recommended that the majority of these impairments be measured and/or derived on a link-by-link basis as well as the impacts made by the different optical devices (OXCs, electronic doped fiber amplifiers, etc.) so that the information can be utilized by the network routing algorithm.

Today, many will claim that optical networks are homogeneous with respect to signal quality. Some of the reasons for this claim are as follows:

- Currently deployed optical networks have a single transparent segment and are therefore considered opaque, in other words they have a very small domain of transparency. Currently, network system engineers simplify many of the optical impairments being discussed to a "maximum optical distance" allowed in order to sustain the minimum value of SNR for the network.
- Heterogeneous networks caused by physical optical impairments are overcompensated by utilizing FEC at the end-nodes, which has the ultimate effect of homogenizing the network. Note that this is useful only for digital signals.
- Currently deployed optical networks route signals operate at bit rates less than 10 Gbps. A number of publications state that physical optical impairments play a more significant role at bit rates of 10 Gbps and higher. As bit rates increase so does signal power; 40 Gbps is a given, and 160 Gbps is on the horizon.

These arguments are valid only when the deployed domains of transparency are very small relative to the envisioned next-generation all-photonic networks. Today's carriers often engineer their small domains of transparency to a maximum number of spans and their distances within a transparent network (six spans at 80 km each maximum) and are pre-engineered (maximum distance per given bit rate for a particular BER requirement). However it is envisioned that the future optical network will have a much larger domain of transparency and will therefore require more detailed impairments calculations to determine routes.

Although many carriers will be reluctant to change their current practices for engineering optical networks, they may find it necessary in order to profit from upcoming technologies. There actually exist many motivations for carriers to change the current strategy of pre-engineering the optical network and pursue high-speed all-optical paths over large areas, either within the same domain or in multiple domains:

- Re-use of existing fiber in networks for lower QoS signals while adding new technology for routes requiring a higher level of QoS.
- Engineering the optical network for homogeneity forces designers to evaluate the network based on the lowest common denominator (from a QoS perspective), which does not consider utilizing the higher QoS links for stricter QoS services (signals).
- Many carriers today realize that having the capability to offer differentiated services is a very profitable business compared with a single-QoS service.

12.5.3.1 Definitions of physical layer impairments

Many impairments in the optical plane can degrade optical signal quality. They are divided into two categories: (i) linear impairments and (ii) nonlinear impairments [41]. Linear impairments are independent of signal power, in contrast to nonlinear impairments, whose values change with power change.

Linear impairments
- *Amplifier-induced noise (ASE).* The only link-dependent information needed by the routing algorithm is the noise of the link, denoted as link noise, which is the sum of the noise of all spans on the link. Therefore, the ASE constraint is the sum of all the link noise of all links.
- *Polarized Mode Dispersion (PMD).* This is the fiber-induced noise. Efforts are being made today to provide PMD compensation devices, which may relieve the network from PMD constraint.
- *Chromatic Dispersion (CD).* This is also fiber-induced noise, which has the effect of pulse broadening. In today's deployed networks, CD is usually compensated for in compensation devices based on DCF (dispersion compensation fiber).

Nonlinear effects

The authors of ref. 41 believe that it is unlikely that these impairments can be dealt with explicitly in a routing algorithm due to their complexities. Others advocate that, due to the complexity of nonlinear impairments, it may be reasonable to assume that these impairments could increase the required SNR_{min} by 1 to 2 dB:

- Self-Phase Modulation (SPM);
- Cross-Phase Modulation (XMP) is dependent on channel spacing;

12.5 All-photonic Grid Network Services

- Four-Wave Mixing (FWM) becomes significant at 50 GHz channel spacing or lower – solution;
- Stimulated Raman Scattering effects (SRS) will decrease OSNR;
- Stimulated Brillouin (SBS) produces a loss in the incident signal.

Another important impairment parameter is *linear cross-talk*, which occurs at the OXCs and filters. Cross-talk occurs at the OXCs when output ports are transmitting the same wavelength and leaking occurs. Out-of-band and in-band cross-talk adds a penalty at the receiver on the required OSNR to maintain a given value of BER. In dense networks, per-link cross-talk information needs to be summed and added to the OSNR margin.

The authors of ref. 41 proposed the following link-dependent information for routing algorithms considering optical layer impairments:

- PMD – link PMD squared (square of the total PMD on a link);
- ASE – link noise;
- link span length – total number of spans in a link;
- link cross-talk (or total number of OXCs on a link);
- number of narrow filters.

When an all-photonic connection is not possible to set up due the optical layer limits, a cross-layer connection consisting of OEO boundary needs to be found [42].

12.5.4 REQUIREMENTS FOR AN ALL-PHOTONIC END-TO-END GRID SERVICE

It is assumed that the first phase of establishing a network Grid service, the service agreement with an end-user, has been achieved, and that this takes care of most policy matters such as AAA, pricing for the different QoS levels, etc.

The network Grid service shall provide the following operations for Grid applications:

- verify if the destination address is reachable via all-photonic connection;
- verify if the all-photonic connection to the destination can meet the minimum requested BER;
- verify if an end-to-end connection to the destination is available;
- Sign up for a push notification service from Grid monitoring services to monitor possible violations of SLAs.

Potential input parameters of interest for such a service may include destination addresses, QoS requirement (wavelength, minimum BER, restoration times, and priority and pre-emption, etc.), bandwidth, duration, and protocols.

12.5.5 OPEN ISSUES AND CHALLENGES

Multiple control planes (GMPLS, Just-In-Time (JIT), etc.) may exist, crossing multiple domains for one end-to-end connection within a Grid VO (virtual office). Each provider will have an agreement with its individual Grid members (GUNI agreements), and these providers must also have agreements with each other (G-NNI agreements). Some Grid providers might not even be aware of Grid members. A transit domain might just interact with other service providers.

This leads to the following open issues:

- Will only the access (GUNI) provider to an individual Grid member, i.e., be involved in that user's GUNI agreement?
- Will unsolicited GUNI notification reach a Grid member only from their prospective access (GUNI) provider?
- Will a Grid network service have an instantiation for each client or for each Grid/VO?
- Will there be a common policy repository that includes the individual and common "rules" for each VO/Grid?
- If a Grid has a quality monitoring service running, will it be responsible for the entire Grid, or will there be an instance per client connection or service/GUNI agreement?
- Will the Grid monitoring service get feeds (quality monitoring information) from each of the domains as necessary?
- New network provisioning problems include advanced resource allocation, VO topology reconfiguration, inter-domain routing with incomplete information, etc.

To answer above challenges, many research/development projects are under way, many based on global collaboration [22,36].

12.6 OPTICAL BURST SWITCHING AND GRID INFRASTRUCTURE

An optical network is built by interconnecting optical switches with Dense Wavelength-Division Multiplexing (DWDM) fibers. In an optical network, the transmission is always in the optical domain but the switching technologies differ. A number of optical switching technologies have been proposed: Optical-to-Electrical-to-Optical (OEO) switching, Optical Circuit Switching (OCS) switching (a.k.a. photonic/lightpath/wavelength-routed switching), Optical Burst Switching (OBS), and Optical Packet Switching (OPS). Most of today's optical networks, such as SONET, operate using OEO switching, in which the optical signal is terminated at each network node, then translated to electronics for processing and then translated back to the optical domain before transmission.

The other common method of optical switching today is OCS, in which static, long-term lightpath connections are set up manually between the source–destination pairs. In OPS, the data is transmitted in optical packets with in-band control information. The OPS technology can provide the best utilization of the resources; however, it requires the availability of optical processing and optical buffers. Unfortunately, the technology for these two requirements is still years away. Given the state of the optical networking technology, the OBS architecture is a viable solution for the control plane in an optical Grid network. OBS combines the best features of

12.6 Optical Burst Switching and Grid Infrastructure

packet switching and circuit switching. The main advantages of OBS in comparison with other optical switching technologies are that [43]: (a) in contrast to the OCS networks, the optical bandwidth is reserved only for the duration of the burst; (b) unlike the OPS network it can be bufferless. In the literature, there are many variants of OBS [44], but in general some main characteristics can be identified.

12.6.1 INTRODUCTION TO OBS

An OBS network consists of core nodes and end-devices interconnected by WDM fibers, as shown in Figure 12.4. An OBS core node consists of an OXC, an electronic switch control unit, and routing and signaling processors. An OXC is a nonblocking switch that can switch an optical signal from an input port to an output port without converting the signal to electronics. The OBS end-devices are electronic IP routers, ATM switches, or frame relay switches, equipped with an OBS interface (Figure 12.4). Each OBS end-device is connected to an ingress OBS core node. The end-device collects traffic from various electronic networks (such as ATM, IP, frame relay, gigabit Ethernet). It sorts the traffic per destination OBS end-device address and assembles it into larger variable-size units called *bursts*. The burst size can vary from a single IP packet to a large dataset at the millisecond timescale. This allows for fine-grain multiplexing of data over a single wavelength and therefore efficient use of the optical bandwidth through sharing of resources (i.e., lightpaths) among a number of users. Data bursts remain in the optical plane end to end, and are typically not buffered as they transit the network core. The bursts' content, protocol, bit rate, modulation format, and encoding are completely transparent to the intermediate routers.

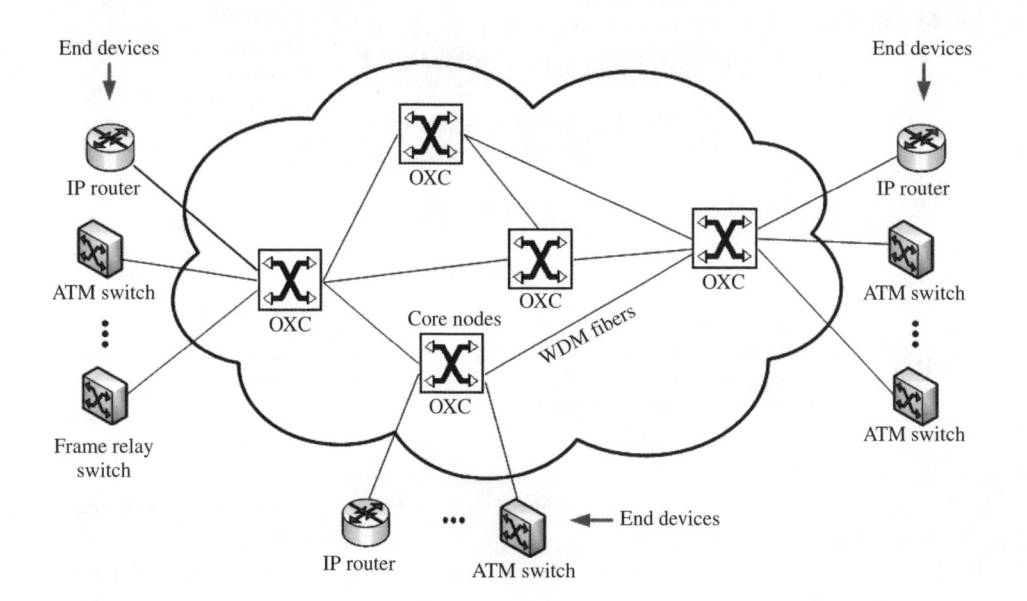

Figure 12.4. OBS network.

For each burst, the end-device also constructs a Burst Control Packet (BCP), which contains information about the burst, such as the burst length and the burst destination address. This control packet is immediately sent along the route of the burst and is electronically processed at each node. The function of the control packet is to inform the nodes of the impending data burst and to set up an end-to-end optical path between the source and the destination. Upon receipt of the control packet, an OBS core node schedules a free wavelength on the desired output port and configures its switching fabric to transparently switch the upcoming burst.

After a delay time, known as the *offset*, the end-device also transmits the burst itself. The burst travels as an optical signal over the end-to-end optical path set up by its control packet. This optical path is torn down after the burst transmission is completed. Figure 12.5 shows a generic model of an edge-OBS node and its functionality.

The separation of the control information and the burst data is one of the main advantages of OBS. It facilitates efficient electronic control while it allows for a great flexibility in the format and transmission rate of the user data. This is because the bursts are transmitted entirely as an optical signal, which remains transparent throughout the network.

12.6.1.1 Connection provisioning

There are two schemes that can be used to set up a connection, namely *on-the-fly connection setup* and *confirmed connection setup*. In the on-the-fly connection setup scheme, the burst is transmitted after an offset without any knowledge of whether

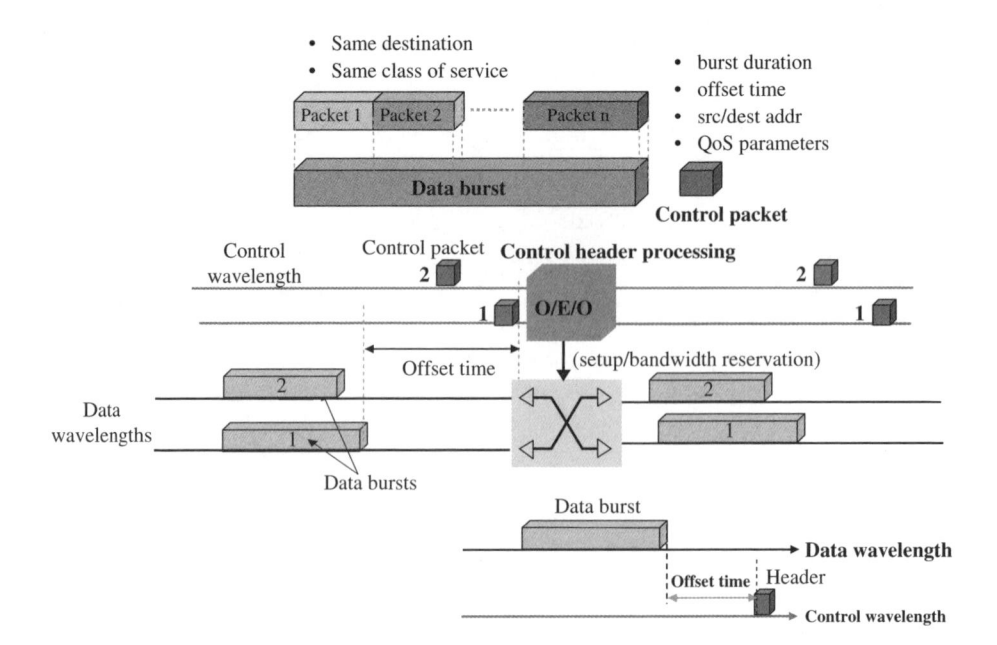

Figure 12.5. Generic model of an OBS end-device.

12.6 Optical Burst Switching and Grid Infrastructure

the connection has been successfully established end to end. In the confirmed connection setup scheme, a burst is transmitted after the end-device receives a confirmation from the OBS network that the connection has been established. This scheme is also known as *Tell And Wait* (TAW).

An example of the on-the-fly connection setup scheme is shown in Figure 12.6. End-devices A and B are connected via two OBS nodes. The vertical line under each device in Figure 12.6 is a time line and it shows the actions taken by the device. End-device A transmits a control packet to its ingress OBS node. The control packet is processed by the control unit of the node and, if the connection can be accepted, it is forwarded to the next node. This processing time is shown by a vertical shaded box.

The control packet is received by the next OBS node, processed, and, assuming that the node can accept the connection, forwarded to the destination end-device node. In the meantime, after an offset delay, end-device A starts transmitting the burst, which is propagated through the two OBS nodes to the end-device B. As can be seen in this example, the transmission of the burst begins before the control packet has reached the destination. In this scheme it is possible that a burst may be lost if the control packet cannot reserve resources at an OBS node along the burst's path. The OBS architecture is not concerned with retransmissions, as this is left to the upper networking layers. Also, it is important that the offset is calculated correctly. If it is too short, then the burst may arrive at a node prior to the control packet, and it will be lost. If it is too long, then this will reduce the throughput of the end-device.

An example of the confirmed connection setup scheme is shown in Figure 12.7. The end-device A transmits a control packet, which is propagated and processed at each node along the path as in the previous scheme. However, the transmission of

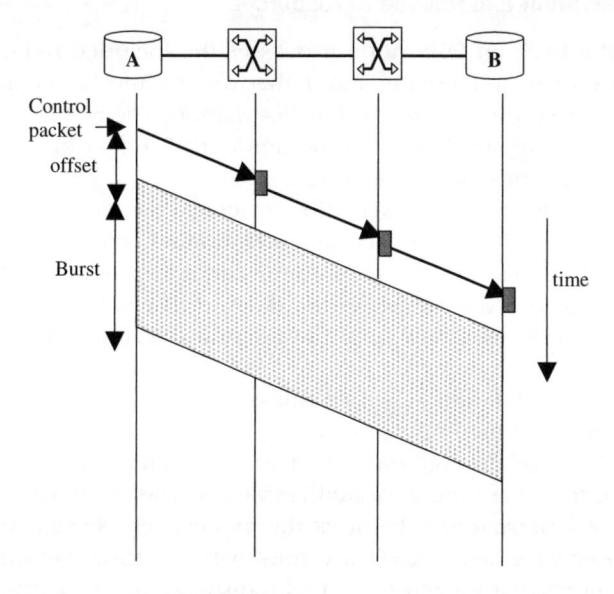

Figure 12.6. The on-the-fly connection setup scheme.

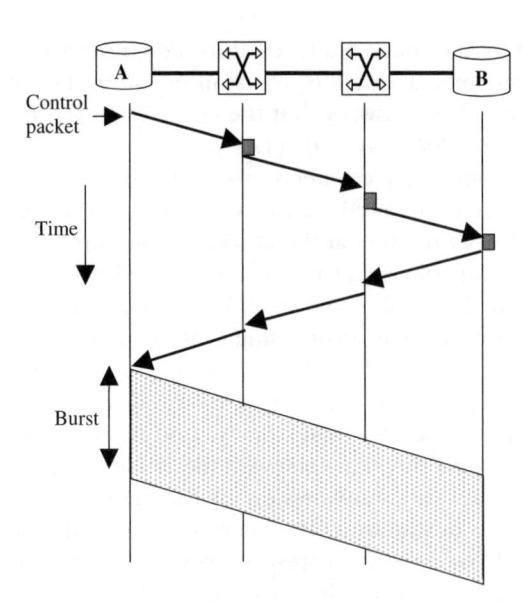

Figure 12.7. The confirmed connection setup scheme.

the burst does not start until A receives a confirmation that the connection has been established. In this case, there is no burst loss and the offset can be seen as being the time it takes to establish the connection and return a confirmation message to the transmitting end-device.

12.6.1.2 Reservation and release of resources

Upon receipt of a BCP, an OBS node processes the included burst information. It also allocates resources in its switch fabric that will permit the incoming burst to be switched out on an output port toward its destination. The resource reservation and release schemes in OBS are based on the amount of time a burst occupies a path inside the switching fabric of an OBS node.

There are two OBS resource reservation schemes, namely *immediate reservation* and *delayed reservation*. In the immediate reservation scheme, the control unit configures the switch fabric to switch the burst to the correct output port immediately after it has processed the control packet. In the delayed reservation scheme, the control unit calculates the time of arrival t_b of the burst at the node, and it configures the switch fabric at t_b.

There are also two different resource release schemes, namely *timed release* and *explicit release*. In the timed-release scheme, the control unit calculates when the burst will completely go through the switch fabric. When this time occurs, it instructs the switch fabric to release the allocated resources. This requires knowledge of the burst duration. An alternative scheme is the explicit release scheme, in which the transmitting end-device sends a release message to inform the OBS nodes along the path of the burst that it has finished its transmission. The control unit instructs the switch fabric to release the connection when it receives this message.

12.6 Optical Burst Switching and Grid Infrastructure

Combining the two reservation schemes with the two release schemes results in the following four possibilities: immediate reservation/explicit release, immediate reservation/timed release, delayed reservation/explicit release, and delayed reservation/timed release (see Figure 12.8). Each of these schemes has advantages and disadvantages.

For example, when timed release is implemented, the OBS core node knows the exact length of the burst. Thus, it can release the resources immediately upon burst departure. This results in shorter occupation periods and thus higher network throughput than in the explicit release. The difficulty, however, is that the timed-release schemes require complicated scheduling and their performance greatly depends on whether the offset estimates are correct. On the contrary, the immediate reservation/explicit release scheme requires no scheduling. It is easier to implement, but it occupies the switching fabrics for longer periods than the actual burst transmission. Therefore, it may result in a high burst loss.

In the OBS literature, the three most popular OBS variants are Just-In-Time (JIT) [44], Just-Enough-Time (JET) [45], and horizon. They mainly differ based on their wavelength reservation schemes. The JIT protocol utilizes the immediate reservation scheme while the JET protocol uses the delayed reservation scheme. The horizon reservation scheme can be classified as somewhere between immediate and delayed. In horizon, upon receipt of the control packet, the control unit scheduler assigns the wavelength whose deadline (horizon) to become free is closest to the time before the burst arrives.

12.6.2 GRID-OBS AS A CONTROL PLANE FOR GRID NETWORKING

In general, given the state of the optical technology, OBS is a viable near-term optical switching solution because it achieves good network resource utilization and it does not require optical buffers or optical processing. In this section, we identify why the OBS architecture might be a good candidate for the control plane in the specific context of Grid networking.

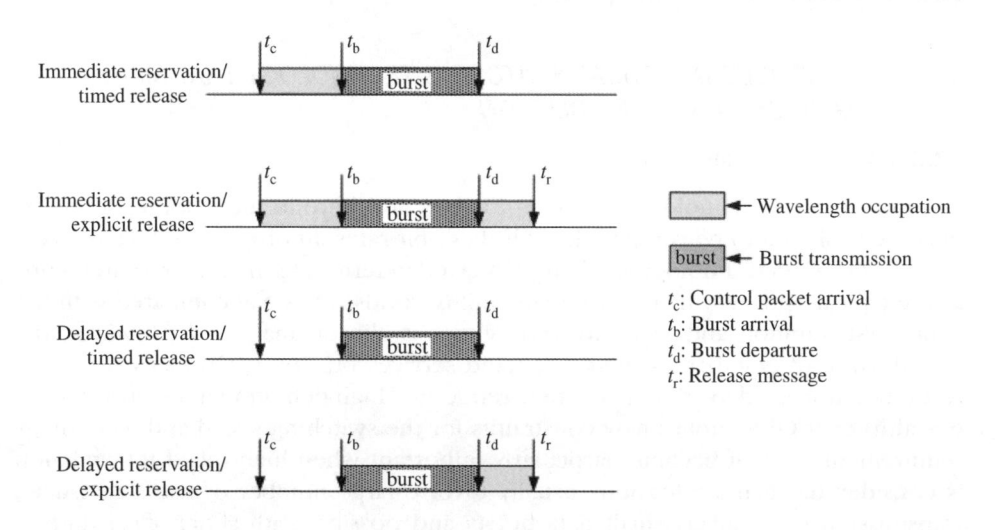

t_c: Control packet arrival
t_b: Burst arrival
t_d: Burst departure
t_r: Release message

Figure 12.8. Reservation and release of resources.

The variable size of data bursts in OBS allows for a flexible, close mapping to the user/application Grid requests. In other words, the variable-size bursts provide a flexible granularity that can support users/applications with different needs from the Grid. Users/applications that require a shorter duration connections will generate small bursts that may last only a few milliseconds whereas users/applications that require a larger bandwidth connection can generate a large enough burst that will hold the resources for longer time, i.e., similar to a long-lived all-optical lightpath. This fine-grain bandwidth granularity allows for the efficient transmission of Grid jobs with different traffic profiles.

The dynamic nature of OBS, i.e., connections are set up and torn down for the transmission of each burst, allows for a better sharing and utilization of the networking resources than in a optical circuit-switched network. The statistical multiplexing achieved by the bursts allows a large number of Grid users/ applications to access the resources.

Another advantage for Grid networking is the fast connection provisioning time in OBS. In most OBS variants, in order to minimize the connection setup time, the signaling of connections is accomplished using the on-the-fly connection setup scheme from Figure 12.6. In this *one-way* signaling scheme, the burst is transmitted after an offset without any knowledge of whether the optical path has been successfully established end to end. Note that the connection setup time can be even further decreased if it is implemented in hardware rather than software [45].

The separation of the control and data plane in OBS is yet another advantage for Grid networking. In OBS, the control packet is transported prior to its corresponding data burst and it is electronically processed at each node along the route between the source and the destination. The OBS technology can be adapted so that it can interact with the Grid middleware for resource reservation and scheduling. Therefore, the Grid application/user can include Grid protocol layer functionalities, such as intelligent resource discovery, authentication information, etc., in the information contained in the burst control packet.

12.6.3 ADVANCES IN OPTICAL SWITCHING TECHNOLOGY THAT MAKE GRID-OBS A VIABLE SOLUTION

12.6.3.1 At OBS core node

As future optical technology moves to 40 Gbps and beyond, networking solutions must be designed to be compatible with these bit rates, in order to reduce the cost per bit [43]. OBS technology is relatively relaxed in terms of switching requirements, as the typical optical switch setup times (milliseconds) are small compared with the data burst duration and therefore throughput is almost unaffected. However, the introduction of new bandwidth-on-demand services [46] (e.g., Grid services: high-resolution home video editing, real-time rendering, high-definition interactive TV and e-health) over OBS implies new constraints for the switching speed and technology requirements, which become particularly important when high-speed transmission is considered. Such applications usually involve large number of users that need transmission of relatively small data bursts and possibly with short offset time. A flexible OBS network must be able to support the small data bursts generated by the

12.6 Optical Burst Switching and Grid Infrastructure

aforementioned types of applications and services. For example, a burst of 300 ms duration transmitted at 10 Gbps can be switched by a MEMS-based switch typically within 20 ms. Considering only the switching time, the throughput of the system is 93.7%. If the same burst is transmitted at 160 Gbps then its duration is 18.75 ms and routing through the same switch would decrease the system's throughput to less than 50%. This becomes more severe when smaller bursts with a short offset time are treated by the OBS switch. For this reason, the deployment of fast switching technology is essential for future high-speed OBS networks where the evolving bandwidth on demand services is supported.

It should be noted, though, that the Burst Control Packet/header (BCP) requires intensive and intelligent processing (i.e., QoS, routing and contention resolution algorithms) which can be performed only by specially designed fast electronic circuits. Recent advances in the technology of integrated electronic circuits allow complicated processing of bursty data directly up to 10 Gbps [47]. This sets the upper limit in the transmission speed of the BCP. On the other hand, the optical data bursts (which do not need to be converted to the electronic domain for processing) are those that determine the capacity utilization of the network.

The optical bursts (data burst) can be transmitted at ultra-high bit rates (40 or 160 Gbps), providing that the switching elements can support these bit rates. Faster bursts indicate higher capacity utilization of the existing fiber infrastructure and significantly improved network economics. The deployment of fast switching assists the efficient bandwidth utilization but provides an expensive solution when it scales to many input ports. On the other hand, there is no additional benefit for long bursts of data, if fast switching is utilized. Therefore, one possible solution can be a switch architecture that utilizes a combination of fast (e.g., based on semiconductor optical amplifier) and slow (e.g., MEMS-based) switches. The switch architecture is shown in Figure 12.9.

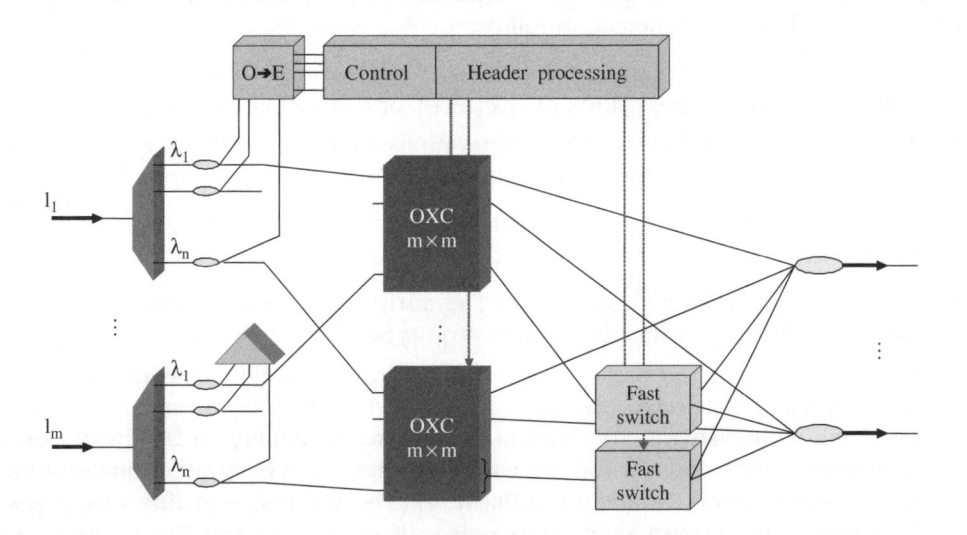

Figure 12.9. Architecture that combines slow switching (OXCs) and fast switching elements.

The general idea is based on the use of MEMS-based OXCs, which have a number of output ports connected to a fast optical switches. When a BCP appears, the control mechanism must first recognize if the BCP belongs to a burst with slow switching requirements (usually long burst) or a burst with fast switching requirements (usually short burst). In the first case the OXC is reconfigured so that when the long burst arrives it is automatically routed to the appropriate output port. In the other case the short bursts are routed directly to the fast switch (through predefined paths) and switched immediately to the next node. This architecture requires all the switching paths inside the OXC to be initially connected to the fast switch ports and special design constraints must be considered to avoid collision. The benefit of the proposed scheme is that it reduces the requirements on fast switching and therefore only smaller and cost-efficient matrices are required.

The fast switching mechanism can be based on the use of fast active components, such as semiconductor optical amplifiers. Switching is achieved by converting the signal's wavelength and routing it to an output port of a passive routing device (Arrayed Waveguide Grating, AWG). This solution is scalable but the bit rate is dependent on the utilized conversion technique. However, almost bit rate-transparent wavelength conversion schemes have been proposed, and fast switching of asynchronous bursty data at 40 Gbps has been demonstrated, with technology scalable to more than 160 Gbps [48]. This solution provides switching in nanoseconds and therefore can almost eliminate the required offset time for the short data bursts, offering increased throughput.

12.6.3.2 At OBS edge node

To facilitate on-demand access to Grid services, interoperable procedures between Grid users and optical network for agreement negotiation and Grid service activation have to be developed. These procedures constitute the Grid user optical network interface (G-OUNI). The G-OUNI functionalities and implementation will be influenced by number of parameters, as follows:

- Service invocation scenarios: the Grid user can request Grid services from the optical network control plane either directly or through Grid middleware [3].
- 2-Optical transport format, which determines transmission format of signaling and control messages as well as data from the Grid user to the optical network.

In the Grid-enabled OBS network with heterogeneous types of services and user demands the G-OUNI needs to provide the following functionalities:

- *Subwavelength bandwidth allocation.* The attribute "flexible" is used to indicate that G-OUNI will in principle support various bandwidth services.
- *Support for claiming existing agreements.* G-OUNI must facilitate the incorporation of information that relates to an existing agreement. This covers the support of a lambda time-sharing mechanism to facilitate scheduling of bandwidth over predefined time windows for the Grid users/service (i.e., lambda time-sharing for efficient/low-cost bandwidth utilization). The G-OUNI signaling would also be required to support ownership policy of bandwidth and the transport of authentication and authorization-related credentials.

12.6 Optical Burst Switching and Grid Infrastructure

- *Automatic and timely light-path setup.* Grid users, through G-OUNI, can automatically schedule, provision, and set up lightpaths across the network.
- *Traffic classification, grooming, shaping, and transmission entity construction.* At the transport layer (physical layer) the G-OUNI must be able to map the data traffic to a transmission entity (i.e., optical burst). In the case of in-band signaling the G-OUNI will provide a mapping mechanism for transmission of control messages (e.g., control wavelength allocation).

In a Grid-enabled OBS network, in which network resources are treated the same as Grid resources, the edge router must be able to perform G-OUNI functionality through mapping user jobs into the optical domain in the form of variable-length optical bursts. Therefore, the main characteristics of a G-OUNI-enabled edge OBS router are wavelength tuneability, traffic aggregation, variable-length optical burst construction, data burst and BCP transmission, and support for UNI functionality by interfacing with the control plane. Figure 12.10 shows functional architecture of an edge OBS router.

This architecture comprises the following units:

- input interfaces to accept user jobs through the gigabit Ethernet links;
- traffic aggregation and optical burst assembly unit to generate optical bursts and their associated BCPs;
- tuneable laser source and its controller to facilitate wavelength assignment for data bursts and BCPs.
- user–network signaling and control interface (UNI) to obtain the required information from control plane (i.e., data burst and BCP wavelengths, BCP information and burst transmission parameters such as offset time).

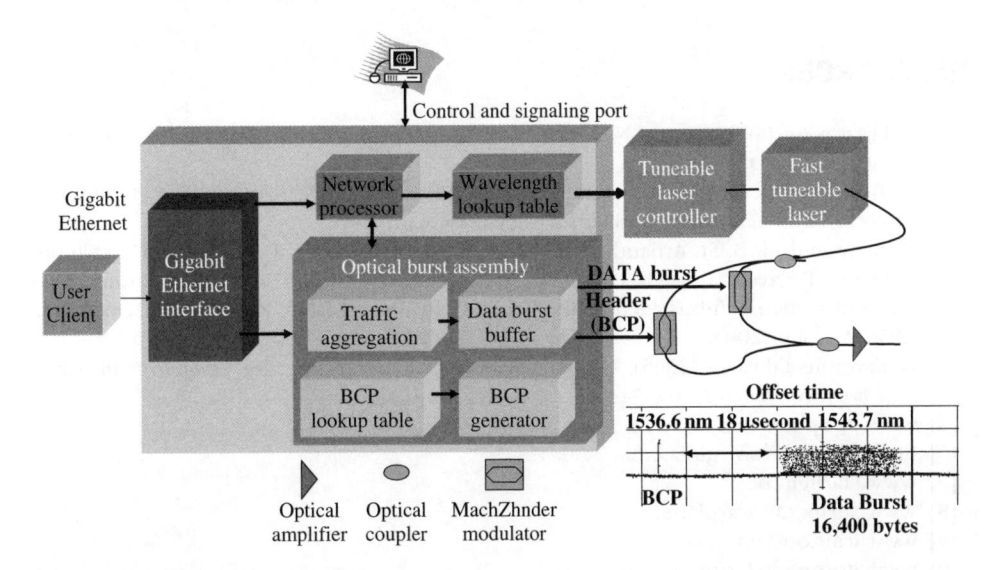

Figure 12.10. Functional architecture of a tuneable edge optical burst switching interface.

In this architecture, Grid user jobs from user clients enter into the edge router through a gigabit Ethernet input interface. The incoming data is aggregated with the help of a network processor in aggregation buffers based on type of the associated Grid jobs well as Grid resource requirements. Before transmission of each aggregated data burst a BCP is transmitted in front of the data burst. In addition, the tuneable laser is set to emit suitable wavelengths for each BCP as well as each data burst.

12.6.4 GRID-OBS USE SCENARIO

In this section, a typical Grid network scenario using OBS technology will be described. On the way there, the Grid service/application sends the request for the Grid service through the UNI (edge router) by using burst control signal on a dedicated wavelength. The request is distributed through the network for the resource discovery (both network and Grid resources) by the core OBS routers using optical multicast or broadcast. After source discovery and allocation, an acknowledgment message determines the data transmission parameters such as allocated lightpath and the time duration that each lightpath is available. Consequently, the user sends the data burst (Grid job) through the allocated lightpath's time window.

Once the job has been done, the results have to be reported back (if there are any results for the user/sender). On the way back, based on the type of results as well as their requirements in term of the network resources, the same reserved path can be used or a new path can be reserved with new OBS signaling.

In such a Grid networking scenario, the control of the OBS routers must support the functionality of the Grid protocol architecture (i.e., collective layer, resource layer, connectivity layer) [49]. This control architecture will ensure that resource allocation/sharing, data aggregation, and routing of the application data bursts will fulfill Grid service requirements.

REFERENCES

[1] A. Jajszczyk (2005) "Optical Networks – the Electro-Optic Reality," *Optical Switching and Networking*, 1(1), 3–18.
[2] G. Bernstein, B. Ragagopalan, and D. Saha (2003) *Optical Network Control: Architecture, Protocols, and Standards*, Addison-Wesley-Longman.
[3] D. Simeonidou, B. St. Arnaud, M. Beck, B. Berde, F. Dijkstra, D.B. Hoang, G. Karmous-Edwards, T. Lavian, J. Leigh, J. Mambretti, R. Nejabati, J. Strand, and F. Travostino (2004) "Optical Network Infrastructure for Grid," Grid Forum Draft, Grid Forum Document-I.036, October 2004.
[4] G. Karmous-Edwards (2005) "Global E-Science Collaboration", *IEEE Computing in Science and Engineering*, 7(2), 67–74.
[5] www.optiputer.net.
[6] http://cheetah.cs.virginia.edu/.
[7] www.starlight.net.
[8] www.surfnet.nl/starplane.
[9] www.icair.org/omninet.
[10] www.dotresearch.org.
[11] www.ultralight.net.

References

[12] B. Mukherjee (1997) *Optical Communications Networks*, McGraw Hill.
[13] F. Dijkstra and C. de Laat (2004) "Optical Exchanges," GridNets 2004 conference proceedings, October 2004.
[14] S.V. Kartalopoulos (2003) *DWDM: Networks, Devices, and Technology*, John Wiley & Sons Ltd.
[15] H. Takara, T. Ohara, K. Mori, K. Sato, E. Yamada, Y. Inoue, T. Shibata, M. Abe, T. Morioka, and K.-I. Sato (2000) "More than 1000 Channel Optical Frequency Chain Generation from Single Supercontinuum Source with 12.5 GHz Channel Spacing," *Electronic Letters* 36, 2089.
[16] E. Mannie (ed.) (2001) "Generalized Multi-Protocol Label Switching (GMPLS) Architecture," IETF RFC 3945, October 2004.
[17] ITU-T (2001) "Architecture for the Automatically Switched Optical Network (ASON)," Recommendation,G.8080/Y.1304, November 2001.
[18] T. Dimicelli, "Emerging Control Plane Standards and the Impact on Optical Layer Services", www.oiforum.com/public/downloads/DiMicelli.ppt.
[19] T. Takeda (ed.) "Framework and Requirements for Layer 1 Virtual Private Networks," draft-ietf-l1vpn-framework-01.txt.
[20] http://www.ietf.org/html.charters/l1vpn-charter.html).
[21] www.canarie.ca/canet4/uclp/UCLP_Roadmap.doc.
[22] Global Lambda Integrated Facility, http://www.glif.is.
[23] www.glif.is/working-groups/controlplane/.
[24] High Voltage Electron Microscope at Korea Basic Science Institute, http://hvem.kbsi.re.kr/.
[25] GLObal Ring network for Advanced application Development (GLORIAD) Project, http://www.gloriad-kr.org.
[26] G. Swallow, J. Drake, H. Ishimatsu, and Y. Rekhter (2004) "Generalized Multi-Protocol Label Switching (GMPLS) User Network Interface (UNI), RSVP-TE Support for the Overlay Model," draft-ietf-ccamp-gmpls-overlay-05.txt, October 2004.
[27] J.Y. Wei (2002) "Advances in the Management and Control of Optical Internet," *IEEE Journal on Selected Areas in Communication*, 20, 768–785.
[28] http://dragon.east.isi.edu/dragon_projsumm.htm.
[29] L.L. Smarr, A.A. Chien, T. DeFanti, J. Leigh, and P.M. Papadopoulos (2003) "The OptIPuter," *Communications of the ACM*, 46, 68–77.
[30] M. Médard and S. Lumetta (2003) "Network Reliability and Fault Tolerance," *Wiley Encyclopedia of Engineering* (ed. by J.G. Proakis), John Wiley & Sons Ltd.
[31] J. Mambretti, J. Weinberger, J. Chen, E. Bacon, F. Yeh, D. Lillethun, B. Grossman, Y. Gu, and M. Mazzuco (2003) "The Photonic TeraStream: Enabling Next Generation Applications Through Intelligent Optical Networking at iGrid 2002," *Journal of Future Generation Computer Systems*, 19, 897–908.
[32] R. Grossman, Y. Gu, D. Hanley, X. Hong, J. Levera, M. Mazzucco, D. Lillethun, J. Mambretti, and J. Weinberger (2002) "Photonic Data Services: Integrating Path, Network and Data Services to Support Next Generation Data Mining Applications," Proceedings of NSF Workshop on Next Generation Data Mining (NGDM) '02, Baltimore, MD, November 1–3, 2002. http://www.cs.umbc.edu/NGDM02.
[33] G. Allen and E. Seidel (2003) "Computational Astrophysics and the Grid," in *The Grid: Blueprint for a New Computing Infrastructure*, 2nd edn, Morgan Kaufmann.
[34] G. Allen, K. Davis, K.N. Dolkas, N.D. Doulamis, T. Goodale, T. Kielmann, A. Merzky, J. Nabrzyski, J. Pukacki, T. Radke, M. Russell, E. Seidel, J. Shalf, and I. Taylor (2003) "Enabling Applications on the Grid: A Gridlab Overview," *International Journal of High Performance Computing Applications*, special issue on "Grid Computing: Infrastructure and Applications", August 2003.
[35] NSF Workshop on Optical Networks, April 2004.

[36] Global Grid Forum, http://www.ggf.org.
[37] Y. Zhu, A. Jukan, M. Ammar and W. Alanqar (2004) "End-to-End Service Provisioning in Multi-granularity Multi-domain Optical Networks", IEEE ICC 2004, Paris, France.
[38] H. Zang, J. Jue, and B. Mukherjee (2000) "A Reviewof Routing and Wavelength Assignment Approaches for Wavelength-Routed Optical WDM Networks," *Optical Networks Magazine*, 1(1).
[39] Y. Xin and G.N. Rouskas, "Multicast Routing Under Optical Layer Constraints," Proceedings of IEEE Infocom 2004, March 2004, Hong Kong.
[40] P. Mehrotra, G. Karmous-Edwards, and D. Stevenson (2003) "Defining Optical Plane QoS Parameters for OBS Networks", Workshop for Optical Burst Switching (WOBs).
[41] A. Chiu (ed.) (2002) "Impairments and Other Constraints on Optical Layer Routing," draft-ietf-ipo-impairments-02.txt, August 2002.
[42] http://www.glif.is/publications/press/20040525.php.
[43] Y. Chen, C. Qiao and X. Yu (2004) "Optical Burst Switching: A New Area in Optical Networking Research," *IEEE Network Magazine*, 18(3), 16–23.
[44] T. Battestilli and H. Perros (2003) "An Introduction to Optical Burst Switching", *IEEE Communication Magazine*, 41(8), S10–S15.
[45] S.R. Thorpe, D.S. Stevenson and G. Karmous-Edwards (2004) "Using Just-in-Time to Enable Optical Networking for Grids," GridNets Workshop (co-located with Broadnets 2004), October 2004.
[46] E. Breusegem, M. Levenheer, J. Cheyns, P. Demeester, D. Simeonidou, M. O'Mahoney, R. Nejabati, A. Tzanakaki, and I. Tomkos (2004) "An OBS Architecture for Pervasive Grid Computing", Proceedings of Broadnets 2004, October 2004.
[47] J. Gaither (2004) "300-pin msa Bit-error Rate Tester for the ml10g Board and Rocketphy Transceiver", XILINX Application note: XAPP677 Virtex-II Pro Family, January 2004.
[48] D. Klonidis, R. Nejabati, C. Politi, M. O'Mahony, and D. Simeonidou (2004) "Demonstration of a Fully Functional and Controlled Optical Packet Switch at 40gb/s", Proceedings of 30th European Conference on Optical Communications PD Th4.4.5.
[49] Workshop on Optical Control Planes for the Grid Community, April 23 and November 12 2004. http://www.mcnc.org/mcnopticalworkshop/nov04.

Chapter 13

Network Performance Monitoring, Fault Detection, Recovery, and Restoration

Richard Hughes-Jones, Yufeng Xin, Gigi Karmous-Edwards, John Strand

13.1 INTRODUCTION

Many emerging Grid-based applications require guarantees of high performance and high availability of computing and network resources [1]. Being a geographically distributed large-scale system consisting of processors, storage, and software components interconnected with wide-area networks, Grid services are provisioned on a dynamic basis whereby service components may join or leave a Grid infrastructure at any time.

Network administrators are usually interested in overall utilization, traffic patterns, and trends, e.g., for capacity planning, as well as visual tools and alarms for operational support. Network applications developers may want to observe their application's traffic and competing traffic. Network protocol designers may want to see the contents of every packet in the protocol. Grid user experiencing "poor network" performance may want tools to monitor and probe the network paths and monitor the effects of TCP tuning.

Grid Networks: Enabling Grids with Advanced Communication Technology Franco Travostino, Joe Mambretti,
Gigi Karmous-Edwards   © 2006 John Wiley & Sons, Ltd

As well as making information available for human use, network monitoring provides vital data about the operational state and performance of the network resources for the Grid middleware itself. By analysis of recent measured historic network performance, the middleware can make simple projections of the future potential performance. This network resource information, together with data about CPU and storage capability, will enable efficient scheduling of the Grid for both "batch" and real-time computing.

Thus, network monitoring and the ability to monitor the distributed computing environment at many levels and with various detail is most important to the smooth operation of the Grid.

Furthermore, a Grid infrastructure may suffer from multiple failures modes: computer crash, network failure, unplanned resource down-time, and process fault, etc. Once the complexity of fault management is tamed, a Grid in fact holds promise of great reliability, in that its constituent services can be efficiently migrated and rebalanced within a virtual organization. Since many Grid implementations build upon Web Services, the reliability strengths in a Grid will directly depend upon the Web Services ones. Clearly, Web Services started as a document-based exchange to access backend servers. As new functionalities are added and Web Services are increasingly part of everyday processes, it is a fair question to ask whether Web Services are on a trajectory to become increasingly similar to distributed object systems [2], whose difficulties in fault management are well known.

This chapter first examines network monitoring, discussing the network characteristics in Section 13.2 followed by the methods of instrumentation and analyzing the data in Section 13.3. The fault management functions follow with an introduction in Section 13.4, Section 13.5 is dedicated to fault detection techniques and Section 13.6 discusses fault recovery and restoration.

13.2 MONITORING CHARACTERISTICS

Chapter 7 introduced and discussed the nomenclature and schemata being developed by the Network Measurements Working Group (NMWG) of the Global Grid Forum (GGF) for describing network measurements taken in Grid environments. This chapter examines the characteristics and network monitoring techniques in more detail. Following their terminology, *network entity* is a general term that includes nodes, paths, hops autonomous systems, etc.

A *network characteristic* is an intrinsic property that is related to the performance and reliability of a network entity. It is a characteristic that is the property of the network, or of the traffic on it, not an observation of that characteristic. An example of a characteristic is the round-trip delay on a path.

Measurement methodologies are techniques for measuring, recording, or estimating a characteristic. Generally, there will be multiple ways to measure a given characteristic, but it should be clear that all measurement methodologies under a particular characteristic should be measuring "the same thing," and could be used mostly interchangeably. As an example, consider the round-trip delay characteristic. It may be measured directly using ping, calculated using the transmission time of a TCP packet and receipt of a corresponding ACK, projected from combining separate one-way delay measurements, or estimated from link propagation data and queue lengths. Each of these techniques is a separate measurement methodology for calculating the round-trip delay characteristic.

An *observation* is an instance of the information obtained about a characteristic by applying the appropriate measurement methodology. Following RFC 2330 [3] an individual observation is called a singleton, a number of singletons of the same characteristic taken together form a sample, and the computation of a statistic on a sample gives a statistical observation.

A network measurement consists of two elements, the characteristic being measured and the network entity to which the methodology was applied together with the conditions under which the observation was performed. Because network characteristics are highly dynamic, each reported observation must be attributed with timing information indicating when the observation was made. In addition, attributes about the nodes and paths such as the protocol (e.g., TCP over IPv6), QoS, applicable network layer, "hoplist", end-host CPU power, and Network Interface Card (NIC) [4] might all be important in interpreting the observations, and need to be included.

Figure 7.6 shows the nomenclature and hierarchy of network characteristics developed by the GGF Network Measurements Working Group (NMWG). It shows the relationships between commonly used measurements and allows measurements to be grouped according to the network characteristic they are measuring. Any number of these characteristics may be applied to a network entity but some characteristics, such as route or queue information, are sensible only when applied to either paths or nodes. It is worth noting that many network characteristics are inherently hop-by-hop values, whereas most measurement methodologies are end to end and, for Grid environments, may cross multiple administrative boundaries. Therefore, what is actually being reported by the measurements may be the result for the smallest (or "bottleneck") hop.

To enable real-time performance in Grid networks together with compliance monitoring, tracking, and comparison, the major characteristics used to measure the end-to-end performance of networks are bandwidth, delay, and data loss. For certain multimedia applications, jitter is also important. The following sections discuss these characteristics. High network availability and short restoration time after a failure are often a part of the Service Level Agreement (SLA) for those applications with dependability requirements.

13.2.1 THE HOPLIST CHARACTERISTIC

A path is a unidirectional connection from one node to another node that is used as a measurement endpoint. Network paths can represent any connection between nodes, including an end-host to end-host connection though the Internet, a router-to-router link that uses several different link layer technologies, as well as a simple Ethernet link between two Ethernet switches. A path may consist of several hops; for example, a path between two hosts might pass though several layer 3 routers, each of which would be a hop. The hoplist characteristic records the hops in sequence. "Traceroute" is an example of a tool that reports the hops traversed at layer 3.

Knowledge of both wide-area and local campus topology is important for understanding potential bottlenecks and debugging end-to-end problems.

13.2.2 THE BANDWIDTH CHARACTERISTIC

For network measurements, in line with the IETF IP Performance Metrics (IPPM) "bulk transfer capacity" [5], bandwidth is defined as data per unit time. There are four characteristics that describe the "bandwidth of a path":

- *Capacity*. Capacity is the maximum amount of data per time unit that a hop or path can carry. For a path that includes a bottleneck, the capacity of the bottleneck hop will give the upper bound on the capacity of a path that includes that hop.
- *Utilization*. The aggregate of all traffic currently flowing on that path. Recording the amount of data that traversed the path over a period of time provides a singleton observation of the average bandwidth consumed. Selection of the period of time for the observation requires care: too long an interval can mask peaks in the usage and, if the time is too short, considerable CPU power may be consumed, the traffic counters may not be reliably reflect the traffic, and the observed bandwidths may vary dramatically.
- *Available bandwidth*. The maximum amount of data per time unit that a hop or path can provide given the current utilization. This can be measured directly or estimated from capacity and utilization.
- *Achievable bandwidth*. The maximum amount of data per time unit that a hop or path can provide to an application, given the current utilization, the protocol used, the operating system and parameters (e.g., TCP buffer size) used, and the end-host performance capability. The aim of this characteristic is to indicate what throughput a real user application would expect. Tools such as iperf [6] observe the achievable bandwidth. On a path consisting of several hops, the hop with the minimum transmission rate determines the capacity of the path. However, while the hop with the minimum available bandwidth may well limit the achievable bandwidth, it is possible (and even likely on high-speed networks) that the hardware configuration or software load on the end-hosts actually limits the bandwidth delivered to the application.

Each of these characteristics can be used to describe characteristics of an entire path as well as a path's hop-by-hop behavior. The network layer at which the bandwidth is measured is also important. As an example, consider the capacity at layer 1, this would give the physical bit rate, but as layer number increases, the capacity would apparently decrease by taking into account the framing and packet spacing at the link layer (layer 2) and then the protocol overheads for layers 3 and 4.

13.2.3 THE DELAY CHARACTERISTIC

Delay is the time taken for a packet to travel between two nodes, and may be divided into one-way and round-trip delay. More formally following RFC 2330 [3], RFC 2679 [7] and RFC 2681 [8], delay is the time between the first bit of an object (e.g., a frame or packet) passes an observational position (e.g. a host's NIC) and the time the last bit of that object or a related object (e.g. a response frame or packet) passes a second observational point. The two observation points may be located at the same position on the path, for example when measuring round-trip delays.

Observations of the one-way delays are useful in providing detailed understanding of the network as the outgoing and return paths can be asymmetric, have different QoS settings, or have different utilization or queues. In addition, some applications, such as file transfer, may depend more on performance in the direction in which the data flows. Round-trip delays may be formed from the corresponding one-way delays, but they are easier to measure than one-way delays since only one observation point and one clock is needed. It is worth noting that methodologies that measure absolute one-way delays on gigabit or multi-gigabit links require clocks to be accurately synchronized to the order of microseconds or better, which is usually achieved using global positioning systems.

The variance in one-way delay is commonly referred to as jitter. In the IP world, IP Packet Delay Variation (IPDV) is defined for a selected pair of packets in a stream of packets as the difference between the delay of the first packet and the delay of second of the selected packets. Jitter may be derived from a sample of one-way delay observations or be measured by observing the inter-arrival times of packets sent at fixed intervals.

13.2.4 THE LOSS CHARACTERISTIC

A singleton packet loss is the observation that a packet sent from a source node does not traverse the path and arrive at the destination node. Packet loss may be measured over one-way or round-trip paths; ping is a common methodology for observing round-trip packet loss. Loss is usually presented as a statistic, the fraction of packets lost. A loss pattern can be used to represent more detailed observations than a simple fraction.

The sensitivity to the loss of individual packets, or the frequency or the pattern of packet loss, is strongly dependent on the transport protocol and the application itself. For streaming applications (audio/video), packet loss results in reduced quality of sound and images. For data transfers using TCP, packet loss can cause severe degradation of the achievable bandwidth.

13.2.5 THE CLOSENESS CHARACTERISTIC

It is possible to combine multiple characteristics into a single number that is indicative of the "distance" or "closeness" of two nodes in a network determined by traffic performance, rather than according to the actual physical connection. For example, Ferrari [9] proposes closeness as a function combining available bandwidth with round-trip delay. As envisioned in Chapter 3, this information could be used by the Grid middleware, such as a resource scheduler or a data replication manager, to choose the optimum location on the Grid for a computation requiring large amounts of data, or to determine if a request for bandwidth on demand can be satisfied, or to choose the best site to locate a replica of some dataset. In practice, a Grid application itself might chose a particular "closeness" characteristic that weighs bandwidth, latency, long transfers, or short transfers according to that particular application's needs.

13.3 NETWORK MONITORING INSTRUMENTATION AND ANALYSIS

The network measurements and monitoring that might be used in distributed Grid computing include both active and passive techniques [10], and may be grouped into a framework that covers four broad areas:

- the monitoring of traffic flows and patterns;
- regular lightweight observations;
- detailed network investigations;
- monitoring at the application level.

The tools and techniques within this framework aim to support the needs of the Grid community, which includes the network providers, the network operations staff at Grid computing centers, network researchers, and Grid users themselves. The areas are presented below and followed by a discussion of how they might be used to support the Grid environment.

13.3.1 MONITORING OF TRAFFIC FLOWS AND PATTERNS

Many network nodes, such as routers, switches, and multiplexers, maintain registers that record information about the network traffic, such as counting the number of packets and bytes transmitted and received by each of the interfaces. The Simple Network Management Protocol (SNMP) provides an interface via the Management Information Base (MIB) to the traffic statistics collected by that device. Many frameworks, both commercial and open source, e.g., Cricket [11] or Multi Router Traffic Grapher (MRTG) [12], use SNMP to retrieve this data on a regular basis and then store it in some form of archive.

This noninvasive network monitoring typically provides visualization in the form of historic time-series graphs presenting average utilization, traffic patterns, and trends. However, an equally important facility for operational staff at Grid computer centers is the ability to generate alarms when conditions on the network exceed some limit or some incident occurs. Many network problems, e.g., the loss of a trunk link, will cause a series of changes, and useful alarm systems contain heuristics that filter out duplicate alarms associated with one incident. This operational requirement for "intelligent" alarms is also required for the lightweight monitoring, described below. SNMP and the frameworks are not without some problems, for example configuration and control are usually restricted to network administrators and a SNMP requester must use the correct community string to allow access to the SNMP data.

As well as layer 2 information, packets, and byte counts, many hardware units provide information on the optical and electrical properties at layer 1. This could include, for example, the transmitted and received optical power levels, or the bit error rates corrected by forward error correction systems and, for optical systems, the number of "colors" available and those in use.

A clear use of this information is in capacity planning for network managers and administrators, but using this information to determine the available bandwidth can help user applications verify or understand their observed network performance. Some frameworks provide web or proxy access to allow network administrators/users to query the MIBs on demand, and this can be most useful when commissioning links or improving application performance by pinpointing bottlenecks or potential problems. Even a simple use such as sending a series of bytes and verifying that they correctly arrived and passed each point in the network can be a great help. Often this may be the only way when dealing with layer 2 networks distributed over large physical distances.

13.3.2 LIGHTWEIGHT MONITORING

To gauge the overall operation and performance of their network connectivity, many network administrators find it useful to make automatic, periodic observations of certain network characteristics and to record this information in some form of database. Usually, the monitoring is performed by dedicated hosts with sufficient I/O to ensure that the observation of the characteristic represents the performance of the network at that time and not artifacts due to loading of the monitoring system. It is also very important to connect these systems to the network at a point that is representative of the campus or Grid computing facility. The network group in the European Union DataGrid Project set up a framework to monitor the Grid sites [13,14], which is being developed into the EGEE Network Performance Monitoring toolkit e2emonit [15]. Some network monitoring frameworks also provide methods of requesting that measurements be made on demand along with suitable authentication and authorization techniques [16].

Simple web interfaces also exist for some common tools, for example traceroute, which allow users to determine hoplists from many locations around the world [17]. Another example of a web-based tool is the Network Diagnostic Tool (NDT) [18], which is a client–server program that provides network configuration and performance testing using the path between the web client and the node running the server. NDT servers can be installed at strategic points of the network being used.

The most commonly monitored characteristics are round-trip delay, packet loss, and TCP-achievable bandwidth using tools such as ping and iperf. An example of this type of monitoring is the Internet End-to-end Performance Monitoring (IEPM) [19,20] framework developed at SLAC by Les Cottrell and Connie Logg. This work started in 1995 with the PingER project, which used ping and has considerable historic data. Current work includes regular observations of the round-trip delay by ping, and TCP-achievable bandwidth using thrulay [21] and iperf as well as that from tools using packet dispersion techniques [22] such as pathload [23] and pathchirp [24].

The project also makes measurements of the TCP-achievable bandwidth for disk-to-disk transfer rates using BbFTP [25] and BBCP [26].

Visualization of this lightweight monitoring data is usually in the form of time-series plots showing the observation plotted against time together with histograms of the projections. Web-based interfaces are used that allow the user to select the range of time for the display and the end sites. Figure 13.1 gives an example from the IEPM framework web visualization display. The plots show a break in service in the middle of the plots, and reasonable tracking of the achievable TCP bandwidth observed with different methodologies.

As well as time-series analysis, the data may be interpreted by investigating the correlation of observations of the same characteristic by different methodologies, e.g., the round-trip delay reported from ping and THRUput and delay (thrulay) or the achievable TCP throughput reported from the pathload, pathchirp, thrulay, and iperf methodologies. There is also considerable interest in studying the correlations of observations of different characteristics over the same paths. An informative analysis of historic network data by researchers at SLAC [27] is shown Figure 13.2. This shows a scatterplot of the disk-to-disk transfer rates plotted against the measured memory-to memory achievable TCP throughput that was observed closest in time to the disk transfer for many different paths. The horizontal bands indicate limitations caused by the disk subsystem while the vertical bands identify limitations due to the network connectivity. Clearly, the network performance is very important, but it is also clear that often it is not the limiting item.

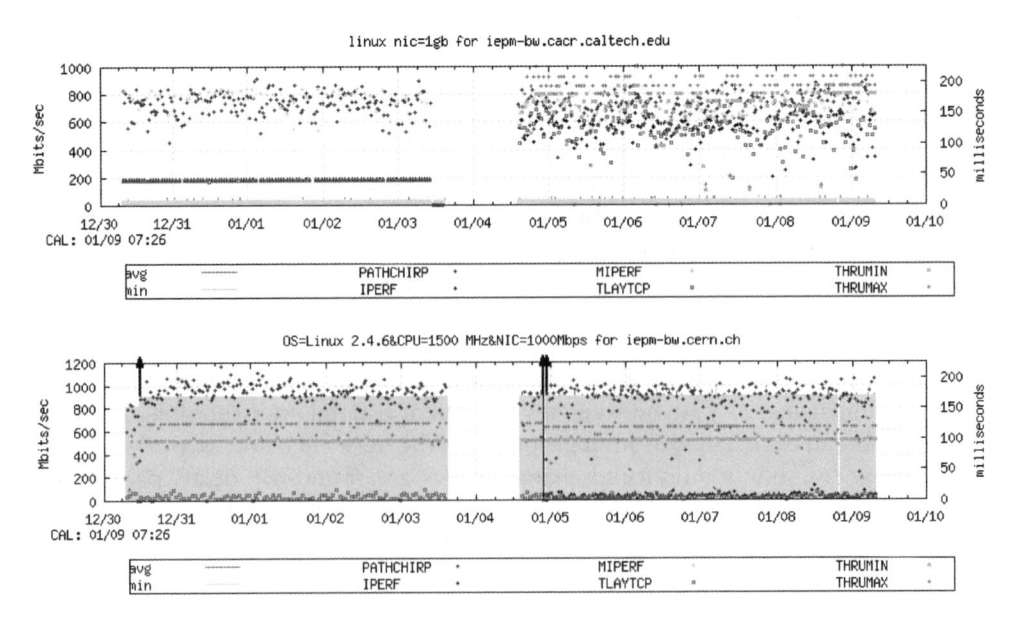

Figure 13.1. Example of the visualization of the round-trip delay (shaded) and achievable TCP bandwidth observed with different methodologies. Top plot: SLAC to Caltech rtt ~12 ms; lower plot: SLAC to CERN rtt ~170 ms.

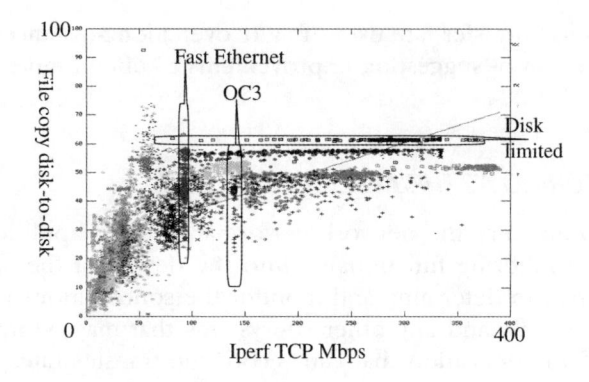

13.3 Network Monitoring Instrumentation and Analysis

Figure 13.2. Disk-to-disk transfer rates plotted against the achievable TCP bandwidth observation that was made closest in time to the disk transfer. Ideally the correlation would be a linear relation. However, for rates over about 60 Mbps the iperf TCP throughput is greater than the disk transfer rate.

13.3.3 DETAILED NETWORK INVESTIGATIONS

When new network equipment is being commissioned, or to investigate problems reported by network users, or to understand unusual changes that have been noticed from alarms or the visualization of traffic pattern monitoring or lightweight monitoring, more detailed network measurements will be required. Often, these will be quite intrusive and may involve considerable loads on the network; however, they are not undertaken regularly. An example of this type of investigation would be the location or isolation of packet loss in an end-to-end network which could be done by sending long streams of UDP or TCP packets at line speed using tools such as UDPmon [28] or iperf. Detailed measurements are also required to determine and characterize the performance and behavior of network hardware components, including links, multiplexers, routers, switches, and end-host interfaces.

Nonintrusive or passive observations may be made using optical splitters with one output feeding an interface dedicated to monitoring, or by using special-purpose devices enabling line-rate capture of network traffic, even on fully loaded 10 Gbit Ethernet and OC192/STM-64 links [29]. This approach of making use of optical or electrical splitters in conjunction with dedicated interfaces also allows more detailed examination of traffic flows by recording packet, transport, and application-level protocol headers, for example to investigate web server delays [30]. One of the challenges in applying this type of methodology on high-speed networks, 1 Gbps and above, is the large amounts of data (~10 Mbytes/s just for the header information for bidirectional monitoring at 1 Gbit) that need to be recorded and processed in real time.

Instrumentation of the protocol stacks on end-hosts using tools such as Web100 [31] could be used to investigate the behavior of network protocols themselves, e.g., the sharing of the available bandwidth with certain topologies, traffic patterns, or network technologies. This methodology can also be used to investigate the reasons why users are getting the performance they report when using real applications on Grid networks. An example of this type of work might be the understanding of the observed disk-to-disk transfer rate using TCP/IP over a long-distance high-bandwidth network, with the aim of suggesting improvements to obtain more efficient end-to-end transfers.

13.3.4 MONITORING AT THE APPLICATION LEVEL

The ideal way to measure the network performance of an application is to incorporate network monitoring functionality into the design of the application. With careful design, one can determine and monitor the contributions from the application code, the network, and any other subsystems that may be involved. GridFTP is an example of an application that can record the transfer rates achieved by the user; and NetLogger [32] is an example of a toolkit and framework for monitoring application performance.

Of course, it is not always possible to extend network monitoring functionality into existing programs, nor is it practical to create the environment to run the real application in a network test configuration. In these cases, simple test programs that use and instrument the application-level protocol may be written, so that the effect of the network hardware or network protocol may be investigated.

Owing to its distributed nature, raw data collected in a Grid infrastructure must be analyzed and interpreted to compute the performance metrics in a reliable way. Two fundamental functions of data analysis are historic data summary retrieval and future prediction. These are often coupled with data visualization modules in the Grid system. Chapter 7 has further discussion of how monitoring mechanisms fit within the Grid architecture.

Grid network infrastructure is thought to be capable of continuously tracking the fulfillment of SLAs as well as the available network capacity (see Chapter 7). In turn, this information is communicated within other services in the Grid infrastructure (e.g., Globus MDS).

13.4 GENERAL CONSIDERATIONS ON AVAILABILITY

System *availability* is decided by its *reliability*, the continuous availability, and *maintainability*, the ability to recover the service after the service is down [33]. Therefore, Grid systems of high availability have to provide resource and service discovering capability along with fault tolerance and fault restoration capability. This relates back to the earlier sections on how to evaluate and report the behavior and effectiveness of the system and elements. On one hand, Grid service unavailability has a negative impact on system performance. On the other hand, any high-availability technique will incur extra resource consumption and management overhead. Reliability monitoring and data analysis are usually integrated in the performance monitoring and data analysis subsystem.

High-availability systems are achieved through fault-tolerant or survivable system design so that services can be restored to normal operations (with some probability of degraded performance) using redundant resource in face of component failures. A typical system of high performance and availability consists of *redundant components* (computers, bandwidths ...) and *intelligent fault-tolerant software* solutions providing system-level load balancing, fast fault detection, and recovery and restoration using the off-shelf hardware components. An efficient software solution not only provides fast fail-over capability to guarantee system performance, but also reduces the requirement on the reliability and redundancy of system components by optimally using the redundant resources in face of failure(s) or performance degradation. The topic of the following sections is the cost-efficient fault-tolerant software solutions for Grid networks with the focus on the general fault management architecture and methodologies.

A simple solution could be that the Grid middleware provides only a performance and reliability monitoring service and fault tolerance is explicitly achieved in the application layer by discovering and redirecting to redundant resource and services [34]. However, it is more desirable to provide transparent fault tolerance capability in the Grid middleware layer to reduce the application development difficulties and improve application performance.

As a Grid middleware managing distributed compute and network resources, a fault-tolerant Grid platform should have a complete spectrum of fault management functionalities, including performance and reliability monitoring, monitoring data analysis, fault detection, and fault recovery. In the past, monitoring and fault recovery of distributed computing systems were independent from the underlying network fault management [35]. They assume that the network links between client and servers as well as primary server and backup servers are always on. This assumption is apparently not true, especially for WANs, which may suffer many types of failures.

Furthermore, without explicit knowledge of network reliability and fault tolerance, a Grid platform will not be able to distinguish if a service failure is due to the node crash or network partition, which will reduce system performance and availability dramatically. Therefore, it is necessary to look at the available fault tolerance and fault management techniques in both distributed computing and networking contexts so that an integrated fault management component in the Grid platform can be designed. Based on these techniques, a Grid platform will be able to provide different types of fault tolerance services for applications. A Fault-Tolerant Grid platform (FT-Grid) with brief consideration on network availability is proposed in ref. 36.

13.5 FAULT DETECTION

Given the accrual of past and recent performance data, it is possible to tie fault detection to a drop in performance data (e.g., latency, throughput) of such magnitude and duration that it affects the component of a Grid infrastructure with the strictest performance requirements.

In addition, selected modules can elect to maintain liveness information (i.e., "are you there now?") in a peer-wise fashion. They typically do so by way of "heartbeat" messages. For the heartbeat handshake to be a trustworthy indication of well-being for the overall service, it is crucial that the heartbeat be generated as close as possible to the service logic and its context. Conversely, the use of a lower-level heartbeat (such as the "keep alive" function built into the TCP protocol) would not yield any valuable proof of an application making forward progress.

Lastly, it is worth recalling that fault detection is often, in fact, nothing more than a mere fault suspicion. Over an asynchronous network, for instance, it is impossible for a module to distinguish whether the network is experiencing packet loss or the peering module is being unusually slow. In fault detection, a false positive bears a negative impact on the progress of the overall system, with fault detections reciprocated and resources unduly partitioned.

Grid systems are typical asynchronous redundant distributed systems in which it is impossible to obtain consensus because accurate failure detection is impossible. For such systems, unreliable fault detection proves an efficient solution without timing synchronization assumption. System-level fault detection always incurs extra management overhead (periodic fault detection message exchange and processing). It is very important to design a fault detection scheme with good trade-off between management overhead, detection accuracy, and timeliness. Scalability and flexibility are also major design concerns for large-scale Grid systems.

Unreliable fault detection in Globus toolkit used heartbeat to monitor the health status of Grid processes and hosts [37]. It is made up of a local monitor for each host/process generating the periodic heartbeat and a data collector. It supports a range of application-specific fault detection services, i.e., applications have control over which entities are monitored, and how often, fault criteria, and fault report policies.

A network consists of a number of nodes (hosts, routers, etc.) interconnected by (physical or logical) links. The nodes usually have self-fault detection capability and their status information can be queried via the SNMP tools. Today's Internet consists of multiple layers in which the network topology and link have different meaning and features in different layers. In the IP layer, link failure detection is done by periodic Hello message exchange between peer routers via the Interior Gateway Protocols (IGPs) such as OSPF or IS-IS.

It is proposed to speed up Hello exchange to be in the subsecond range to accelerate the IGP (or BGP) convergence (see Chapter 10). To avoid the increasing route flaps, it is proposed that bad news should travel fast but good news should travel slowly, i.e., a new path should be quickly calculated when a link goes down, but not as fast as when a link comes up [38]. The SONET/SDH layer has a built-in fast fault detection mechanism (loss of signal, loss of frame, alarm indication signal, etc.). The newly developed (backbone) Generalized MultiProtocol Label Switching (GMPLS) network protocol suites contain a Link Management Protocol (LMP) [39] that provides control channel management (data plane Traffic Engineering (TE)), link property correlation, link connectivity verification, and fault management for data plane channels (links) via exchange channel status messages.

Clearly, a scalable distributed fault detection service hierarchy can be drawn from the above discussion to integrate computer/process and network fault detection.

13.6 RECOVERY AND RESTORATION

Temporal redundancy mechanisms, like the Automatic Repeat reQuest (ARQ) algorithm in TCP sliding window transmission control, is preferable when dealing with transient failures. However, for permanent failures such as fiber cut and node crash,

13.6 Recovery and Restoration

spatial redundancy with replicated system resource or service connections is essential [40]. A permanent failure is usually called a fault because it results in total service losses (zero performance) if no fault tolerance and recovery are in place.

A network that is able to deal with permanent failures is a survivable network. A basic requirement of a survivable network is that its topology must be at least two-connected, i.e., there are at least two disjoint paths between any pair of nodes so that a backup path can be used for any end-to-end connection when its primary path is broken due to either a node failure or a link failure. In reality, a single-link failure is regarded as the major network failure mode because the probability of multiple-link failure is very low and node (routers and end hosts) failures are usually dealt with the local redundant hardware.

The techniques used for network survivability design can be broadly classified into two categories: *preplanned protection* and *dynamic restoration*. Compared with protection, restoration has more efficient resource utilization, but typically takes more time. Another advantage of dynamic restoration is the better scalability in term of the fault management overheads, as backup connections for the disrupted services need only to be discovered and maintained after the network failure(s).

However, dynamic restoration requires longer restoration time and does not guarantee the re-establishment of the disrupted services, since the backup resources may not be available at the time when a failure happens. Therefore, protection is often used for the premier long-term service connections in circuit-switched or connection-oriented data networks. Existing restoration mechanisms focus on the recovery of disrupted existing connections in the circuit-switched or connection-oriented data networks. For the connectionless networks, such as IP networks, the restoration is achieved by dynamic routing through global routing table update upon a network failure, where all new incoming connections whose initial routes traverse the failure will be dropped before this slow restoration process (up to minutes in the worst case) is completed.

While survivability is not only desirable but required in order to support dependable applications and services, there are certain costs associated with designing a survivable network. These costs reflect the additional network resources that have to be provisioned to restore any connections disrupted by a failure, including hardware resources (fibers, wavelengths, switches, transceivers, etc.) as well as software resources (e.g., diverse routing algorithms, restoration and protection switching protocols, etc.).

An efficient network survivability scheme should be designed with following objectives:

(1) The restoration process is fast such that interrupted services can be restored in a short time.
(2) The overall network performance degradation during the failure and restoration is as low as possible due to the reduced network capacity.
(3) The fault management overhead is low in terms of extra control message exchange and processing.

It should be noted that network intervention (whether it is protection or restoration) is not the sole asset with which to recover from failures. In Grids, service migration is another important dimension to recovery. Researchers [41] have explored the coupling of service migration and GMPLS path restoration.

13.6.1 PROTECTION FOR CIRCUIT SWITCHED NETWORKS

The protection techniques can be further classified as path based or link based. In path-based schemes, upon the occurrence of a failure, a fault notification is sent to the source (and destination) of each affected connection; each such source–destination pair then takes the necessary actions to switch its connection from the primary to the backup path. In link-based schemes, on the other hand, the nodes at the two ends of a failed link are responsible for rerouting all affected traffic around the failure; the source and destination of an affected connection need not become aware of the failure at all. Usually, path-based schemes require less spare capacity than link-based ones. However, path-based recovery may take longer to restore service since it requires the source and destination of a connection to take action, whereas, with link-based schemes, the nodes closest to the failure are the ones involved in the recovery process. Path-based protection is the preferable technique for circuit-switched networks, such as MPLS and optical networks [42]. Readers are also referred to ref. 40 for more information.

Regardless of the path-based protection scheme used (dedicated or shared path), in order for the network to be able to recover from any single-link failure, each connection must be assigned both a primary and a backup path, and these paths must not share a common link. Therefore, the protected connection routing problem is equivalent to finding two link-disjoint paths in the network topology for each connection. Here we look at two major protection schemes for protecting connections (lightpaths) in wavelength-routed WDM networks from any single-link failure. Assuming no wavelength conversion, the objective is to minimize the network resource (wavelength) required to provision a set of lightpath requests [43].

A. *Dedicated Path Protection (DPP)*. This is a 1+1 path protection scheme in which there is no wavelength sharing among backup paths and a service demand (lightpath) is transported along the primary and backup paths simultaneously, and the receiver selects the best signal. Therefore, each lightpath (primary and backup) will be assigned an individual wavelength.

B. *Shared Path Protection (SPP)*. This protection scheme allows a group of backup paths with common links to share a wavelength whenever (1) their corresponding primary lightpaths do not share a common link; or (2) the backup lightpaths do not share a common link with any primary lightpath. To survive from a single-link failure, all the primary paths that share the same spare capacity must be mutually link-disjoint. Obviously, shared path protection requires less spare capacity than the dedicated protection scheme. After the routes are found for a set of lightpaths (primary and backup), the wavelength assignment can be modeled as the NP-hard vertex coloring problem and efficient heuristic algorithms can used to find an approximate solutions. In this study two possible approaches are considered to solve the routing and wavelength assignment problem for primary and backup paths in shared path protection. The main

13.6 Recovery and Restoration

difference in the two approaches is in whether the backup paths are considered jointly with, or separately from, the primary paths when assigning wavelengths.

B.1 *Shared Path Protection with Joint Wavelength Assignment (SPP-JWA)*. In this approach, we first determine the primary and backup paths for each lightpath, by finding a pair of link-disjoint paths. We then assign the wavelength to all the lightpaths altogether according to the two sharing conditions.

B.2 *Shared Path Protection with Separate Wavelength Assignment (SPP-SWA)*. In this approach, after obtaining a pair of link-disjoint primary and backup paths for each lightpath in the request set, we first assign the wavelengths for all the primary lightpaths; a group of primary lightpaths can use a common wavelength as long as they do not have common link on their routes. We then assign wavelengths to a group of backup lightpaths according to the two sharing conditions.

Figure 13.3 demonstrates a simulation result on the wavelength consumption under the above three protection schemes and no protection provided. The simulated network is a backbone IP/WDM network in which the number of OXCs (M) is proportional to the number of IP routers (N) evenly attached on OXCs ($N/M = 20$) and each router is equipped with 4 transceivers. We consider the scenario where every transmitter is request to set up a lightpath with an arbitrary router. The result clearly shows that the shared protection scheme is very cost-efficient in terms of wavelength consumption, while it provides 100% survivability against any single link failure.

Generally, shared protection requires a sophisticated protection switching scheme to configure the backup path for a particular service path upon a failure, whereas

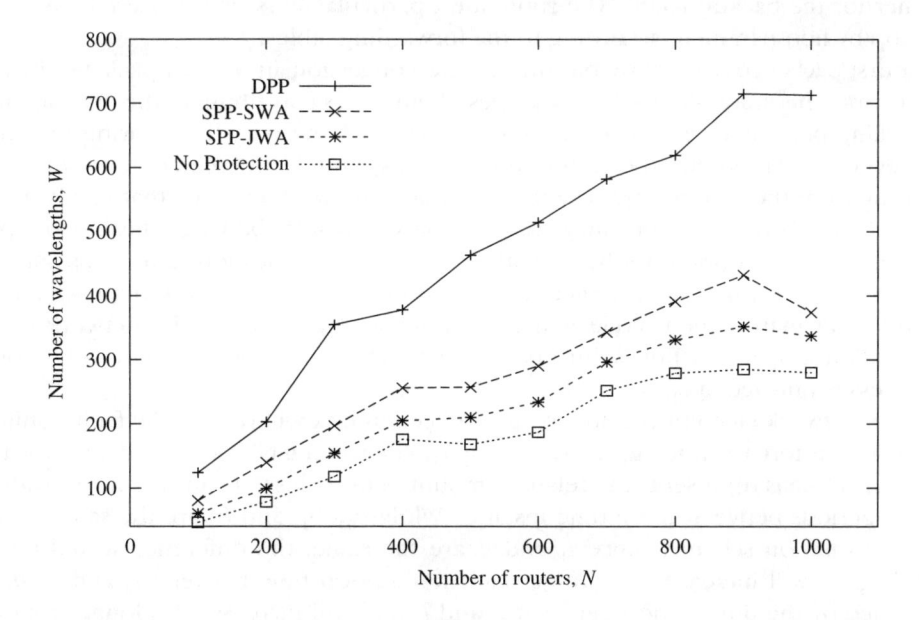

Figure 13.3. Efficiency of protection schemes.

dedicated protection is simpler and faster since the only action required when a failure occurs is for the receiver to switch to the backup path. Because of the big saving on network resource, shared protection is preferred for circuit switched networks. Because of the differentiated performance and management overhead of different protection schemes, QoS based protection mechanism can be deployed [44].

13.6.2 RESTORATION FOR BURST/PACKET-SWITCHED NETWORKS

The dominating packet-switched network is the IP network, in which every IP router maintains a local forwarding table. The entries in the forwarding table contain the next hop information for packets per destination and per Forward Equivalence Class FEC (forward equivalence class). IP routers forward the incoming burst control packets and set up the connections by looking up the next-hop information in their forwarding tables. An overview of IP network fault tolerance can be found in ref. 45.

With recent development in optical network technologies, Optical Burst Switching (OBS) and Optical Packet Switching (OPS) have emerged as promising architectures [46]. A complete introduction to OBS and OPS networks can be found in Chapter 12. In such a network, optical burst/packet routing and forwarding are accomplished in a similar way as in the IP network, e.g., OSPF protocol can be used to calculate the forwarding table. In the following, we introduce a fault management architecture with fast restoration techniques for OBS networks from refs 47 and 48.

In this architecture, the routing protocol uses a two-shortest-disjoint-path algorithm to support alternative routing. The resulting forwarding table at an OBS node contains two next-hop entries per destination, one for the primary route and the other for the backup route. The route for a particular burst is discovered based on a hop-by-hop paradigm according to the forwarding tables.

Burst/packet loss is inevitable due to the congestion in a burst/packet-switched network. This study shows that two types of burst loss contribute to the overall burst blocking performance: (1) *restoration loss*, which is the burst loss during the fault detection, localization, and notification periods; and (2) increased *congestion loss* arising from the diverted traffic and the reduced network capacity that results from the failure. However, there may exist complex trade-offs between these two types of loss. A fast restoration scheme with shorter fault notification time (thus smaller restoration loss) may incur larger congestion loss, and vice versa. Burst loss during a full fault management cycle is demonstrated in Figure 13.4. The network starts operation at t_0 and a link failure occurs at t_1. The fault detection and localization process is finished at t_2.

The network element conducting the deflection task will receive the fault notification and restore the interrupted services at t_3. The routing tables get updated globally at t_4. The y-axis represents the relative amount of burst loss, b_i, during the individual time periods between these time instants. While b_0, b_1, and b_4 are the same for all the restoration schemes since t_1 and t_2 are the same, the difference would be b_2 and b_3. b_2 will increase with longer fault notification time between t_2 and t_3. b_3 is decided by the time period between t_1 and t_4 and will increase with longer routing table update time (t_4). Furthermore, during the period between t_3 and t_4, network

13.6 Recovery and Restoration

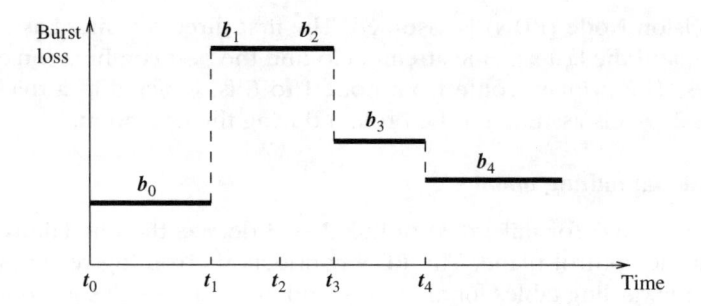

Figure 13.4. Burst losses during a link failure.

congestion could increase when extra traffic is deflected to the alternative route(s). Therefore, the total burst loss during the time period between t_2 and t_4 is determined decided by the fault notification time and the network congestion condition.

While the fault detection and localization time may normally be a constant, the fault notification time makes the major difference among different restoration schemes. Generally, it is preferable that the node that makes the rerouting decision be close to the faulty link so that the fault notification time is reduced. This also leads to light fault management overhead as only a small amount of fault notification message exchange is needed. In the following, we demonstrate several fast restoration schemes with an example network (Figure 13.5), in which a centralized routing computation entity

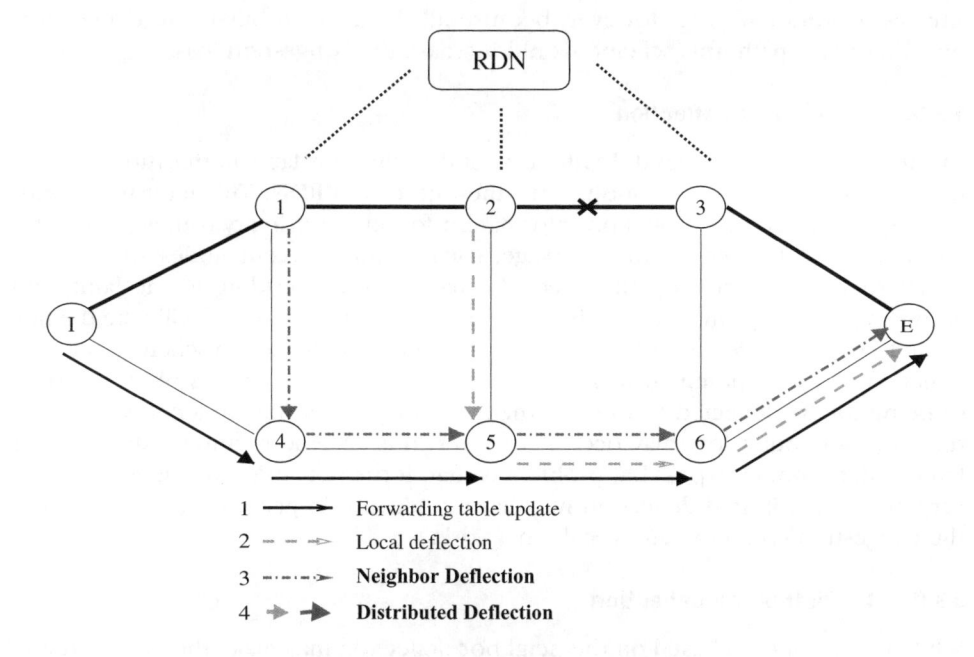

Figure 13.5. Fast restoration schemes.

Routing Decision Node (RDN) is assumed. The first three schemes basic standalone mechanisms, and the last scheme attempts to find the best combination of the above two schemes. The primary route from node I to E is depicted as a thick black line and the link 2 → 3 is assumed to be broken during the operation.

13.6.2.1 Global routing update

When the headend 2 (or tailend 3) of link 2 → 3 detects the link failure, it informs the RDN via the control plane. The RDN conducts the routing recomputation and updates the forwarding tables for all nodes, and new bursts will subsequently follow the new routes. For example, new bursts will follow the route I → 4 → 5 → 6 E. This solution is optimal and the existing routing protocol can handle it well. However, global routing table updating is a slow process (in seconds or even minutes) due to the long round-trip time for the signal transmission and processing between the OBS nodes and the routing entity. As a result, a large amount of bursts will be lost before the forwarding tables are updated.

13.6.2.2 Local deflection

This is similar to the traditional deflection routing usually seen in a congestion resolution scheme. When the headend 2 detects the link failure, it will automatically pick up the alternative next hop in the forwarding table for every new burst whose next hop on its primary route passes the faulty link. In the example, new bursts from I to E will follow the alternative route I → 1 → 2 → 5 → 6 → E. This would be the fastest restoration scheme since new bursts will be deflected to an alternative good link right after the link failure is detected locally. Therefore, it will incur the smallest restoration loss. However, because all the affected bursts are deflected to one alternative path, this scheme would increase the congestion loss.

13.6.2.3 Neighbor deflection

In this scheme, the headend 2 will also send a different fault notification message to all its adjacent nodes in addition to the one to the RDN. This fault notification message contains the destination information for all the primary routes passing the faulty link. After receiving this message, each of the adjacent nodes will pick up an alternative next hop for the affected bursts that are heading to the faulty link according to their primary route. In the example, bursts from I to E will take the new route I → 1 → 4 → 5 → 6 → E. Compared with the local deflection scheme, neighbor deflection has the potential to make the rerouted traffic more distributed instead of being totally deflected to one alternative path. In this way, less congestion and therefore less burst loss may occur. However, this scheme requires extra one-hop fault notification. One possible problem is that, if the network traffic load is already very heavy, distributed deflection may have a negative impact as it may deteriorate the congestion condition all over the network.

13.6.2.4 Distributed deflection

While the restoration based on the neighbor deflection may make the deflected load more balanced, it suffers from an extra one-hop fault notification delay that will result

13.6 Recovery and Restoration

in greater overall burst loss because of bigger restoration loss. Therefore, a more efficient algorithm is a combination of local deflection and neighbor deflection, i.e., the affected bursts are deflected locally until the adjacent nodes receive the fault notification. At that time the affected bursts will be deflected in a distributive way. We name this scheme distributed deflection.

13.6.2.5 α Distributed deflection

One interesting observation from scheme 2 is that the capacity of the links between the headend (node 2) of the faulty link and its adjacent nodes (node 1) will not be utilized if affected bursts are deflected at adjacent nodes. Therefore, we define a *distribution ratio*, α, to determine the portion of affected bursts that will be deflected at the adjacent nodes. That is, after the adjacent nodes receive the fault notification, α affected bursts will be deflected distributively and $(1-\alpha)$ affected bursts will be forwarded to the headend node of the faulty link to be deflected locally.

With a different value of $\alpha \in [0,1]$, we have a different variance of the distributed restoration scheme. When $\alpha = 0$, it is equivalent to scheme 1, local deflection-based restoration. When $\alpha = 1$, it becomes scheme 3, the distributed deflection-based restoration. We use α distributed deflection to denote the generalized distributed deflection mechanism. We also note that using α introduces only a tiny amount of management complexity in the adjacent nodes. We expect that there exists an optimal value of α that makes the affected bursts deflected in a most balanced way such that the minimum burst loss can be achieved.

13.6.2.6 Class-based QoS restoration scheme

Here are defined three restoration classes with different priority values for all bursts. When a burst is generated at the edge node, it will be assigned a priority value in its control packet according to its restoration QoS requirement. Bursts in the highest class will pick up best from the local and neighbor deflection restoration schemes during different restoration periods. The local deflection will be chosen during the fault notification period because of its shorter fault notification time. And the local and neighbor deflection scheme with shorter alternative route length (number of hops in this paper) during the deflection period will be chosen because of its lower backup capacity requirement (and possible lower average burst loss probability). Bursts in the middle class will be restored via neighbor deflection, and the bursts in the lowest class will be dropped until the global forwarding table update is finished. In this way, the lower class bursts do not compete for the bandwidth with the higher class bursts in the deflected alternative routes

Figure 13.6 depicts the simulation results on a NSF topology. The y-axis represents the overall burst loss probability under medium traffic during the three network operational phases around a link failure. In the x-axis, 1 represents the period before the link failure, 2 represents the period before a global forwarding table update and only the proposed fast restoration schemes are in place, and 3 represents the period after the global forwarding table update. We observe that the burst loss probability is very low for both phases 1 and 3, though it is actually a little bit higher in phase 3 due to the reduction in the network capacity.

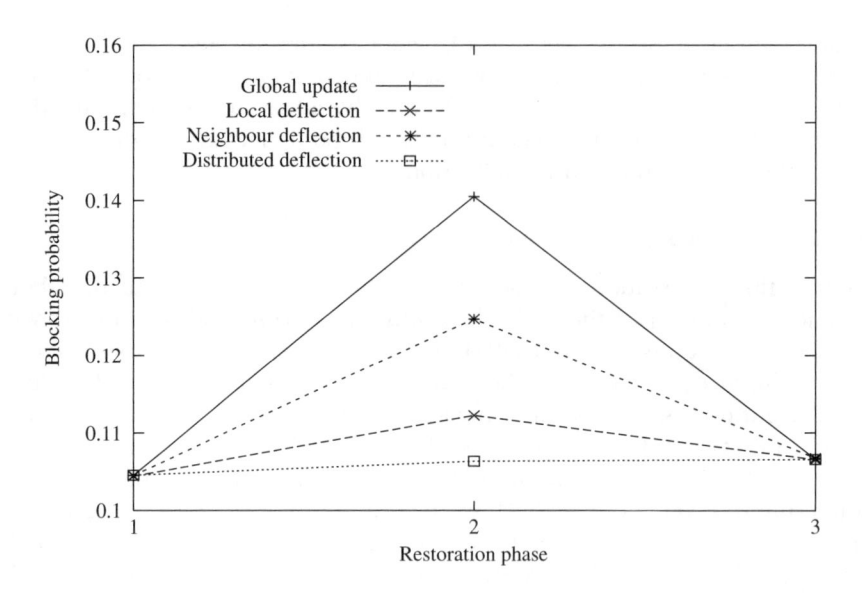

Figure 13.6. Restoration performance.

However, the loss probability could increase significantly in phase 2. Relying only on the forwarding table update would incur very high burst loss in this phase. Nonetheless, the loss probability increases only moderately when the proposed fast restoration schemes are used in this phase. Among the four restoration schemes, distributed deflection shows the best performance (almost no extra burst loss) followed by local deflection and neighbor deflection. Global routing update incurs the highest burst loss. Specifically, the improvements from using the three fast restoration schemes over the global forwarding table update are 24.3%, 20.1%, and 10.5%, respectively.

The α distributed deflection and class-based QoS restoration schemes are built based on the performance differentiation of above four basic restoration schemes, therefore they can be directly built into the fault management module to provide differentiated restoration service. We note that the above schemes are studied for OBS networks, but they could be directly extended to other packet-switched networks.

13.7 INTEGRATED FAULT MANAGEMENT

Except for the unreliable fault detector, the current Globus toolkit does not provide other fault tolerant mechanisms. There exist a number of different fault handling mechanisms for Grid applications to respond to system/component failures [37]:

- fail-stop: stop the application;
- ignore the failure: continue the application execution;
- fail-over: assign the application to new resources and restart;
- migration: replication and reliable group communication to continue the execution.

Fail-over and migration are the two choices for fault-tolerant computing and can be achieved either in a deterministic way, in which the backup compute resource and network route are precomputed along with the primary, or in a dynamic way, in which the backup is dynamically discovered after the failure occurrence and detection. We note that this could be coupled with the connection protection and restoration mechanism in the network context discussed in Section 13.6.

At the system level, performance monitoring, data analysis and fault detection, fault recovery will form a feedback loop to guarantee the system performance and availability. The complexity comes from the fact that performance/reliability data can be collected in different timescales from different layers of the Grid networks (layer 1/2/3 and application). Therefore, a Grid platform needs to provide different levels of fault tolerance capability in terms of timescale and granularity. A large portion of the fault management activities are conducted within the Grid infrastructure and transparent to the end-user applications.

From the point view of applications, we identify the following failure scenarios and fault handling mechanisms to explicitly integrate the Grid service migration with the network protection or restoration:

- *Fail-over or migrate within the same host(s)*. The application requires dedicated or shared backup connections for the primary connections (either unicast or multicast). The applications can also specify the network restoration time requirement.
- *Fail-over or mitigate to different host(s)*. This mechanism is more generic in the sense that the application requires redundant Grid resource/service. In the case that the primary Grid resource fails, the Grid middleware can reset the connections to the backup Grid resource/service and jobs can be migrated directly from the failed primary resource to the backups.

Further studies are under way for the aforementioned cases in terms of dynamic connections setup and network resource allocation.

REFERENCES

[1] F. Cappello (2005) "Fault Tolerance in Grid and Grid 5000," IFIP Workshop on Grid Computing and Dependability, Hakone, Japan, July 2005.
[2] K. Birman, "Like it or Not, Web Services are Distributed Objects!", white paper.
[3] V. Paxson, G. Almes, J. Mahdavi, and M. Mathis (1998) "Framework for IP Performance Metrics." RFC 2330, May 1998.
[4] R. Hughes-Jones, P. Clarke, and S. Dallison (2005) "Performance of Gigabit and 10 Gigabit Ethernet NICs with Server Quality Motherboards," grid edition of *Future Generation Computer Systems*, 21, 469–488.
[5] M. Mathis and M. Allman (2001) "A Framework for Defining Empirical Bulk Transfer Capacity Metrics," RFC 3148, July 2001.
[6] Source code and documentation of iperf is available at http://dast.nlanr.net/Projects/Iperf/.
[7] G. Almes, S. Kalidindi, and M. Zekauskas (1999) "A One-way Delay Metric for IPPM." RFC 2679, September 1999.
[8] G. Almes, S. Kalidindi, and M. Zekauskas (1999) "A Round-trip Delay Metric for IPPM." RFC2681, September 1999.

[9] T. Ferrari and F. Giacomini (2004) "Network Monitoring for Grid Performance Optimization." *Computer Communications*, 27, 1357–1363.

[10] S. Andreozzi, D. Antoniades, A. Ciuffoletti, A. Ghiselli, E.P. Markatos, M. Polychronakis, and P. Trimintzios (2005) "Issues about the Integration of Passive and Active Monitoring for Grid Networks," Proceedings of the CoreGRID Integration Workshop, November 2005.

[11] Source code and documentation of cricket is available at http://people.ee.ethz.ch/~oetiker/webtools/mrtg/.

[12] Multi Router Traffic Grapher software and documentation, http://people.ee.ethz.ch/~oetiker/webtools/mrtg/.

[13] "Grid network Monitoring: Demonstration of Enhanced Monitoring Tools," Deliverable D7.2, EU Datagrid Document: WP7-D7.2–0110-4-1, January 2002.

[14] "Final Report On Network Infrastructure And Services", Deliverable D7.4, EU Datagrid Document Datagrid-07-D7-4-0206-2.0.doc, January 26, 2004.

[15] EGGE JRA4: "Development of Network Services, Network Performance Monitoring – e2emonit," http://marianne.in2p3.fr/egee/network/download.shtml.

[16] Internet2 end-to-end performance initiative. http://e2epi.internet2.edu/web traceroute. Home page for web based traceroute http://www.traceroute.org/.

[17] A large collection of traceroute, looking glass, route servers and BGP links traceroute may be found at http://www.traceroute.org/.

[18] J. Boote, R. Carlson, and I. Kim. NDT source code and documentation available at http://sourceforge.net/projects/ndt.

[19] R.L. Cottrell and Connie Logg (2002) "A new high performance network and application monitoring infrastructure." Technical Report SLAC-PUB-9202, SLAC.

[20] IEPM monitoring page, http://www.slac.stanford.edu/comp/net/iepm-bw.slac. stanford.edu/slac_wan_bw_tests.html.

[21] Source code and documentation of thrulay is available at http://people.internet2.edu/~shalunov/thrulay/.

[22] C. Dovrolis, P. Ramanathan, and D. Moore (2001) "What do Packet Dispersion Techniques Measure?," IEEE INFOCOM.

[23] Source code and documentation of pathload is available at http://www.cc.gatech.edu/fac/Constantinos.Dovrolis/pathload.html.

[24] Source code and documentation of pathchirp is available at http://www.spin.rice.edu/Software/pathChirp/.

[25] Source code and documentation of bbftp is available at http://doc.in2p3.fr/bbftp/.

[26] A. Hanushevsky, source code and documentation of bbcp is available at http://www.slac.stanford.edu/~abh/bbcp/.

[27] L. Cottrell (2002) "IEPM-BW a New Network/Application Throughput Performance Measurement Infrastructure," Network Measurements Working Group GGF Toronto, February 2002, www.slac.stanford.edu/grp/scs/net/talk/ggf-feb02.html.

[28] R. Hughes-Jones, source code and documentation of UDPmon, a tool for investigating network performance, is available at www.hep.man.ac.uk/~rich/net.

[29] Endace Measurement Systems, http://www.endace.com.

[30] J. Hall, I. Pratt, and I. Leslie (2001) "Non-Intrusive Estimation of Web Server Delays," Proceedings of IEEE LCN2001, pp. 215–224.

[31] Web100 Project team. Web100 project, http://www.web100.org.

[32] B. Tierney, NetLogger source code and documentation available at http://www-didc.lbl.gov/NetLogger/.

[33] "Telecommunications: Glossary of Telecommunication Terms", Federal Standard 1037C, August 7, 1996.

[34] X. Zhang, D. Zagorodnov, M. Hiltunen, K. Marzullo, and R.D. Schlichting (2004) "Fault-tolerant Grid Services Using Primary-Backup: Feasibility and Performance," Proceedings of the IEEE International Conference on Cluster Computing (CLUSTER), San Diego, CA, USA, September 2004.

[35] N. Aghdaie and Y. Tamir (2002) "Implementation and Evaluation of Transparent Fault-tolerant Web Service with Kernel-level Support," Proceedings of the IEEE Conference on Computer Communications and Networks, October 2002, Miami, FL, USA.

[36] H. Jin, D. Zou, H. Chen, J. Sun, and S. Wu (2003) "Fault-tolerant Grid Architecture and Practice," *Journal of Computing Science and Technology*, 18, 423–433.

[37] P. Stelling, C. DeMatteis, I. Foster, C. Kesselman, C. Lee, and G. von Laszewski (1999) "A Fault Detection Service for Wide Area Distributed Computations," *Cluster Computing*, 2(2), 117–128.

[38] O. Bonaventure, C. Filsfils, and P. Francois (2005) "Achieving Sub-50 Milliseconds Recovery upon BGP Peering Link Failures," Co-Next 2005, Toulouse, France, October 2005.

[39] J. Lang, "Link Management Protocol (LMP)," IETF RFC 4204.

[40] M. Médard and S. Lumetta (2003) "Network Reliability and Fault Tolerance", *Wiley Encyclopedia of Engineering* (ed. by J.G. Proakis), John & Wiley & Sons Ltd.

[41] L. Valcarenghi (2004) "On the Advantages of Integrating Service Migration and GMPLS Path Restoration for Recovering Grid Service Connectivity", 1st International Workshop on Networks for Grid Applications (gridNets 2004), San Jose, CA, October 29, 2004.

[42] D. Papadimitriou (ed.) "Analysis of Generalized Multi-Protocol Label Switching (GMPLS) based Recovery Mechanisms (including Protection and Restoration)," Draft-ietf-ccamp-gmpls-recovery-analysis-05.

[43] Y. Xin and G.N. Rouskas (2004) "A Study of Path Protection in Large-Scale Optical Networks," *Photonic Network Communications*, 7(3), 267–278.

[44] J. L Marzo, E. Calle, C. Scoglio, and T. Anjali (2003) "Adding QoS Protection in Order to Enhance MPLS QoS Routing," Proceedings of IEEE ICC '03.

[45] S. Rai and B. Mukherjee (2005) "IP Resilience within an Autonomous System: Current Approaches, Challenges, and Future Directions", *IEEE Communications Magazine*, 43(10), 142–149.

[46] L. Xu, H.G. Perros, and G.N. Rouskas (2001) "A Survey of Optical Packet Switching and Optical Burst Switching", *IEEE Communications Magazine*, 39(1), 136–142.

[47] Y. Xin, J. Teng, G. Karmous-Edwards, G.N. Rouskas, and D. Stevenson, "Fault Management with Fast Restoration for Optical Burst Switched Networks," Proceedings of Broadnets 2004, San Jose, CA, USA, October 2004.

[48] Y. Xin, J. Teng, G. Karmous-Edwards, G.N. Rouskas, and D. Stevenson (2004) "A Novel Fast Restoration Mechanism for Optical Burst Switched Networks", Proceedings of the Third Workshop on Optical Burst Switching, San Jose, CA, USA, October 2004.

Chapter 14

Grid Network Services Infrastructure

Cees de Laat, Freek Dijkstra, and Joe Mambretti

14.1 INTRODUCTION

Previous chapters have presented a number of topics related to Grid network services, architecture, protocols, and technologies, including layer-based services, such as layer 4 transport, layer 3 IP, Ethernet, wide-area optical channels, and lightpath services. This chapter examines new designs for large-scale distributed facilities that are being created to support types of those Grid network services. These facilities are being designed in response to new services requirements, the evolution of network architecture standards, recent technology innovations, a need for enhanced management techniques, and changing infrastructure economics.

Currently, many communities are examining requirements and designs for next-generation networks, including research communities, standards associations, technology developers, and international networking consortia. Some of these new designs are implemented in early prototypes, including in next-generation open communication exchanges. This chapter describes the general concept of such an exchange, and presents the services and functions of one that would be based on a foundation of optical services.

14.2 CREATING NEXT-GENERATION NETWORK SERVICES AND INFRASTRUCTURE

If current communications infrastructure did not exist and an initiative was established to create a new infrastructure specifically to support Grid network services, a fundamentally different approach would be used. This new infrastructure would not resemble today's legacy communications architecture and implementations. The new communications infrastructure would be easily scalable and it would incorporate many options for efficiently and flexibly providing multiple, flexible high-quality services. It would not be a facility for providing only predetermined services, but would provide capabilities for the ad hoc creation of services. It would incorporate capabilities for self-configuration and self-organization. The design of such a communications infrastructure would be derived from many of the key concepts inherent in Internet architecture. For example, it would be highly distributed.

Even without the Grid as a driving force to motivate the creation of communications infrastructure with these characteristics, architectural designs based on these principles are motivated by other dynamics. Major changes are required in the methods by which communications services and infrastructure are designed and provisioned. These changes are motivated by the explosive growth in new data services, by the additional communities adopting data services, and by the number and type of communication-enabled devices. Soon services will be required for 3 billion mobile general communication devices.

Today, there is a growing need to provide seamless, ubiquitous access, at any time from any location and any device. In addition to general communication devices, many more billions of special-purpose communications-enabled devices are anticipated: sensors, Radio Frequency Identification (RFID) tags, smart dust, nano-embedded materials, and others. All of these requirements are motivating a transformation of the design of core network infrastructure. The legacy architectural models for services design and provisioning cannot meet the needs of these new requirements. A fundamentally new approach is required. The guiding principles for that new approach will be based on those that informed Internet architecture. The next section presents some of these principles, which are key enablers for next-generation communication services and facilities.

14.2.1 END-TO-END PRINCIPLE

In previous chapters, the important of the end-to-end principle in distributed infrastructure design was discussed. Its basic premise is that the core of such distributed infrastructure, including networks, should be kept as simple as possible, and the intelligence of the infrastructure should be placed at the edge. This principle is now almost universally accepted by all networking standards bodies. Recently, the ITU formally endorsed the basic principles of the Internet design through its Study Group 13 recommendation for a Next-Generation Network (NGN) [1]. Balancing the need to maintain the end-to-end principle and to create new types of "intelligent" core facilities is a key challenge.

14.2.2 PACKET-BASED DATA UNITS

After 35 years of increasing Internet success, communications standards organizations have now almost universally endorsed a architecture founded on the communication of packet-based data. The IETF has demonstrated the utility of this approach. The IEEE has established initiatives to create layer 2 transport enhancements to better support packet-based services. The ITU-T NGN recommendation specifies "A packet-based network able to provide telecommunication services and able to make use of multiple broadband, QoS-enabled transport capabilities." The NGN Release 1 framework indicated that all near-term future services are to be based on IP, regardless of underlying transport.

14.2.3 ENHANCED FUNCTIONAL ABSTRACTION

Earlier chapters have stressed the importance of abstracting capabilities from infrastructure. The design of the Internet has been based on this principle. Attaining this goal is being assisted by the general recognition that legacy vertical infrastructure stacks, each supporting a separate service, must be replaced by an infrastructure that is capable of supporting multiple services. This design direction implies a fundamentally new consideration of basic architectural principals. The ITU-T has begun to formally adopt these concepts. One indication of this direction is inherent in the ITU-T NGN recommendation that "service-related functions" should be "independent from underlying transport related technologies." These recommendations indicates that the design should enable "unfettered access for users to networks," through open interfaces. [1]

Also, the ITU-T is beginning to move toward a more flexible architectural model than the one set forth in the classic 7 layer Open Systems Interconnect (OSI) basic reference model [2]. This trend is reflected in the ITU-T recommendations for the functional models for NGN services [3]. This document notes that the standard model may not be carried forward into future designs. The number of layers may not be the same, the functions of the layers may not be the same, standard attributes may not be the same, future protocols may not be defined by X.200, e.g., IP, and understandings of adherence to standards may not be the same.

14.2.4 SELF-ORGANIZATION

Another important architectural design principle for any future network is self-organization.

The Internet has been described as a self-organizing network. This description has been used to define its architecture in general, because many of its individual protocols are designed to adapt automatically to changing network conditions without having to rely on external processes. This term has also been used to describe the behavior of individual protocols and functions, such as those that automatically adjust to changing conditions. Because the Internet was designed for an unreliable infrastructure, it can, therefore, tolerate uncertain conditions better than networks that depend on a highly reliable infrastructure.

The principle of self-organization for networks is particularly important for scalability in a ubiquitous communication infrastructure containing an extremely large number of edge devices, especially mobile edge devices. The only method for ensuring reliable services in this type of global distributed environment is through an architecture that enables edge devices to be self-sufficient, to be context aware, self-configuring, adjustable to fault conditions and highly tolerant of instability. These attributes enable individual services, objects, and processes to automatically adjust to changing circumstances. Recently, progress has been made in identifying fundamental issues in requirements and potential solutions for autoconfiguration for components in large-scale dynamic IP networks [4].

Another motivation for developing architecture for self organizing networks, such as ad hoc networks, is to enable communication services in areas with minimal infrastructure or unstable infrastructure.

14.2.5 DECENTRALIZATION

A complementary principle to self-organization is decentralization. Another major attribute of the Internet is that it is highly decentralized, which is one reason that it is a highly scalable, cost-effective network. With continued rapid growth in communication services, service edge capabilities, traffic volume, individual traffic streams, and communications devices, especially mobile devices, it is important to endorse decentralization as a key principle for architectural designs. Centralized processes become constraining factors in large-scale distributed environments.

14.2.6 DISTRIBUTED SERVICE CREATION

Another key principle for next generation communications architecture is ensuring capabilities for distributed service creation. Traditionally, service providers have carefully defined services, and have then designed the technology and infrastructure to support those services, with an understanding that they will remain in place for long periods of time. However, in an environment of rapid change, in which new services can be conceptualized continually, this approach is no longer feasible. The architecture must anticipate and support continuous changes in services.

Also, the architecture should be designed to provide support for distributed service creation. It should provide the tools and functionality for services creation to broad communities and allows those communities to create, manage, and change their own services.

14.3 LARGE-SCALE DISTRIBUTED FACILITIES

These design principles are inherent in the basic architecture of the Internet, and they are reflected in Grid design principles. Currently, these principles are used to design new types of large-scale distributed infrastructure to provide support for Grid services. The communications infrastructure for these environments does not resemble a traditional network. Within this environment, it is possible either to accept predefined default services or to create required services through dynamic, ad hoc processes. These environments are used to develop and implement customized Grid services and specialized communication services. For example, in some Grid environments, communications services are used as a large-scale distributed backbone to support specialized Grid processes.

This architectural approach provides highly distributed communities with direct access to resources, toolsets, and other components, including complete resource management environments, that allow them to create new services and to change existing services. Network services, core components, resources such as lightpaths, layer 2 channels, cross-connections, and even individual elements can be identified and partitioned into separate domains and provided with secure access mechanisms that enable organizations, individuals, communities, or applications to discover, interlink and utilize these resources. Accessible resources can even include basic optical components, which can be managed and controlled by highly distributed edge processes.

14.4 DESIGNS FOR AN OPEN SERVICES COMMUNICATIONS EXCHANGE

The architecture described here is implemented within many Grid environments. As these environments were developed, it was recognized that an important component within this type of large scale distributed facility is an open, neutral communications exchange point. The design of such exchanges constitutes a significant departure from that of traditional carrier communications exchanges.

Today, almost all data networks exchange information at Internet exchange facilities. Almost all of these facilities have an extremely limited range of services, and limited peering capabilities – primarily they provide capabilities for provider peering at layer 3 and, for some under restrictive policies, layer 2. Also, traditional exchanges are oriented almost exclusively to provider-to-provider traffic exchanges.

The international advanced networking research community is developing new designs for data communication exchange points that provide for many more types of services, including Grid services and layer 1 peering capabilities. These designs are currently being developed and implemented in prototype by research communities.

An open Grid service exchange provides mechanisms for nonprovider services exchange through subpartitioning of resources. For example, it can allow segments of resources along with selected control and management processes for those resources to be allocated to specific external entities.

14.4.1 THE DESIGN OF AN OPEN GRID SERVICES EXCHANGE

A basic design objective for an open Grid services exchange is the creation of a facility that fully supports interactions among a complete range of advanced Grid services, within single domains and across multiple domains. Because Grid services are defined as those that can be constructed by integrating multiple resources, advertised by common methods, this type of exchange provides capabilities for resource discovery, assembly, use, reconfiguration, and discard.

Unlike existing exchanges, this type of exchange will provide methods for discovering a complete range of services that can be utilized and manipulated as Grid resources, including options for integration with other distributed Grid resources. This capability provides far more options for fulfilling requirements of services and applications than are possible with traditional exchanges.

This type of exchange fulfills the design principles described earlier, especially those related to decentralization. Using such an exchange, there is no need to rely on a carrier cloud for transport. Grid collaborations will be able to manage dedicated communications resources and services within an environment that they can customize directly in accordance with their requirements.

14.4.2 PROVISIONING IMPLICATIONS

The infrastructure design used for traditional exchanges is specifically oriented to a narrow set of limited services. As a result, such exchanges require major physical provisioning to add new services, enhance existing services, or to remove obsolete services. The new type of exchange described here will provide a flexible infrastructure with many component resources that can be dynamically combined and shaped into an almost unlimited number of services.

The abstraction layer made possible by the Grid service-oriented approach to architecture, combined with new types of network middleware and agile optical components, allows provisioning to be undertaken through high-level signaling, protocols, and APIs rather than through physical provisioning.

14.4.3 EXCHANGE FACILITY CHARACTERISTICS

These exchanges have the following characteristics.

- They are based on a services-oriented network model, and they provide facilities to support that model, incorporating capabilities that provide direct access to network services. Resources are advertised through standard, common services models.
- They are provider and services neutral. They allow any legitimate communication service entity to utilize internal resources. Because they are based on a services-oriented architecture, services provisioning, access, and control can be undertaken by multiple service providers, using distributed management and control systems.
- They provide a significantly expanded set of services, both a wide range of standard, advanced, and novel communication services and other data-oriented services, including Grid services.
- They support ad hoc dynamic services and resource provisioning.
- They are highly decentralized. They enable multiple capabilities for autonomous peerings, including ad hoc peerings, among client constituencies, both service providers and other communities such as individual organizations.

- They enable peering to take place at any service or communications level, including those for data transport services at layer 1.
- They provide capabilities for extensive integration and service concatenation at multiple levels, e.g., layer 1, layer 2, and layer 3, and among multiple services and resources.
- They enable such integration across multiple independent domains.
- They allow for direct peer-to-peer connections from specifically designated sites, and enable those sites to established unique shared network services.

Currently, a new architecture for next-generation open exchange points based on these concepts is being designed, developed, implemented in prototype, and in some cases being placed into production. The most advanced designs for these exchange points incorporate innovative methods of providing a services foundation based on dynamic lightpath provisioning. Unlike traditional exchanges points, these exchanges will provide accessible layer 1 transport services as foundation blocks for all other services. This capability will be a key foundation service within such exchanges.

14.5 OPEN GRID OPTICAL EXCHANGES

Because optical services layers within these exchanges will provide for key foundation capabilities, it is appropriate to term these types of exchanges "open grid optical exchanges." As noted, these facilities represent a significant departure from traditional exchanges.

14.5.1 TRADITIONAL INTERNET EXCHANGES

There are currently three types of Internet exchanges [5,6]:

(1) *LAN-based Internet exchanges.* These are the most common exchanges, and they typically implement layer 2 switches at their core, although some exchanges are distributed. This is the only stateless exchange. Blocking is possible if multiple peers want to send traffic to the same peer at the same time.
(2) *ATM-based Internet exchanges.* These are statefull exchanges, with Permanent Virtual Circuits (PVCs) at their core. If Variable Bit Rate (VBR) circuits are used, there is no guaranteed congestion-free transmission. The use of Constant Bit Rate (CBR), on the other hand, results in very inefficient use of resources and poor scalability.
(3) *MPLS-based Internet exchanges.* MPLS exchanges are both stateful and packet based. Though the service may be congestion free, it requires a lookup at the IP routing table at the ingress edge Label Switching Router (LSR). This approach results in a relatively expensive operation for very-high-bandwidth datastreams.

Because of the properties of these types of exchanges, they cannot provide many services required by Grids. These services are not sufficient to support even a few very high-bandwidth flows that do not require routing. Either it is technically impossible (for LAN-based Internet exchanges) or it yields unnecessary and costly overhead (for ATM- and MPLS-based Internet exchanges). Also, such exchanges provide no capabilities for external entities to implement and customize services dynamically.

14.5.2 RATIONALE FOR AN OPEN OPTICAL EXCHANGE

This section describes the rationale and model for an exchange that can be termed an "open Grid optical exchange." There are multiple reasons for developing this type of exchange, including those that are derived from the design principles described in earlier sections. This section also describes the basic interfaces and services that such an optical exchange may provide.

The case for an open Grid optical exchange is derived directly from a number of key requirements, including the need for direct access to services at lower layers. A growing requirement has been recognized for transport services that allow data traffic to remain at lower layers (layer 2, layer 1) and not be converted, for example for routing through other layers. Also, there are reasons based on requirements for service quality, and others that relate to taking advantage of multiple new optical innovations. First, optical connection-oriented solutions can ensure high quality, e.g., they can be guaranteed congestion free. This feature eliminates any need to impose complex QoS technologies. In addition, connection-oriented links allow users to use more efficient, but TCP-unfriendly, transport protocols. Such flows would be problematic in a common layer 3 exchange. Although other technologies also attempt to offer high-quality services with connectionless transport, such as Multi-Protocol Label Switching (MPLS) and Optical Burst Switching (OBS), these services do not match the performance of optical services for many required Grid services.

Second, there is an economic argument for this type of exchange – connection-oriented technologies will always remain an order of magnitude cheaper because there is no need to look into headers at the datastream itself. Even as traditional facility components are becoming less expensive, optical components are also continuing to fall in price, and the relative price ratio is at least remaining constant (although there is an argument that optical component prices are falling faster). Many new technologies are being developed [7], such as Optical Add-Drop Multiplexers (OADM)s and advanced photonic devices that will provide powerful new capabilities to these environments. Furthermore, the long-term forecast for optical network components is that they will become increasingly less expensive, including such key devices as optical shift registers [8]. If current research projects develop optical routers, this type of facility would be able to take immediate advantage of their capabilities.

Third, it is more efficient to keep traffic that does not need any high-level functionality at the lowest layer possible. In particular, this will apply to relatively long (minutes or more) and high-bandwidth (gigabps or more) datastreams between a limited number of destinations. Examples of these requirements are data replication processes and instrumentation. Grids where huge datastreams must be collected at a central place (like the eVLBI project [9]). The packets in these streams do not need routing. One knows the data has to go from this source to that destination. In general, this is called the service approach: apply only the required services, nothing more.

Fourth, even with a limited number of destinations, it is very likely that data-intensive traffic streams will traverse multiple network domains. Considering scalability issues, it is likely that the most cost-effective means for peering among the domains will be at optical peering locations.

14.5.3 THE CONCEPT OF AN OPTICAL EXCHANGE

The next sections of this chapter describe the concept of an optical exchange, and they provide an example of how such an exchange may be implemented. Different applications require different transport services, including layer 1 services [10], ranging from layer 0 to layer 3. There could be many types of optical exchanges. For example, one could be a trivial co-location site that provides no more functionality other than rack space and the ability for providers to have bilateral peerings with other providers at the same co-locations. Others can be highly complex with extensive services. Instead of describing four or more different potential exchanges, a generic architecture for an exchange is described, consisting of a "black box" with advertised interfaces and services (Figure 14.1).

The "black box" interfaces can implement different services and protocols. For example, one interface may be used to carry undefined traffic over 32 wavelengths using Dense Wavelength Division Multiplexing (DWDM), while an other interface may carry only one signal at 1310 nm, which carries SDH-framed traffic. A third interface may be LAN-PHY-based Ethernet, where traffic must reside in the 145.146.100.0/24 subnet.

The basic functionality of any exchange is to transport traffic from one organizational entity to another, usually a provider. For an optical exchange, this requirement implies that the main function is providing cross-connections among interfaces, forming circuits. The different interfaces require the optical exchange to extract a given signal from a provider circuit and inject it in another provider circuit using some form of multiplexing.

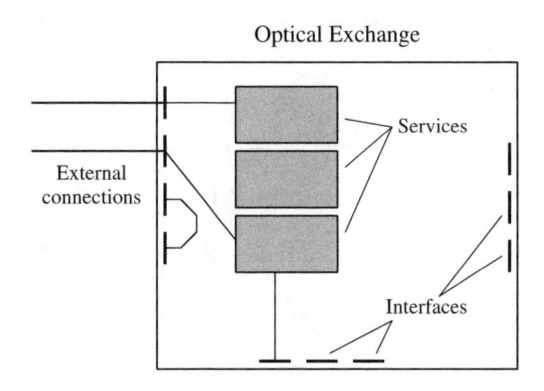

Figure 14.1. Components of an optical exchange.

For example, if one provider uses DWDM with 50 GHz spacing, and another uses 100 GHz spacing, it will probably be necessary to first demultiplex (demux) and then multiplex (mux) the signal again. If a signal enters at one layer and leaves at a different layer, the optical exchange effectively acts as an "elevator," lowering or raising traffic to different layers. For example, if a signal comes in as a wavelength using DWDM, and is injected in a vLAN, the signal is elevated from layer 1 to layer 2.

The first type of service can be described as providing cross-connections. The second type of service would be providing capabilities for muxing and demuxing bitstreams. Other more complex services may also be required, such as aggregation of traffic using a switch and optical multicast and store-and-forward services. Any given optical exchange may provide only a subset of all possible services. However, at such an exchange, if other services are not offered by the exchange itself, they may be offered by a service provider that is connected at the exchange, as Figure 14.2 shows.

An important point is that there is no technical obligation to place one set of services in the optical exchange domain (the *internal services*) and another set in a services domain (the *external services*). The decision as to which service should be an internal and which external service is a no technical one determined by other, external, considerations.

14.5.4 INTERFACES AND PROTOCOLS WITHIN AN OPTICAL EXCHANGE

As noted, an optical exchange may accept one or more types of interfaces. The list below describes common interfaces.

At layer 0, only single-mode fibers are considered. Specifically ignored here are multimode fiber and electrical (UTP) carriers, because currently within optical exchanges such as NetherLight [11] and StarLight [12], there is a trend toward single-mode fiber and away from multimode fiber and UTP. There are three likely causes for this trend. First, optical cross-connects with MEMS switches absorb light at 800 nm, and thus cannot deal with multimode fiber. Secondly, DWDM use is increasing, and

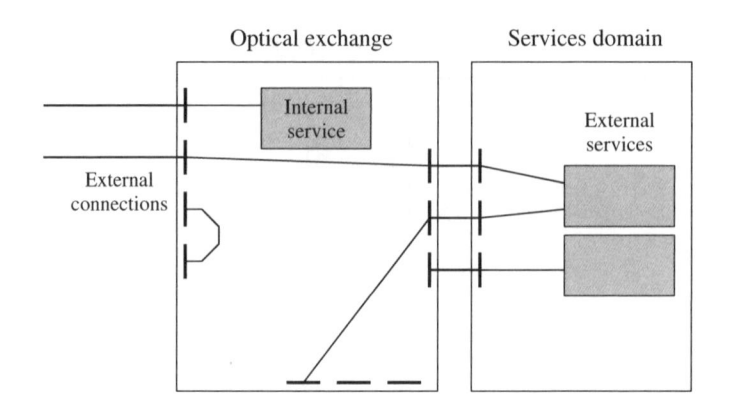

Figure 14.2. Internal and external services.

14.5 Open Grid Optical Exchanges

DWDM is possible only with single-mode fiber. Third, single-mode has a wider range of operation (a few hundred kilometers) than multimode (about 2 km).

At layer 1, each fiber may either carry a single bitstream within the 1260–1675 nm range or multiple bitstreams of data using separate DWDM wavelengths (lambdas).

Each bitstream (lambda) can use one of the following sampling and framings:

(1) a bitstream, at a certain wavelength, with up to 10.1 GHz or up to 40 GHz sampling, without other known properties;
(2) a bitstream, at a certain wavelength, with SONET/SDH framing, either OC-48 (2.5 Gbps), OC-192 (10 Gbps), or OC-768 (40 Gbps);
(3) a bitstream, at a certain wavelength, with 1 Gbps Ethernet;
(4) a bitstream, at a certain wavelength, with LAN PHY Ethernet;
(5) a bitstream, at a certain wavelength, with WAN PHY Ethernet.

Ethernet is a major service that scales from the local to the WAN scale. Each fiber typically carries one wavelength, or it carries 32 or 64 wavelengths though DWDM. The speed is typically either 1 Gbps or 10 Gbps LAN-PHY variant.

From the regional to global scale, typically a single SONET/SDH or WAN-PHY signal is used per carrier. The SONET speed may be OC-48 (2.5 Gbps), OC-192 (10 Gbps), or OC-768 (40 Gbps). The reason that TDM (with SONET or SDH) is used on a worldwide scale is that providers of transoceanic links require customers to comply with SONET/SDH framing. Note that WAN-PHY is compatible with SONET framing and also with SONET optical equipment [13]. Thus, WAN-PHY can be used on transoceanic links.

There are many ways to multiplex signals in one carrier using DWDM, because of the variety of wavelength bands and channel spacings. Similarly, not all vendors encapsulate Ethernet in the same way over a SONET/SDH connection. An optical exchange must have this type of information to determine the correct service to use. Of course, if no demultiplexing is required, then the optical exchange does not need this information.

A major issue when connecting two fibers together is to correctly tune the power levels. Especially with single-mode fibers, a single speck of dust can ruin signal strength. When making and breaking connections between two carriers dynamically, using an optical cross-connect, the signal strength should have a known, predefined value when leaving the optical exchange domain.

The protocol used on each circuit either is unknown to the optical exchange or may be specified by:

(1) IP over Ethernet (note that if all interfaces are of this type, the optical exchange would be reduced to a LAN-based Internet exchange);
(2) Ethernet implemented with 802.1q (vLAN tagging) [14].

Most end-to-end connections are full-duplex, even if the application does not require dedicated bandwidth in both directions. Most applications do not support a one-way connection, but a relatively easy way to optimize resource use is to support asynchronous connections. At a minimum, an optical exchange must be aware if an interface is always full-duplex or supports one-way connections.

14.5.5 OPTICAL EXCHANGE SERVICES

This section describes possible services for an optical exchange. However, not all of these services have to be offered by the optical exchange itself: some may be outsourced to a services provider connected to the optical exchange, or may not be offered at all. The abbreviations are used in the service matrix below.

(a) *Cross (connect)*. Given two interfaces of equal type, be able to make a cross-connect between these interfaces. Typically, this should be done in a user- or traffic-initiated way by a software control plane. However, here this limitation is not imposed.

(b) *Regenerate*. Amplify or attenuate the power levels, to match a certain output power level; amplify and reshape; or reamplify, reshape, and retime (3R).

(c) λ *convert*. Wavelength conversion, by regenerating the signal or by physically altering the wavelength. Regenerating, using tunable transponders, may allow network engineers to monitor the Bit Error Rate (BER) of a signal, but requires the regenerator to understand the modulation and possible framing of the signal.

(d) *WDM mux/demux*. Multiplex wavelengths of different color into a single carrier, and demultiplex different signals in a single carrier into many separate fibers. This process does not need to convert the wavelengths itself. An advanced demultiplexer may first demux the signal into sets of wavelengths, called wavebands, before the wavebands are demuxed into individual wavelengths [15]. Also, not all multiplexers are compatible, since different optical bands and channel spacings may be used.

(e) *(Optical) multicast*. The ability to duplicate an optical signal as-is. Of course, this can only be done one-way. Possible uses include visualization.

(f) *TDM mux/demux*. The ability to extract an Ethernet signal from a SONET or SDH carrier or to insert one or more Ethernet connections in a SONET or SDH carrier. It should be noted that not all SONET/SDH multiplexers are compatible.

(g) *SONET (switching)*: The ability to combine and switch SONET or SDH circuits, without knowing the contents of the circuits.

(h) *Aggregate*. There may be different needs for an Ethernet switch in an optical exchange. First, it allows aggregation of traffic, although this may cause congestion.

(i) *(Ethernet) conversion*. A second use for an Ethernet switch is the conversion between different types of framing. The most useful conversion is perhaps the ability to convert between LAN-PHY and WAN-PHY.

(j) *vLAN encap/decap*. Ethernet encapsulation. A third use for an Ethernet switch is the encapsulation of traffic in a vLAN trunk [14]. This allows the combining of different, separable datastreams on a single link.

The combination of DWDM support, wavelength conversion, and cross-connects will effectively provide for an OAD) facility. Multiple services may be combined in a single device.

14.5.6 EXTERNAL SERVICES

It is expected that an optical exchange itself does not offer any services on layer 3. However, service domains connected to an optical exchange may provide layer 3 services, for example:

- *Layer 3 exit/entry.* The ability to exit to the layer 3 Internet (layer 3 exit/entry), coming from a dedicated connection (lambda) or vice versa. This service may include not only the establishment of a physical link, but also negotiation of IP addresses to use. Care should be taken when offering this service, because it may allow easy creation of bypasses on the regular Internet, causing BGP ripples if such bypasses are continuously created and destroyed.
- *Store-and-forward.* One way to reduce blocking chances is to transport large chunks of data on a hop-by-hop basis. A location near an optical exchange would be the ideal place for a terabyte storage facility.

This list of external services is not complete. For example, MPLS and ATM services were not considered in this list.

14.5.7 SERVICE MATRIX

Services transition from one type of interface to another or to the same type of interface, or sometimes to multiple interfaces. Table 14.1 is a service matrix, showing the conversion from an interface listed on the left to an interface on top.

14.5.8 BLUEPRINT FOR AN OPTICAL EXCHANGE

An OXC, essentially an automated patch panel, is the simplest and least expensive type of device available. In addition, since it is a true optical device, ignorant of transmission speed, a 10-Gbps port costs as much as 1-Gbps port. Therefore, one vision for an optical exchange is to place an OXC at the core, with each interface connected to it. This applies to both external interfaces as well as interfaces to the different service devices. Figure 14.3 shows an example of such an exchange. Of course, it is possible that an exchange can contain multiple OXCs. (Note that in this discussion, references to "optical device" designate OOO devices only, not OEO devices.) It is expected that optical exchange, along with network connections (lambdas) in between, and Grid services on the edges will together form a global layer 1 data network.

14.5.9 MONITORING IN A MULTILAYER EXCHANGE

A disadvantage of handling traffic at layer 0 is that is not possible to monitor error counters. Typically only the signal strength can be measured, which does not give an indication of the bit error rate. Also, providers prefer to keep traffic at layer 0 within a domain, but utilize OEO conversion at the edges. If errors occur, this enables network engineers to at least pinpoint the administrative domain where the

Table 14.1 Service matrix

From \ To	WDM (multiple λ)	Single λ, any bitstream	SONET/SDH	1-Gbit/s Ethernet	LAN-PHY Ethernet	WAN-PHY Ethernet	VLA-tagged Ethernet	IP over Ethernet
WDM (multiple λ)	Cross-connect, multicast, regenerate, multicast	WDM demux	WDM demux*	WDM demux*	WDM demu*	WDM demux*	WDM demux*	WDM demux*
Single λ, any bitstream	WDM mux	Cross-connect, λ conversion, regenerate multicast	N/A*	N/A*	N/A*	N/A*	N/A*	N/A*
SONET/SDH	WDM mux	N/A*	SONET switch†	TDM demux*	TDM demux[6]	SONET switch	TDM demux*	TDM demux*
1-Gbit/s Ethernet	WDM mux	N/A*	TDM mux	Aggregate, Ethernet conversion†	Aggregate, Ethernet conversion	Aggregate, Ethernet conversion	Aggregate, VLAN encapsulation	Layer 3 entry*
LAN-PHY Ethernet	WDM mux	N/A*	TDM mux[6]	Aggregate, Ethernet conversion	Aggregate, Ethernet conversion, +	Ethernet conversion	Aggregate, VLAN encapsulation	Layer 3 entry*
WAN-PHY Ethernet	WDM mux	N/A*	SONET switch	Aggregate, Ethernet conversion	Ethernet conversion	Aggregate, Ethernet conversion†	Aggregate, VLAN encapsulation	Layer 3 entry*
VLAN-tagged Ethernet	WDM mux	N/A*	TDM mux	Aggregate, VLAN decap	Aggregate, VLAN decap	Aggregate, VLAN decap	Aggregate, VLAN decap and encap†	N/A
IP over Ethernet	WDM mux	N/A*	TDM mux	Layer 3 exit*	Layer 3 exit*	Layer 3 exit*	N/A	Store and forward, layer 3 entry/exit†

NA/, Not available. However, it may be possible to go from one interface to the other by applying two services in series, using an intermediate interface or protocol.

* Only possible if the interface type is correct. For example, it may be possible to demux a DWDM carrier to LAN-PHY Ethernet, but only, of course, if one of the wavelengths already contains LAN-PHY Ethernet.

* Note that, unlike WAN-PHY, it is not possible to place LAN-PHY in an OC-192 channel because LAN-PHY has a higher bit rate. However, it is technically possible, albeit uncommon, to put LAN-PHY in an OC-256 or OC-768 channel, so it is listed here.

† These functions may also be possible between one interface and another interface of the same type: cross-connect, regeneration, and λ conversion.

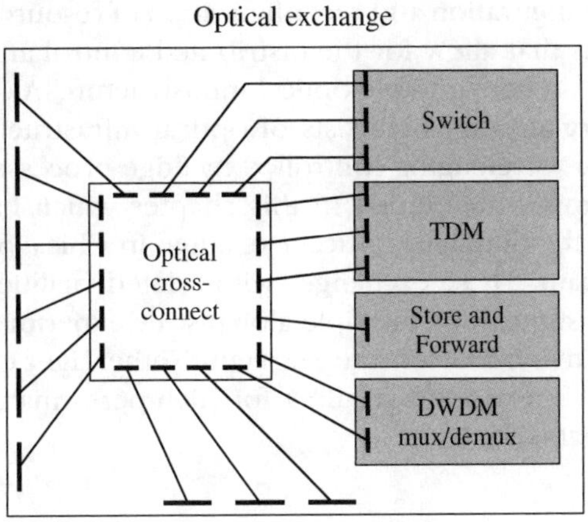

14.6 Prototype Implementations

Figure 14.3. Possible implementation of an optical exchange with an optical cross-connect at the core and providing miscellaneous services.

error originates. If the optical boundaries stretch between domains, this may not be possible. There are multiple solutions to this problem:

(1) Keep the OEO conversion at domain boundaries, at the expense of higher costs.
(2) Make sure that power levels are correct at domain boundaries, and be able to measure error rates, for example using optical multiplexing and a specific device.
(3) Expose error counters as Web Services to allow other domains to monitor the end-to-end quality of a circuit. This assumes that the domain publishing the error counters does not have an incentive to tamper with this data. This is generally the case if it is used only for QoS monitoring and not for billing purposes.
(4) A provider or optical exchange may offer two types of services for a circuit: a cheap one at layer 0, where the client should expose it error counters, or an expensive circuit at layer 1.

14.6 PROTOTYPE IMPLEMENTATIONS

The type of optical exchanges described here are already being developed and implemented. An international organization has been established to design and develop a global facility, termed the Global Lambda Integrated Facility (GLIF), and incorporates the architectural design principles and services that are based on open Grid optical exchanges [16]. The GLIF is based on an architecture that provides for a set of core basic resources within which it is possible to create multiple different types of specialized networks and services. GLIF is an infrastructure that created a closely integrated environment (networking infrastructure, network engineering, system integration, middleware, applications) and designed support capabilities based on dynamic configuration and reconfiguration of resources. The GLIF architecture provides options that allow for the distributed control and adjustment of layer 1 paths on a dynamic, reconfigurable optical infrastructure. A central service component for the GLIF organizations consists of optical infrastructure that can support dynamically allocated wavelengths controlled by edge processes.

Two optical exchanges mentioned in this chapter which have been operational for several years are the StarLight optical exchange in Chicago and the NetherLight exchange in Amsterdam. These exchanges with related facilities such as CA*Net4 in Canada are currently supporting multiple global-scale experimental Grid networking initiatives. This initiative builds on the StarLight/NetherLight/CA*Net4 development of layer 2 and layer 1 services that allow for high-performance, flexible transport among multiple global fabrics.

REFERENCES

[1] ITU-T (2004) "General Overview of NGN," Y.2001, December 2004.
[2] ITU-T "Information Technology – Open Systems Interconnection – Basic Reference Model: the Basic Model," ITU-T X.200.
[3] ITU-T (2004) "General Principles and General Reference Model for Next Generation Networks," Y.2011, October 2004.
[4] K. Manousaki, J. Baras, A. McAuley, and R. Morera (2005) "Network and Domain Autoconfiguration: A Unified Approach for Large Dynamic Networks," *IEEE Communications*, August, 78–85.
[5] G. Huston, (1999) "Interconnection, Peering, and Settlements", Proceedings of Inet'99, June 1999.
[6] I. Nakagawa, H. Esaki, Y. Kikuchi and K. Nagami (2002) "Design of Next Generation IX Using MPLS Technology", *IPSJ Journal*, 43, 3280–3290.
[7] M. Koga and K.-I. Sato (2002) "Recent Advances in Photonic Networking Technologies," Proceedings of International Conference on Optical Internet, July 2002.
[8] B. Tian, W. van Etten, and W. Beuwer (2002) "Ultra Fast All-Optical Shift Register and Its Perspective Application for Optical Fast Packet Switching", *IEEE Journal of Selected Topics in Quantum Electronics*, 8, 722–728.
[9] Consortium for Very Long Baseline Interferometry in Europe, http://www.evlbi.org/.
[10] C. de Laat, E. Radius, and S. Wallace (2003) "The Rationale of the Current Optical Networking Initiatives", iGrid2002, special issue, *Future Generation Computer Systems*, 19, 999–1008.
[11] NetherLight, http://www.netherlight.net.
[12] StarLight, http://www.startap.net/starlight.
[13] C. Meirosu, P. Golonka, A. Hirstius, S. Stancu, B. Dobinson, E. Radius, A. Antony, F. Dijkstra, J. Blom, C. de Laat, "Native 10 Gigabit Ethernet Experiments between Amsterdam and Geneva," High-Speed Networks and Services for Data-Intensive grids: the DataTAG Project, Special Issue, Future Generation Computer Systems, Volume 21, No. 4 (2005).
[14] IEEE (2001) "VirtualBridged Local Area Networks," IEEE standard 802.1q.
[15] X. Cao, V. Anand, and C. Qiao (2005) "Native 10 Gigabit Ethernet experiments over long distances," *Future Generation Computer Systems*, 21(4), 457–468.
[16] Global Lambda Integrated Facility, http://www.glif.is.

Chapter 15

Emerging Grid Networking Services and Technologies

*Joe Mambretti, Roger Helkey,
Olivier Jerphagnon, John Bowers,
and Franco Travostino*

15.1 INTRODUCTION

Grids can be conceptualized as virtual environments, based on multiple heterogeneous resources. Such resources are not limited to those usually associated with traditional information technology infrastructure. Grid architecture is designed to potentially incorporate almost any resource into its virtual environments, as long as that resource is communications enabled.

Given the extensibility of Grid environments to almost any type of component with a capability for data communications, emerging communications technologies are providing new opportunities for the creation of Grid services. These technologies can be used within distributed communications infrastructure, on which it is possible to place many types of overlay network services. Even highly specialized overlay networks can be implemented world-wide to support new Grid services.

This chapter presents an overview of some of these networking technology trends and it indicates how new types of infrastructure may be used to enhance Grid communication capabilities. This chapter organizes topics into several categories: edge technologies, access technologies, core technologies, and technologies that can be used to support services within multiple domains. To some degree, these distinctions are artificial because individual technologies are increasingly multifunctional, and they can be used at many points within an infrastructure.

15.2 NEW ENABLING TECHNOLOGIES

Multiple innovations in communications technologies are emerging from research laboratories and are being introduced into commercial markets at an extremely fast rate. These innovations will be key enabling technologies for services within next-generation Grid environments. These technologies will enable Grid services, especially communication services, to be much more abstracted from specific physical infrastructure than they are today, and they will provide for greater reach than today's technologies. They will also provide for pervasive communication services at much higher levels of scalable performance, at a lower overall cost, and with enhanced manageability.

Many of the most powerful recent innovations are at the component and device level. Common themes for the design and development of these technologies is that they will have small form factors, use minimal power, be cost-effective to mass produce through advanced manufacturing processes, be made of designed materials that will incorporate functionality, and be embedded into many more objects and placed virtually everywhere. The current generation of electronic integrated circuits consumes more power and generates more heat than is optimal. Also, extending the current generation of electronic circuits is becoming increasingly challenging because of the physical limitations of the materials that are being used. As a consequence, fundamentally different designs and materials are being explored, including circuits based on minute optical channels.

In the future, a large percentage of devices will be produced with internal circuits based on lightpaths, some enabled by Photonic Integrated Circuits (PICs), instead of traditional electronic circuits. These elements are being designed for embedding in both traditional communication products and products that have not previously had communications capabilities. In the future, many consumer and industrial items that have not had any type of communications capability will be manufactured with a variety of inherent functions enabling them to interact with their environments, generally through specialized radio frequencies. They will communicate externally through meshes of communication services. These communications functions will enable these items to be "context aware," to detect and to react to external conditions.

New small-form-factor, low-power-consumption components will also enhance capabilities for access channels to wide-area networks. In addition, these component and device innovations are beginning to eliminate complex, expensive, and burdensome functions in core networks, such as multiple electronic optical conversions, which currently occur at many levels in traditional networks.

15.3 EDGE TECHNOLOGIES

Sharp distinctions, and even modest differentiations, among edge, access, and core facilities are beginning to disappear as abstractions for networking capabilities and virtualization allow for individual capabilities and services to be ubiquitous. Also, the Grid is by definition a distributed infrastructure that can be designed to provide services to any diverse location. Nonetheless, to provide an overview for emerging communications technologies, it is useful to conceptualize these domains to some degree as separate facilities. At the level of basic infrastructure, there are foundation facilities that can be effectively integrated and utilized by the communication services that support distributed Grids.

One of the fastest growing areas in networking consists of many new communication devices being designed and developed for the network edge. Because the network edge can be anywhere, especially within a Grid context, perhaps the "edge" can be broadly defined as any location that contains a device capable of communications intended primarily to serve its own requirements as opposed to supporting other traffic. For example, this definition would encompass many areas related to mobile communications, personal computers, consumer products, especially appliances and related electronic devices, personal area networks, home networks, automobile communication systems, sensors, Radio Frequency Identification (RFID) systems, and related tagging devices designed for embedding into many types of physical objects (manufactured and biologically based), many types of premises security devices, and related technologies.

The word "primarily" is important in this definition because of the trend toward making many devices capable not only of serving their own communication requirements but also of serving as a node in a broader mesh network. Increasingly communication components, particularly those using radio frequency communications, are being designed to provide functional options for supporting not only the direct requirements of the device in which it is embedded, but also functions for transmitting pass-through traffic from other nearby devices. Using these capabilities, multiple distributed small devices, even very minute devices, can function as nodes on networks with a mesh topology.

15.4 WIRELESS TECHNOLOGIES

One of the most powerful revolutions in communications today comprises services based on an expanding field of wireless technologies, especially those related to mobile communications. Although personal communication is driving many of these developments, another major force consists of device-to-device and device-to-system communications.

Traditionally, personal communications have been defined in terms of a particular service (e.g., voice) associated with a specific device (e.g., phone). This approach to communication is rapidly disappearing. Increasingly, personal communications is defined as a capability to access a wide range of general communication services available through any one of multiple devices. A common expectation of personal communications is that any service will be available on any communications device from any location.

Furthermore, the utility of ubiquitous device-to-device and device-to-system communications is being realized. Many functions related to devices that have traditionally been manual are being automated through individual device communications using embedded components. Processes that have required mechanical manipulation are being migrated to communication services that are available seven days a week, 24 hours a day. These processes need not be simple interactive communications. Devices with embedded communications capabilities can be "context aware," i.e., communications content and services can be changed adaptively depending on detected local conditions.

Wireless Grid initiatives have attracted considerable interest because they are directly complementary to the Grid architectural model [1]. A number of Grid research and development projects have been established that take advantage of wireless networking capabilities, especially their potential for self-organizing, self-configuring, ad hoc, adjustable networking. This research direction is exploring powerful capabilities for creating customizable networks on demand.

15.4.1 DEVICE-LEVEL WIRELESS TECHNOLOGIES

Currently, devices are being designed for multifunctional wireless services, including mobile services. Instead of being limited to a particular communication service based on a given technology, e.g., cellular or packet, they can be used seamlessly across multiple types of services. Chipsets are being developed to seamlessly support multiple communication types, e.g., cell spectrum and digital.

There are multiple wireless standards, which are primarily designed for functional attributes, such as those related to spectrum, data rate, range, and power utilization. For example, small consumer items use low-rate, low-power-consumption, and small-range standards, such as Bluetooth (which has a maximum data rate of 1 Mbps, a distance of 10 m), or 802.15.4 (which can have a maximum data rate of 250 kbps or a minimum of 20 kbps and a maximum distance of 30 meters). Devices requiring additional bandwidth may use the developing IEEE standard 802.15.3a (also labeled ultra wide band, UWB), which is attempting to meet the needs of Wireless Personal Area Networks (WPANs), for example possibly 110 Mbps within 10 m and 200 Mbps within 4 m.

15.4.2 IEEE 802.11

A particularly important standard is IEEE 802.11, which is one of the most widely implemented short-range radio wireless specifications [2]. Although this standard is primarily used for local area communication, efforts have been initiated to integrate this approach with wider area wireless capabilities. This architectural direction is part of an overall trend toward developing comprehensive standard architectures instead of multiple discrete suites of protocols used for special-purpose implementations.

The 802.11 standard defines capabilities for Media Access Control (MAC) functions for wireless communications, for example communications between a network radio

15.4 Wireless Technologies

device and a network access node, such as a base station. However, as the next sections will indicate, it is also possible to establish 802.11 networks without such access nodes.

These designs use a physical layer implementation, based on one of several other defined standards, including 802.11b, 802.11a, and 802.11g, which support functions for detecting the wider network and send and receive frames. The 802.11 standard specifies two basic access modes. The primary one used is the Distributed Coordination Function (DCF), which is supported by the traditional Ethernet Carrier Sense Multiple Access Collision Avoidance protocol (CSMA/CA).

The external network is detected by scanning for available access points and best access points from among those that are detected, based on measurements of signal strength. The access points broadcast beacon information that the network radio device can use to determine the strength of the signal. The network radio device associates with the access node in advance of sending data to ensure compatibility for the communications, for example communication rate. Another function relates to security options through authentication, such as through a Wired Equivalent Privacy standard (WEP).

15.4.3 SELF-ORGANIZING AD HOC WIRELESS NETWORKS

The principle of distributed, self-organizing networks has been discussed in earlier chapters. One area of development that is becoming particularly interesting to Grid designers is the current trend toward the development of self-organizing wireless networks based on the 802.11 standard. These networks, which are sometimes called ad hoc networks or multihop wireless networks, provide for distributed, usually mobile, inter-device functionality that enables networks to be created without base stations. These networks are designed to manage circumstances in which some nodes continually appear and disappear from other nodes.

Such networks can be set up quickly and can provide for powerful, flexible communications services in rapidly changing environments in which there is no dependable fixed communications infrastructure. There are a number of performance challenges with these types of networks related to protocol implementation. However, recently several studies have been undertaken to examine these issues and to outline potential directions for solutions [3,4].

15.4.4 IEEE SA 802.11B

One of the most popular standards for wireless communication services is 802.11b, often labeled WiFi, which uses an unregulated part of the radio spectrum, based on three radio frequency channels within 2.4 GHz (ISM – industrial, scientific, and medical). Designed to transfer data on a 25-MHz channel at 11 Mbps as an ideal maximum rate, it has a specified range of 100 m.

802.11b operates in two primary modes. One mode requires an access point, which is often used as an extension to wired networks. Communications services are provided by establishing a link to a common hub. Such access points can employ the Network Address Translation (NAT) protocol, so that one set of IP addresses can be used as a front end to others that cannot be detected behind the access point. Although NAT compromises the Internet end-to-end addressing as noted in Chapter 10, it is an increasingly popular implementation.

The other mode is an ad hoc mode. No access point is required: devices can communicate with each other directly without an access point intermediary. Using this mode, small ad hoc networks of devices can be established.

15.4.5 IEEE 802.11A

Another key standard is 802.11a, also labeled "WiFi," which uses twelve channels (20 MHz) within the 5Ghz spectrum (UNII Unlicensed National Information Infrastructure). This standard is designed to ensure at least 24 Mbps but to also allow for data transfer of 54 Mbps as an option. It has a specified range of 50 meters.

15.4.6 IEEE 802.11G

The 802.11g standard also uses three channels within 2.4 GHz (ISM); however, it is designed to transfer data at the same rate as 802.11a, i.e., 54 Mbps. Its channel bandwidth is 25 MHz, and it has a specified range of 100 m.

15.4.7 SOFTWARE-DEFINED RADIOS AND COGNITIVE RADIOS

The Software-Defined Radio (SDR) [5] represents the convergence between digital radio and component-based software. In a SDR, a radio-enabled hardware platform and a companion set of software modules can become a cost-effective multimode, multiband, multifunctional wireless device. Typically, software modules can be dynamically loaded onto the device for the highest degree of adaptivity. The software modules and their exported APIs provide a capability for hopping frequency bands, switching modulation schemas, adopting custom protocols, and roaming over multiple wireless domains.

Researchers have formulated a vision of Cognitive Radios (CRs) [6] to harness the power of SDRs. A CR is an intelligent communication system that autonomously maintains awareness of the environment and adapts its radio frequency-related operating parameters to achieve greater effectiveness or efficiency in the communication. By way of feedback loops, a CR-enabled network is able to detect the existence of "spectrum holes" in time and space, to predict performance gains, and ultimately to take concerted provisioning actions at transmitters and receivers. These actions may result in fewer losses, longer reach, superior power efficiency, and other benefits that will be continuously benchmarked against requirements and new trade-offs as they present themselves.

It is appropriate to think of CRs as an intelligent, distributed control plane for SDRs. While they clearly have different scope, the concepts of CR and Grid network services are conceptually related, as they are both oriented to distributed resource optimization, dynamic adaptation, and customization.

In a Grid network, it is conceivable that Grid network services will overlay over one or more CR-enabled networks among several other networks. As such, Grid network services would be generating stimuli at the macro level (e.g., an upcoming burst calls for greater throughput). To these, the underlying, CR-enabled network would react with a matching adaptation at the micro level (i.e., with radio frequency scope).

15.4.8 RADIO FREQUENCY IDENTIFICATION

Radio Frequency Identification (RFID) is a quickly growing area of research, development, and implementation, in part because of the need of many communities, private and commercial, to track large numbers of items within a defined environmental context. RFID technology is based on small, often extremely small, transponders, which can be used as embedded tags to identify objects through low-power radio frequency signals. One issue that has been a challenge to RFID developers is that multiple standards organizations are involved with developing RFID architecture, and multiple regulatory bodies have some oversight of RFID implementation activities.

Various types of RFIDs exist today, and others are being created, designed to meet the requirements of a growing number of applications and services.

Passive RFIDs are not internally powered. When certain types of external radio frequency signals are detected by the tag's antenna, an internal electrical current is generated within its integrated electronic circuit. This activation mode powers the interactive communication required for the sending of information, or for the interactive exchange of information. Active RFIDs contain their own power source, usually small batteries.

RFIDs can operate across a broad range of frequencies, e.g., 135 kHz (and less, 30 kHz) on the low end, to almost 6 GHz on the high end, and many frequencies in between. A popular range for many RFID devices is 13.5 MHz, which is used by product labels and smart cards. The high side of the mid-range is around 30 MHz.

15.4.9 SENSORS

Essentially, sensors are devices that detect select physical properties in an environment, measure those properties, and communicate the result. Historically, sensors have had a limited communication range, often simply communicating at the device level and not to objects within an external environment. However, increasingly, sensor communications are being used for enhanced functionality not only to transmit results more broadly, but also to enhance functionality by being configured as a node in a meshed sensor net to provide for more precise results, to increase reliability, and to add additional detection capabilities.

Although most sensor networks have been supported by RF communications, a number of recent projects have integrated them into optically based networks. Currently, multiple major science projects are exploring or implementing this approach as a means of investigating phenomena at remote locations, such as the bottom of the ocean [7–9]. Very small devices based on the Bluetooth standard are also being used in sensor networks [10,11]. Wireless sensor networks also are benefiting from innovations in new methods for context-aware detection, self-organization, self-configuration, and self-adjustment [12,13].

15.4.10 LIGHT-EMITTING DIODES (LEDS)

Light-emitting diodes are widely used solid state semiconductors that convert electricity to light. They are used for display systems, optical communications, consumer light products, traffic lights, interior lighting, and many other applications. Increasingly, they are being used to replace traditional incandescent and fluorescent light sources, because they are cost-effective to manufacture, they have long lifetimes, and they can provide high-quality light with less power than traditional light sources.

However, they have another property that makes them more appealing than traditional sources of light. LEDs can be used for high-speed digital modulation, and, therefore, they can support local and access area communication services at speeds of 1 Gbps and above. Some research laboratories have demonstrated the use of digitally modulated LEDs to support speeds of tens of Gbps for both local areas, around 10–20 m, and access area distances.

15.5 ACCESS TECHNOLOGIES

As edge communications experience hypergrowth, pressure builds for access methods that provide interlinking capabilities between content, services, and related functions that are at the network edge with core communications capabilities. Currently, enterprise and in-home broadband capabilities are growing at much faster rates than access capabilities. Traditionally, access to WANs has been comparatively restrictive.

However, the issue of narrow access capabilities is being addressed. Consequently, a range of new access technologies has become the focus of multiple current research, development, and deployment activities. These initiatives are investigating a broad range of improved access technologies, including broadband wireless, satellite, data communications over power lines, and fiber to the premises technologies.

15.5.1 FIBER TO THE PREMISES (FTTP)

A major driver for access development is the provision of connections to hundreds of millions of domestic premises and small businesses. Next-generation fiber solutions are replacing legacy access technologies used for this type of access. Traditionally, these access points have been served by restrictive legacy technologies, such as copper lines for voice communications, coax for video, and DSL for data.

Some access implementations use electronics along the complete path to the edge premises, supporting layer 2 and layer 3 switching and using technologies derived from standard LAN technologies, such as 802.3ah, which provides for a specification of 100-Mbps full-duplex communications. Another popular approach uses Passive Optical Networking (PON) to avoid such electronic implementations. One of the largest PON-based projects is being undertaken in Japan, where it is being deployed to 30 million premises [14].

15.5.2 WIRELESS ACCESS NETWORKS

The IEEE 802.16 Working Group on Broadband Wireless Access Standards is developing a standard for multiple access services within metropolitan-area networks, especially those based on mobility, for example through the 802.16e specification, which is based on the 2–6 GHz range.

Another standard effort being developed by the IEEE is 802.20, which is designed to operate in licensed bands below 3.5 GHz, at speeds of at least 1 Mbps or higher for distances of up to 15 km.

15.5.3 FREE SPACE OPTICS (FSO)

Free Space Optics (FSO) is a point-to-point line of sight technology than can be useful for avoiding complex land-based provisioning methods, especially in cities requiring building-to-building communications or in geographic locations with difficult topologies. FSOs use point-to-point light signaling versus radio frequency technologies. FSO innovations have continued to advance the data capacity and reach of these technologies.

15.5.4 LIGHT-EMITTING DIODES

As noted above, LEDs can be used to support specialized communications within local-area networks. However, they can also be used for access networks. To some degree, because they can be used to produce a modulated signal, they can be used like FSOs [15]. However, they can be used for more versatile applications, for example ambient lights, as well as localized broadband communications.

15.5.5 BROADBAND OVER POWER LINES (BPL)

The IEEE standards association has established a working group to create standards, termed P1675, that enable power lines to be used as a data distribution channel, e.g., to carry data to premises, a technique called Broadband Over Powerline (BPL). These standards are defining the hardware required to place a high-frequency BPL signal onto electric power lines, such as with an inductive coupler, to convey and maintain the signal as it is sent over the line, and to remove the signal from the line at an edge point. Standards are also addressing issues related to insulation requirements and to components that allow for bypassing electrical elements that would cause attenuation, such as transformers.

15.6 CORE TECHNOLOGIES

Many technologies are transforming core technologies. The most powerful sets of technologies consist of multiple innovations related to optical communications, for example innovations that are significantly improving the flexibility, performance, and cost of optically based transmission. Key drivers consist of DWDM devices, reconfigurable add-drop multiplexers, signal enhancers and shapers, advanced optical cross-connects, tunable amplifiers, wavelength tunable lasers, 2D and 3D MEMS, and nonlinear fibers.

Of particular interest to the Grid community are emerging new technologies related to Photonic Integrated Circuits (PICs) and high-performance optical switches. Neither of these technologies will be limited to deployment to core networks. They will be used through highly distributed optical infrastructure, included at the edge of the network and in some cases within small devices. However, they will provide for particularly powerful capabilities for core network facilities and services.

15.7 PHOTONIC INTEGRATED CIRCUITS (PIC)

Since the mid-1990s, researchers have been making considerable progress in conceptualizing, designing, and implementing capabilities that place photonic logic on integrated circuits, using specialized materials, similar to the way in which electronic logic has been placed on integrated planar electronic circuits. Just as the creation of electronic integrated circuits revolutionized information technology, Photonic Integrated Circuits (PICs) have a substantial potential for driving a major information technology revolution.

These technologies have implications for all network domains – the network edge, access, and core – as well as for numerous edge devices. At the network core, they will contribute to enhance functionality and transparency; at the edge they will replace many traditional electronic devices, and at the edge they will also enable lightwave-based networking to be extended directly into individual edge devices. PICs will provide for substantially enhanced benefits for designers of next-generation communication services, including those related to capacity, flexibility, manageability, and device economics.

15.7.1 HIGH-PERFORMANCE OPTICAL SWITCHES

Optical technologies have played a growing role in Grid computing networks as the exchanged data capacity and distances have increased. DWDM technology and access to dark fiber have provided a large capacity and more control to Grid and network researchers [16]. For high-end computing and e-science applications, which require a large amount of bandwidth with high throughput and short delay, it is more efficient to use optical layer 1 switching to route 10-Gbps links of information. This development has led to the emergence of hybrid IP/optical networks [17–19] and the optical exchange points, such as StarLight [20] and NetherLight, discussed in the Chapter 14, which support optical-based overlay networks such as UltraLight [21].

The capabilities provided by such facilities allows for experiments that would otherwise not be possible. At times, it is actually interesting to "break" the network itself to try new experiments or new architectures, and optical switches become a necessary research tool.

15.7 Photonic Integrated Circuits (PIC)

15.7.2 RECENT ADVANCES IN HIGH PERFORMANCE OPTICAL SWITCHING

The following sections present recent advances in high-performance optical switching, examples of applications, and possible future evolutions that can impact networks for Grids.

Optical switches have existed and been used for many years, but they were limited to small sizes (e.g., 1×2). Several important technologies received significant research attention and range from nanoseconds to milliseconds in switching speed (Figure 15.1) [22]. However, it is only recently, with the advent of 3D Micro-Electro-Mechanical Systems (3D-MEMS), that large optical switches have been able to be built with hundreds of ports and low insertion loss (Figure 15.2). This technology breakthrough relies on switching in free space in three dimensions (3D) instead of two dimensions (2D) and the ability to precisely align and steer beams of light [23].

Large optical switches have been demonstrated with typical insertion loss as low as 1.3 dB [24]. The loss distribution for a 347×347 3D-MEMS switch [25] is shown in Figure 15.3b, where the median insertion loss of the switch is 1.4 dB. Optical loss is minimized by control of lens array focal length uniformity, collimator optical beam pointing accuracy, and MEMS micro-mirror reflectivity. The optical switches are transparent throughout the transmission spectrum of single-mode fibers, having negligible wavelength-dependent losses as well as very low chromatic and polarization mode dispersions [25]. The negligible wavelength-dependent loss is achieved by designing with sufficiently large mirrors to minimize diffraction loss across the whole wavelength band, and with the appropriate design of broadband antireflection coatings on all optical surfaces.

The optical switch produces negligible distortion of high-bandwidth optical signals, as demonstrated by optical loop experiments. In these experiments with a large switch (320×320), Kaman demonstrated sending 10-Gbps signals through 60 nodes with less than 1 dB of degradation [26]. In the other experiment, Kaman demonstrated sending 320 Gbps of data consisting of 32 wavelengths of 10-Gbps data, through 10 nodes over a total of 400 km of fiber without electrical regeneration [27].

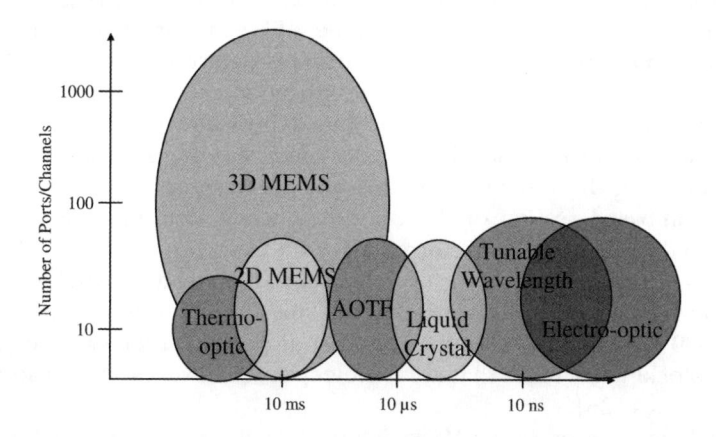

Figure 15.1. Optical switching technologies with speed and size characteristics.

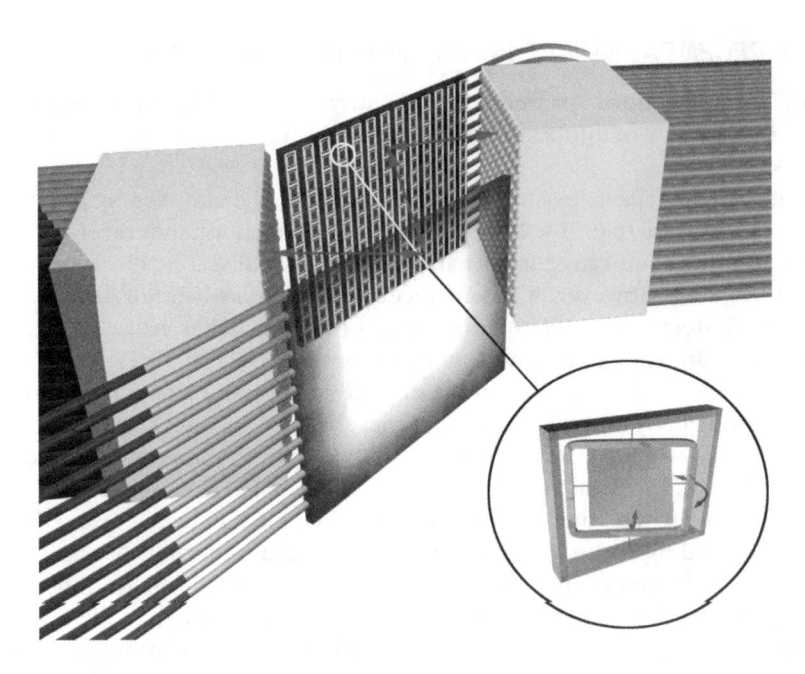

Figure 15.2. Configuration of a large 3D-MEMS switch where optical beams are steered in free-space with micro-mirrors.

These experiments demonstrate the scalability of high-capacity switches to large networks.

15.7.3 OPTICAL SWITCH DESIGN

There is no hard limit to the size of optical switches that can be fabricated, as the maximum port count of an optical switch is proportional to the diameter of the MEMS mirrors [25,28], and the mirrors can be made quite large (Figure 15.2). Size is a practical constraint however. The size of the MEMS mirror arrays and the switching volume between the planes of the MEMS arrays is determined by optical diffraction, the phenomenon causing optical beams to spread as they propagate.

The optical fiber coupling loss introduced by diffraction can be reduced by increasing the size of the optical beams, but increasing the size of the optical beams requires increasing the mirror array area and the distance between mirror arrays. The diffraction-based scaling of the optical switch is deterministic, assuming that the mirror deflection angle and mirror array fill factor are constant [29]. Although there is no theoretical limit to the switch size, the free-space switching core size of nonblocking single-stage optical switches becomes impracticably large at somewhere between 1000 [28] and 5000 input and output ports. Larger switches can also be built using blocking switches that are wavelength selective, or by cascading multiple stages of switches [23].

The switching time of 3D-MEMS switches is in the order of 10 ms, which is fast enough to support most applications [30]. In the simplest configuration, optical

15.7 Photonic Integrated Circuits (PIC)

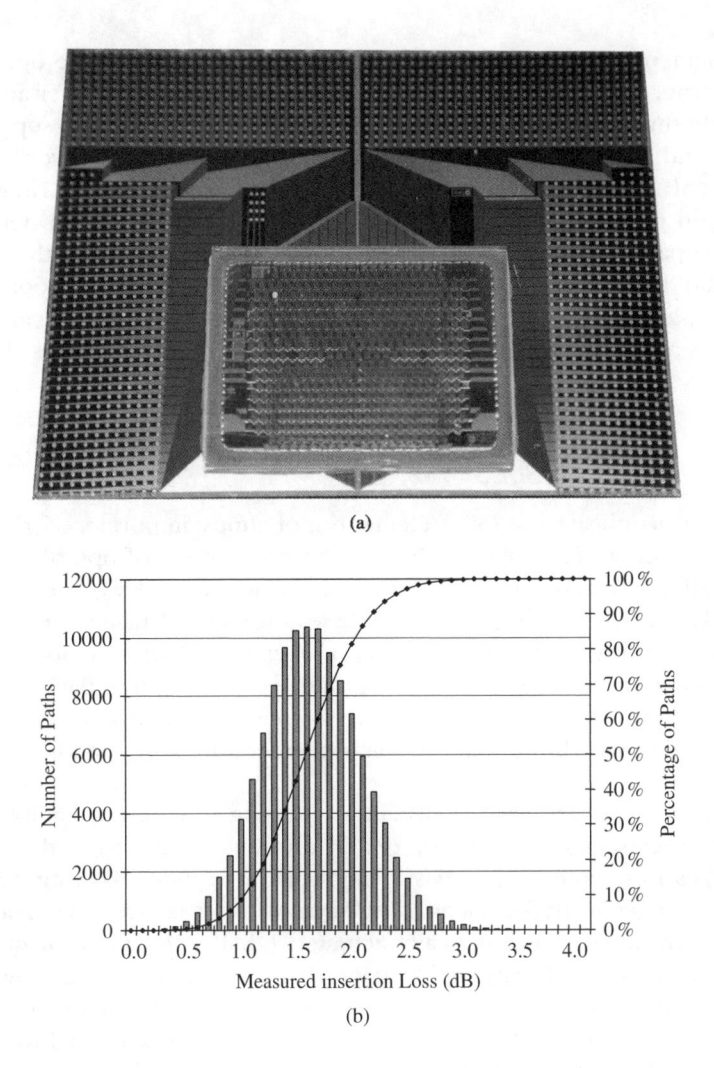

Figure 15.3. (a) Example of a MEMS array with mirrors located in the center sealed under a glass cover. (b) Insertion loss distribution of 347 × 347 3D-MEMS switch.

switches can be used as automated fiber patch panels at exchange points [31]. This simplifies operation and shortens the time it takes to create connectivity between computing users from weeks to minutes or seconds. It allows using network resources in a server configuration much like supercomputers use CPUs today.

15.7.4 OPTICAL SWITCHES IN CORE NETWORKS

Optical switches can also be used in core networks [32] in conjunction with DWDM transport and to reconfigure wavelength connectivity. Typical provisioning times are a second or less and are dominated by the signaling time between the network elements, not the time it takes to actually reconfigure the optical switch. IP and Ethernet services can be managed more efficiently directly over an intelligent layer [18]. In this case it is critical to provide automated protection to support service availability and alleviate the need of a SONET/SDH layer. Mesh network restoration of less than 50 ms is desired as it is supported today by SONET/SDH rings.

Restoration times of a couple of hundred milliseconds have been demonstrated and are compatible with IP/Ethernet networks, as long it is under the link downtime threshold of Ethernet switches and IP routers. Furthermore, coordination of provisioning and protection events across multiple IP/optical layers can be undertaken with a multilayer standardized control plane such as GMPLS, as discussed in Chapter 12.

15.7.5 RELIABILITY ISSUES

The reliability of optical networking elements is of utmost importance for deployment in core networks, an example of which is the deployment of optical switches in a next-generation Japanese telecommunications carrier network [33]. A highly reliable optical package can be formed with a double sealing method, using a hermetic sealing of the MEMS with an optical window to protect the MEMS mirrors during wafer dicing and assembly, and a second sealing interface at the optical package to enclose the entire optical path from input fiber to output fiber. This dual sealing method allows the optical switches to pass stringent temperature and humidity qualification tests.

The MEMS mirrors are passive structures, leading to inherent high reliability of the mirrors themselves. The mirror deflection angles are designed to allow the mirror hinges to operate in the elastic deformation regime, allowing mirrors with bulk silicon hinges to deflect billions of times without failure [34]. The principal failure mechanism of electrostatically actuated MEMS mirrors is dielectric breakdown, which can occur in any high-voltage silicon integrated circuit, owing to the high-voltage and high-density interconnect used to provide control of the MEMS mirrors. Potential dielectric breakdown failure points can be screened with a burn-in process to accelerate dielectric breakdown of components, leading to an extrapolated MEMS failure rate of 10 FIT (10 failures per 10^9 hours) per port, a failure rate that is extremely difficult to achieve in a comparably large high-speed electronic switch.

15.7.6 FUTURE ADVANCES IN HIGH-PERFORMANCE OPTICAL SWITCHES

Future advances in high-performance optical switching will permit Grid networks to provide higher capacity and real-time on-demand optical services. One area of research is to provide faster scalable switching fabrics in support of Optical Burst Switching (OBS) and Optical Packet Switching (OPS), described in Chapter 12. One approach is to optimize MEMS mirror resonant frequency and mirror control to provide submillisecond switching times. Recent results have shown submillisecond switching times on 3D-MEMS switches built for OBS demonstrators [35]. Another approach is to integrate a large number of switching elements in a photonic integrated circuit [36,37].

This method is an attractive approach for integrating other signal processing functionalities needed in OPS such as label processing, wavelength conversion, etc. One fundamental hurdle to migrating from circuit to packet-switching mode is providing enough buffers in the optical domain for congestion and contention packet resolution [38]. Recent research has shown that the buffers do not have to be as large as conventional wisdom suggested; this actually improves jitter and delay performances without affecting throughput [39]. Packet networks can actually be made with very small buffers given a small reduction in network bandwidth utilization. This may reconcile optical technologists and IP network architects to develop next-generation networks.

Another area of development is the integration of other functionalities with optical switching such as DWDM and wavelength conversion to deliver on the promises of all-optical networks. Integration of optical switching and DWDM has been successfully implemented with Reconfigurable Optical Add-Drop Multiplexers (ROADM) for a ring-based network. However, support of mesh configuration is needed to fully leverage the possibilities and advantages of optical networks. This will be beneficial to provide a shared-cost model at layer 1 and to provide lambda connectivity on demand between any two points in a network. Two challenges to address are limited optical reach and wavelength continuity constraints. Advances in wavelength conversion and regeneration are needed as well as in control planes to automate routing and wavelength assignment. Research demonstrations are under initial investigation in laboratories [27] and field testbeds [40] for metropolitan and regional environments.

15.7.7 IMPLICATIONS FOR THE FUTURE

In many ways, advances in optical technologies, and in optical switching in particular, are changing the network operation paradigm from a bandwidth-limited regime to a new paradigm in which bandwidth is no longer a limitation. Network protocols such as TCP were developed at a time when bandwidth was a scarce resource and electronic memory size and processing speed were seen as unlimited. This is no longer the case, and the new circumstances are having a profound implication on network architectures for Grids.

Future networks will have orders of magnitude more throughput than is available in existing networks. A more important point, however, is that future Grid networking services will have substantially more flexibility and potential for customizability than they have today. The future of Grid network services is very promising.

ACKNOWLEDGMENTS

Joe Mambretti developed Sections 15.1–15.4.4 and 15.4.6–15.7, Franco Travostino for Section 15.4.5, and Roger Helkey, Olivier Jerphagnon, and John Bowers Sections 15.7.–15.7.7.

REFERENCES

[1] L. McKnight and S. Bradner (2004) "Wireless Grids: Distributed Resource Sharing by Mobile, Nomadic, and Fixed Devices," *IEEE Internet Computing*, July – August, 24–31.

[2] IEEE (1997) "IEEE Standard for Information Technology – Telecommunications and Information Exchange Between Systems. Local and Metropolitan Area Network – Specific Requirements – Part 11: Wireless LAN Medium Access Control (MAC) and Physical Layer (PHY) Specifications."

[3] C. Chaudet, D. Dhoutaut, and I. Guerin Lassous (2005) "Performance Issues with IEEE 802.11 in Ad Hoc Networking," *IEEE Communications*, July, 110–116.

[4] L. Villasenor, Y. Ge, and L. Lamont (2005) "HOLSR: A Hierarchical Proactive Routing Mechanism for Mobile Ad Hoc Networks," *IEEE Communications*, July, 118–125.

[5] M. Dillinger, K. Madani, and N. Alonistioti (2003) *Software Defined Radio: Architectures, Systems, and Functions*, John Wiley & Sons Ltd.

[6] S. Haykin (2005) "Cognitive Radio: Brain-Empowered Wireless Communications," *IEEE Journal on Selected Areas in Communications*, 23, 201–220.

[7] www.optiputer.net.

[8] F. Zhao and L. Guibas (2004) *Wireless Sensor Networks: An Information Processing Approach*, Morgan Kaufmann.

[9] "Topics in Ad Hoc and Sensor Networks," *IEEE Communications*, October 2005, pp. 92–125.

[10] Bluetooth SIG, "Bluetooth Specification Version 1.1," www.bluetooth.com.

[11] Jan Beutel, Oliver Kasten, Friedemann Mattern, Kay Romer, Frank Siegemund, and Lothar Thiele (2004) "Prototyping Wireless Sensor Network Applications with BTnodes," *Proceedings of the 1st IEEE European Workshop on Wireless Networks*, July 2004, pp. 323–338.

[12] L. Ruiz, T. Braga, F. Silva, H. Assuncao, J. Nogueira, and A. Loureiro (2005) "On the Design of a Self-Managed Wireless Sensor Network," *IEEE Communications*, 43(8), 95–102.

[13] G. Pottie and W. Kaiser (2000) "Wireless Integrated Network Sensors,' *Communications of the ACM*, 43(5), 51–58.

[14] H. Shinohara (2005) "Broadband Access in Japan: Rapidly Growing FTTH Market," *IEEE Communications*, 43(9), 72–78.

[15] M. Akbulut, C. Chen, M. Hargis, A. Weiner, M. Melloch, and J. Woodall (2001) "Digital Communications Above 1 Gb/s Using 890-nm Surface-Emitting Light Emitting Diodes," *IEEE Photonics Technology Letters*, 13(1), 85–87.

[16] National LambdaRail, www.nlr.net.

[17] Super Science Network, http://www.sinet.ad.jp/english/super_sinet.html.

[18] Japan Gigabit Network 2, http://www.jgn.nict.go.jp/e/index.html.

[19] Louisiana Optical Network Initiative, www.loni.org.

[20] StarLight, http://www.startap.net/starlight.

[21] UltraLight, http://ultralight.caltech.edu/web-site/ultralight/html/index.html.

[22] J. Bowers (2001) "Photonic Cross-connects," in *Proceedings OSA Technical Digest, Photonics in Switching*, p. 3.

[23] R. Helkey, J. Bowers, O. Jerphagnon, V. Kaman, A. Keating, B. Liu, C. Pusarla, D. Xu, S. Yuan, X. Zheng, S. Adams, and T. Davis (2002) "Design of Large, MEMS-Based Photonic Switches," *Optics & Photonics News*, 13, 40–43.

[24] V.V.A. Aksyuk, S. Arney, N.R. Basavanhally, D.J. Bishop, C.A. Bolle, C.C. Chang, R. Frahm, A. Gasparyan, J.V. Gates, R. George, C.R. Giles, J. Kim, P.R. Kolodner, T.M. Lee, D.T. Neilson, C. Nijander, C.J. Nuzman, M. Paczkowski, A.R. Papazian, R. Ryf, H. Shea, and M.E. Simon (2002) "238 × 238 Surface Micromachined Optical Crossconnect With 2 dB Maximum Loss", in Technical Digest OFC 2002, postdeadline paper PD-FB9.

[25] X. Zheng, V. Kaman, S. Yuan, Y. Xu, O. Jerphagnon, A. Keating, R.C. Anderson, H.N. Poulsen, B. Liu, J.R. Sechrist, C. Pusarla, R.Helkey, D.J. Blumenthal, and J.E. Bowers (2003) "3D MEMS Photonic Crossconnect Switch Design and Performance", *IEEE Journal on Selected Topics in Quantum Electronics*, April/May, 571–578.
[26] V. Kaman, X. Zheng, S. Yuan, J. Klingshirn, C. Pusarla, R.J. Helkey, O. Jerphagnon, and J.E. Bowers, (2005) "Cascadability of Large-Scale 3-D MEMS-Based Low-Loss Photonic Cross-Connects," *IEEE Photonics Technology Letters*, 17, 771–773.
[27] V. Kaman, X. Zheng, S. Yuan, J. Klingshirn, C. Pusarla, R.J. Helkey, O. Jerphagnon, and J.E. Bowers (2005) "A 32 × 10 Gbps DWDM Metropolitan Network Demonstration Using Wavelength-Selective Photonic Cross-Connects and Narrow-Band EDFAs," *IEEE Photonics Technology Letters*, 17, 1977–1979.
[28] J. Kim, C.J. Nuzman, B. Kumar, D.F. Lieuwen, J.S. Kraus, A. Weiss, C.P. Lichtenwalner, A.R. Papazian, R.E. Frahm, N.R. Basavanhally, D.A. Ramsey, V.A. Aksyuk, F. Pardo, M.E. Simon, V. Lifton, H.B. Chan, M. Haueis, a. Gasparyan, H.R. Shea, S. Arney, C.A. Bolle, P.R. Kolodner, R. Ryf, D.T. Neilson, and J.V. Gates (2003) "1100 × 1100 port MEMS-Based Optical Crossconnect with 4-dB Maximum Loss," *IEEE Photonics Technology Letters*, 15(11), 1537–1539.
[29] R. Helkey (2005) "Transparent Optical Networks with Large MEMS-Based Optical Switches", 8th International Symposium on Contemporary Photonics Technology (CPT).
[30] O. Jerphagnon, R. Anderson, A. Chojnacki, R. Helkey, W. Fant, V. Kaman, A. Keating, B. Liu, c. Pusarla, J.R. Sechrist, D. Xu, Y. Shifu and X. Zheng (2002) "Performance and Applications of Large Port-Count and Low-Loss Photonic Cross-Connect Systems for Optical Networks," Proceedings of IEEE/LEOS Annual Meeting, TuV4.
[31] F. Dijkstra and C. de Laat (2004) "Optical Exchanges," GRIDNets conference proceedings, October 2004.
[32] O. Jerphagno, D. Alstaetter, G. Carvalho, and C. McGugan (2004) "Photonic Switches in the Worldwide National Research and Education Networks," *Lightwave Magazine*, November.
[33] M. Yano, F. Yamagishi, and T. Tsuda (2005) "Optical MEMS for Photonic Switching – Compact and Stable Optical Crossconnect Switches for Simple, Fast, and Flexible Wavelength applications in recent Photonic Networks," *IEEE Journal on Selected Topics in Quantum Electronics*, 11, 383–394.
[34] R. Lingampalli, J.R. Sechrist, P. Wills, J. Chong, A. Okun, C. Broderick, and J. Bowers (2002) "Reliability Of 3D MEMS-Based Photonic Switching Systems, All-Optical Networks," Proceedings of NFOEC.
[35] K.-I. Kitayama, M. Koga, H. Morikawa, S. Hara and M. Kawai (2005) "Optical Burst Switching Network Testbed in Japan, invited paper, OWC3, OFC.
[36] D. Blumenthal and M. Masanovic (2005) "LASOR (Label Switched Optical Router): Architecture and Underlying Integration Technologies", invited paper, We2.1.1, ECOC.
[37] M. Zirngibl (2006) "IRIS – Packet Switching Optical Data Router", invited paper, OFC.
[38] H. Park, E.E. Bwmeister, S. Njorlin, and J.E. Bowers (2004) "40-Gb/s Optical Buffer Design and Simulation", Numerical Simulation of Optoelectronic Devices Conference, Santa Barbara, August 2004.
[39] N. McKeown (2005) "Packet-switching with little or no buffers", plenary session, Mo2.1.4, ECOC.
[40] Optical Metro Network Initiative, http://www.icair.org/omninet/.

Appendix

Advanced Networking Research Testbeds and Prototype Implementations

A.1 INTRODUCTION

The goal of closely integrating network resources within Grid environments has not yet been fully realized. Additional research and development is required, and those efforts are being conducted in laboratories and on networking testbeds. Currently, a multiplicity of advanced research testbeds are being used to investigate and to develop new architecture, methods, and technologies that will enable networks to incorporate Grid attributes to a degree not possible previously. Several research testbeds that may be of particular interest to the Grid community are described in this appendix.

In addition, this appendix describes a number of early prototype facilities that have implemented next-generation Grid network infrastructure and services. These facilities have been used for experiments and demonstrations showcasing the capabilities of advanced Grid network services at national and international forums. They are also being used to support persistent advanced communication services for the Grid community.

The last part of this appendix describes several advanced national networks that are creating large distributed fabrics that implement leading-edge networking concepts and technologies as production facilities. This section also describes a consortium that is designing and implementing an international facility that is capable of supporting next-generation Grid communication services.

A.2 TESTBEDS

In the last few years, these testbeds have produced innovative methods and technologies across a wide range of areas that will be fundamental to next-generation networks. These testbeds are developing new techniques that allow for high levels of abstraction in network service design and implementation, for example through new methods of virtualization that enable functional capabilities to be designed and deployed independently of specific physical infrastructure. These approaches provide services that enable the high degree of resource sharing required by Grid environments, and they also contribute to the programmability and customization of those environments.

Using these techniques, network resources become basic building blocks that can be dynamically assembled and reassembled as required to ensure the provisioning of high-performance, deterministic services. Many of these testbeds have developed innovative architecture and methods that have demonstrated the importance of distributed management and control of core network resources. Some of these innovations are currently migrating to standard bodies, to prototype deployment, and, in a few cases, commercial development.

A.2.1 OMNINET

The Optical Metro Network Initiative (OMNI) was created as a cooperative research partnership to investigate and develop new architecture, methods, and technologies required for high-performance, dynamic metro optical networks. As part of this initiative, the OMNInet metro-area testbed was established in Chicago in 2001 to support experimental research. The testbed has also been extended nationally and internationally to conduct research experiments utilizing international lightpaths and to support demonstrations of lightpath-based services.

OMNInet has been used for multiple investigative projects related to lightpath services. One area of focus has been supporting reliable Gigabit Ethernet (GE) and 10-GE services on lightpaths without relying on SONET for transport. Therefore, to date no SONET has been used for the testbed. However, a new project currently being formulated will experiment with integrating these new techniques with next-generation SONET technologies and new digital framing technology. No routers have been used on this testbed; all transport is exclusively supported by layer 1 and layer 2 services. Through interconnections with other testbeds, experiments have been conducted related to new techniques for layer 3, layer 2, and layer 1 integration.

OMNInet has also been used for research focused on studying the behavior of advanced scientific Grid applications that are closely integrated with high-performance optical networks, based on dynamic lightpath provisioning and supported by reconfigurable photonics components. The testbed initially was provided with 24 10-Gbps lightpaths within a mesh configuration, interconnected by dedicated wavelength-qualified fiber. Each of four photonic nodes sites distributed throughout the city supported 12 10-Gbps lightpaths. The testbed was designed to provide Grids with unique capabilities for dynamic lightpath provisioning, which could be directly signaled and controlled by individual applications. Using these techniques, applications can directly configure network topologies. To investigate real data behavior on this network, as opposed to artificially generated traffic, the testbed was extended directly into individual laboratories at research sites, using dedicated fiber. The testbed was closely integrated with Grid clusters supporting science applications.

Each core site included a Photonic Switch Node (PSN), comprising an experimental Dense Wavelength Division Multiplexing (DWDM) photonic switch, based on two low-loss photonic switches (supported by 2D 8×8 MEMS), an Optical Fiber Amplifier (OFA), and a high-performance layer 2 switch. The photonic switches used were not commercial devices but unique assemblages of technologies and components, including such variable gain optical amplifiers.

OMNInet was implemented with several User–Network Interfaces (UNIs) that collectively constitute API interfaces to allow for communications between higher level processes and low-level optical networking resources, through service intermediaries. These service layers constitute the OMNInet control plane architecture, which was implemented within the framework of emerging standards for optical networking control planes, including the ITU-T Automatically Switched Optical Networks (ASON) standard. Because ASON is a reference architecture only, without defined signaling and routing protocols, various experimental software modules were created to perform these functions.

For example, as a service interface, Optical Dynamic Intelligent Network (ODIN) was created to allow top-level process, including applications, to dynamically provision lightpaths (i.e., connections based on switched optical channels) over the optical core network. Messages functions used the Simple Path Control (SPC) protocol. ODIN functions include discovery, for example determining the accessibility and availability of network resources that could be used to configure a particular topology. This service layer was integrated with an interface based on the OIF optical UNI (O-UNI) standard between edge devices and optical switches. ODIN also uses, as a provisioning tool, IETF GMPLS protocols for the control plane, Optical Network–Network Interface (O-NNI), supported by an out-of-band signaling network, provisioned on separate fiber than that of the data plane. A specialized process manages various optical network control plane and resource provisioning processes, including dynamic provisioning, deletion, and attribute setting of lightpaths. OMNInet was implemented with various mechanisms for protection, reliability and restoration, including highly granulated network monitoring at all levels, including per-wavelength optical protection through specialized software, protocol implementation, and physical-layer impairment automatic detection and response.

OMNInet is developed and managed by Northwestern University, Nortel Research Labs, SBC, the University of Illinois at Chicago (UIC), Argonne National Laboratory, and CANARIE (www.icair.org/omninet).

A.2.2 DISTRIBUTED OPTICAL TESTBED (DOT)

The Distributed Optical Testbed (DOT) is an experimental state-wide Grid testbed in Illinois that was designed, developed, and implemented in 2002 to support high-performance, resource-intensive Grid applications with distributed infrastructure based on dynamic lightpath provisioning. The testbed has been developing innovative architecture and techniques for distributed heterogeneous environments supported by optical networks that are directly integrated with Grid environments.

This approach recognizes that many high-performance applications require the direct ad hoc assembly and reconfiguration of information technology resources, including network configurations. The DOT testbed was designed to fully integrate all network components directly into a single contiguous environment capable of providing deterministic services. No routers were used on the testbed – all services were provided by individually addressable layer 2 paths at the network edge and layer 1 paths at the network core. The DOT testbed was implemented with capabilities for application-driven dynamic lightpath provisioning. DOT provided for an integrated combination of (a) advanced optical technologies based on leading edge photonic components, (b) extremely high-performance capabilities, i.e., multiple 10-Gbps optical channels, and (c) capabilities for direct application signaling to core optical components to allow for dynamic provisioning. The DOT environment is unique in having these types of capabilities.

DOT testbed experiments have included examining process requirements and behaviors from the application level through mid-level processes, through computational infrastructure, through control and management planes, to reconfigurable core photonic components. Specific research topics have included inter-process communications, new techniques for high-performance data transport services, adaptive lightpath provisioning, optical channel-based services for high-intensity, long-term data flows, the integration of high-performance layer 4 protocols and dynamic layer 1 provisioning, and physical impairment detection, compensation and adjustment, control and management planes, among others.

The DOT testbed was established as a cooperative research project by Northwestern University, the University of Illinois at Chicago, Argonne National Laboratory, the National Center for Supercomputing Applications, the Illinois Institute of Technology, and the University of Chicago (www.dotresearch.org). It was funded by the National Science Foundation, award # 0130869.

A.2.3 I-WIRE

I-WIRE is a dark fiber-based communications infrastructure in Illinois, which interconnects research facilities at multiple sites throughout the state. Established early in 2002, I-WIRE was the first such state-wide facility in the USA. I-WIRE was designed specifically for a community of researchers who are investigating and experimenting with architecture and techniques that are not possible on traditional routed networks. This infrastructure enables the interconnectivity and interoperability of computational resources in unique ways that are much more powerful than those found on traditional research networks based on routing.

I-WIRE layer 1 services can be directly connected to individual Grid clusters without having to transit through intermediary devices such as routers. I-WIRE provides point-to-point layer 1 data transport services based on DWDM, enabling each organization to have at least one 2.5-Gbps optical channel. However, the majority of channels are 10 Gbps, including multiple 10 Gbps among the Illinois TeraGrid sites. The Teragrid project, for example, uses I-WIRE to provide 30 Gbps (3 × 10 Gbps optical channels), among StarLight, Argonne, and NCSA. I-WIRE also supports the DOT.

These facilities have been developed by, and are directly managed by, the research community. I-WIRE is governed by a multiorganizational cooperative partnership, including Argonne National Laboratory (ANL), the University of Illinois (Chicago and Urbana campuses, including the National Center for Supercomputing Applications (NCSA), Northwestern University, the Illinois Institute of Technology, the University of Chicago, and others (www.i-wire.org). The I-WIRE project is supported by the state of Illinois.

A.2.4 OPTIPUTER

Established in 2003, the OptIPuter is a large-scale (national and international) research project that is designing a fundamentally new type of distributed cyberinfrastructure that tightly couples computational resources over parallel optical networks using IP. The name is derived from its use of *opt*ical networking, *I*nternet *p*rotocol, comp*uter* storage, processing, and visualization technologies. The OptIPuter design is being developed to exploit a new approach to distributed computing, one in which the central architectural element is optical networking, not computers. This transition is based on the use of parallelism, as it was for a similar shift in supercomputing a decade ago. However, this time the parallelism is in multiple wavelengths of light, or lambdas, on single optical fibers, allowing the creation of creating "supernetworks." This paradigm shift is motivating the researchers involved to understand and develop innovative solutions for a "LambdaGrid" world. The goal of this new architecture is to enable scientists who are generating terabytes and petabytes of data to interactively visualize, analyze, and correlate their data in real time from multiple storage sites connected to optical networks.

The OptIPuter project is reoptimizing the complete Grid stack of software abstractions, demonstrating how to "waste" bandwidth and storage in order to conserve relatively scarce computing in this new world of inverted values. Essentially, the OptIPuter is a virtual parallel computer in which the individual processors consist of widely distributed clusters; its memory consists of large distributed data repositories; its peripherals are very large scientific instruments, visualization displays, and/or sensor arrays; and its backplane consists of standard IP delivered over multiple, dynamically reconfigurable dedicated lambdas.

The OptIPuter project is enabling collaborating scientists to interactively explore massive amounts of previously uncorrelated data with a radical new architecture, which can be integrated with many e-science shared information technology facilities. OptIPuter researchers are conducting large-scale, application-driven system experiments with two data-intensive e-science efforts to ensure a useful and usable

OptIPuter design: EarthScope, funded by the National Science Foundation (NSF), and the Biomedical Informatics Research Network (BIRN), funded by the National Institutes of Health (NIH). The OptIPuter is a five-year information technology research project funded by the National Science Foundation. University of California, San Diego (UCSD), and University of Illinois at Chicago (UIC) lead the research team, with academic partners at Northwestern University; San Diego State University; University of Southern California/Information Sciences Institute; University of California, Irvine; University of Texas A&M; University of Illinois at Urbana-Champaign/National Center for Supercomputing Applications; and affiliate partners at the US Geological Survey EROS, NASA, University of Amsterdam and SARA Computing and Network Services in The Netherlands, CANARIE in Canada, the Korea Institute of Science and Technology Information (KISTI) in Korea, and the National Institute of Advanced Industrial Science and Technology (AIST) in Japan; and, industrial partners (www.optiputer.net).

A.2.5 CHEETAH

CHEETAH (Circuit-switched High-speed End-to-End Transport ArcHitecture) is a research testbed project established in 2004 that is investigating new methods for allowing high-throughput and delay-controlled data exchanges for large-scale e-science applications between distant end-hosts, over end-to-end circuits that consist of a hybrid of Ethernet and Ethernet-over-SONET segments.

The capabilities being developed by CHEETAH would supplement standard routed Internet connectivity by providing additional capacity using circuits. Part of this research is investigating routing decision algorithms that could provide for completely bypassing routers. The decision algorithms within end-hosts would determine whether not whether to set up a CHEETAH circuit or rely on standard routing.

Although the architecture is being used for electronic circuit switches, it can also be applied to all-optical circuit-switched networks. End-to-end "circuits" can be implemented with the CHEETAH using Ethernet (GE or 10GE) signals from end-hosts to Multiservice Provisioning Platforms (MSPPs) within enterprises, which then could be mapped to wide-area SONET circuits interconnecting distant MSPPs. Ethernet-over-SONET (EoS) encapsulation techniques have already been implemented within MSPPs. A key goal of the CHEETAH project is to provide a capability for dynamically establishing and releasing these end-to-end circuits for data applications as required.

The CHEETAH project is managed by the University of Virginia and is funded by the National Science Foundation. (http://cheetah.cs.virginia.edu).

A.2.6 DRAGON

The DRAGON project (dynamic resource allocation via GMPLS optical networks) is another recently established testbed that is also examining methods for dynamic, deterministic, and manageable end-to-end network transport services. The project has implemented a GMPLS-capable optical core network in the Washington DC metropolitan area. DRAGON has a specific focus on meeting the needs of high-end e-science applications.

The initial focus of the project is on means to dynamically control and provision lightpaths. However, this project is also attempting to develop common service definitions to allow for inter-domain service provisioning. The services being developed by DRAGON are intended as supplements to standard routing. These capabilities would provide for deterministic end-to-end multiprotocol services that can be provisioned across multiple administrative domains as well as a variety of conventional network technologies.

DRAGON is developing software necessary for addressing IP control plane end-system requirements related to providing rapid provisioning of inter-domain services through associated policy access, scheduling, and end-system resource implementation functions.

The project is being managed by the Mid-Atlantic Crossroads GigaPOP (MAX), the University of Southern California, the Information Science Institute, George Mason University, and the University of Maryland. It is funded by the National Science Foundation (www.dragon.maxgigapop.net).

A.2.7 JAPAN GIGABIT NETWORK II (JGN II)

Japan Gigabit Network II (JGN II) is a major research and development program which has established the largest advanced communications testbed to date. The JGN2 testbed was established in 2004 to explore advanced research concepts related to next-generation applications, including advanced digital media using high-definition formats, network services at layers 1 through 3, and new communications architecture, protocols, and technologies. Among the research projects supported by the JGN2 testbed are many that are exploring advanced techniques for large-scale science and Grid computing.

The foundation of the testbed consists of dedicated optical fibers, which support layer 1 services controlled with GMPLS, including multiple parallel 10-Gbps lightpaths, supported by DWDM. Layer 2 is supported primarily with Ethernet. Layer 3 consists of IPv6 implementations, with support for IPv6 multicast. The testbed was implemented nationwide, and it has multiple access points at major centers of research.

However, JGN2 also has implemented multiple international circuits, some through the T-LEX 10G optical exchange in Tokyo, to allow for interconnection to research testbeds world-wide, including to many Asia Pacific POPs and to the USA, for example a 10-Gbps channel from Tokyo to Pacific Northwest GigaPoP (PNWGP) and to the StarLight facility in Chicago. The 10G path to StarLight is being used for basic research and experiments in layer 1 and layer 2 technologies, including those related to advanced Grid technologies. The testbed is also used to demonstrate advanced applications, services, and technologies at major national and international forums.

The JGN2 project has multiple government, research laboratory, university, and industrial partners. JGN2 is sponsored by, and is operated by, the National Institute of Information and Communications Technology (NICT) (http://www.jgn.nict.go.jp).

A.2.8 VERTICALLY INTEGRATED OPTICAL TESTBED FOR LARGE SCALE APPLICATIONS (VIOLA)

In Germany, an inter-organizational consortium established in 2004 an optical testbed to explore the requirements of large-scale next-generation applications and to investigate new architecture and technologies that could address these requirements. The applications being examined include those supported by distributed computer infrastructure, such as e-science applications supported by Grids, which require dynamic access to high-bandwidth paths and guaranteed high quality of service, nationally and internationally.

Research activities include (a) testing new optical network components and network architectures, (b) interoperability studies using network technology from multiple producers, and (c) developing and testing software used for reserving and dynamically provisioning transmission capacity at gigabps ranges and interacting with related networking projects in other parts of Europe and the world.

To create the testbed, dedicated wavelength-qualified fiber (dark fiber) was implemented on a wide-area regional testbed. Equipment was implemented that could support lightpaths with multiple 10 Gbps of capacity simultaneously, using DWDM. The testbed was also implemented with high-performance computational clusters, with GE and/or 10GE access paths. Testbed switching technologies include SDH, GE, and 10GE along with new control and management plane technology. Also, new types of testing and measurement methods are being explored.

VIOLA was established by a consortium of industrial partners, major research institutions, universities, and science researchers, including the IMK, Research Center Jülich, RWTH Aachen and Fachhochschule Bonn-Rhein-Sieg, the DFN-Verein (Germany's National Research and Education Network), Berlin, the University of Bonn, Alcatel SEL AG, Stuttgart, Research Center Jülich, Central Institute for Applied Mathematics, Jülich, Rheinische Friedrich-Wilhelms-University, Institute for Computerscience IV, Bonn, Siemens AG, München, and T-Systems International GmbH (TSI), Nürnberg. VIOLA is sponsored by the German Federal Ministry of Education and Research (BMBF) (www.viola-testbed.de).

A.2.9 STARPLANE

Established in 2005, the StarPlane research initiative in The Netherlands is using a national optical infrastructure that is investigating and developing new techniques for optical provisioning. One focal research area is deterministic networking, providing applications with predictable, consistent, reliable, and repeatable services. Applications are provided with mechanisms that can directly control these services. Another research area is the development of architecture that allows applications to assemble specific types of protocol stacks to meet their requirements and to directly control core network elements, in part through low-level resource partitioning. Related to both of these functions are methods that provide for policy-based access control.

To provide for a comprehensive framework for this architecture, this project is designing and implementing a Grid infrastructure that will be closely integrated with these optical provisioning methods. The Grid infrastructure will support specialized applications that are capable of utilizing mechanisms provided to control network resources, through management plane software. The infrastructure will be implemented with this StarPlane management middleware as a standard library of components, including protocols and middleware that will be advertised to applications. The testbed will be extended internationally through NetherLight for interoperable experiments with other testbeds.

StarPlane is managed by the University of Amsterdam and SURFnet. It is funded by the government of The Netherlands.

A.2.10 ENLIGHTENED

The goal of the EnLIGHTened research project is establishing dynamic, adaptive, coordinated, and optimized use of networks connecting geographically distributed high-end computing and scientific instrumentation resources for faster real-time problem resolution. The EnLIGHTened project, established in 2005, is a collaborative interdisciplinary research initiative that seeks to research the integration of optical control planes with Grid middleware under *highly dynamic* requests for heterogeneous resources. Request for the coordinated resources are application, workflow engine, and aggregated traffic driven. The critical feedback loop consists of resource monitoring for discovery, performance, and SLA compliance and information is fed back to co-schedulers for coordinated *adaptive* resource allocation and co-scheduling. In this context, several network research challenges will be theoretically analyzed, followed by proposed novel extensions to existing control plane protocols, such as RSVP-TE, OSPF-TE, and LMP. Taking full advantage of the EnLIGHTened testbed, prototypes will be implemented and experimental testing will be conducted on extensions that make possible national and global testbeds.

Several network research topics within the above context arise with a central theme of *rethinking of behavioral control* of networks:

(1) centralized control – via management plane or middleware and
(2) distributed control – via control plane.

Highlights of the network research include:

(1) network-aware distributed advanced scheduling without preemption;
(2) lightpath group reservation: (a) sequential/parallel and (b) time;
(3) dynamic restoration and congestion avoidance using intelligent deflection routing;
(4) analyzing transport protocols for dedicated optical connections;
(5) resource optimization.

This research is developing an advanced software architecture that will vertically integrate the applications, advanced middleware, and the underlying optical network control plane technologies. The ultimate goal is to provide high-end dynamic applications with the capabilities to make highly dynamic, coordinated, adaptive, and optimized use of globally distributed compute, storage, instrument, and network resources. Monitoring tools will be used for resource discovery and near-real-time performance and availability of the disparate resources. A better understanding of how best to implement this tight feedback loop between monitored resource information and resource coordination and allocation will be one area of focus.

The proposed *vertical integration software architecture* consists of three layers:

(1) the application layer: focused on the capabilities required for application abstraction and service transaction protocols;
(2) the resource management layer: coordinates application requirements with resource discovery and performance monitoring, service provisioning and reconfiguration, and fault tolerance;
(3) the provisioning layer: APIs for on-demand and in-advance resource reservation to interface with the underlying resource control plane.

The system-level mechanisms and algorithms behind the software architecture will be studied and developed to satisfy high performance, resource sharing, resource coordination, and high availability.

This project is exploring a number of research problems regarding data management, load balance, scheduling, performance monitoring, service reconfiguration, and fault tolerance that are still open in this integrated environment. In addition, the following key components are being developed:

- advanced monitoring capabilities, which provide real-time information on the performance and availability of compute, instrument and network resources;
- the possibility of closer interaction with the optical network control plane;
- software that provides unified services to jointly and optimally allocate and control compute and networking resource and enable the applications to autonomously adapt to the available resources;
- methods to ensure security and service level agreement regulation within the distributed resources.

Internationally, the EnLIGHTened project is a partner project to the EU's LUCIFER control plane research testbed. Both projects are collaborating on control plane for Grid computing research as well as extending each other's testbed. The EnLIGHTened project is also collaborating with the Japanese G-Lambda research project.

The EnLIGHTened project is a large collaborative interdisciplinary research and testbed effort having partners from MCNC, RENCI, NCSU, LSU, Cisco, Calient, SURA, AT&T Research, NRL, and NLR, and is supported by the National Science Foundation.

A.2.11 LAMBDA USER CONTROLLED INFRASTRUCTURE FOR EUROPEAN RESEARCH

The Lambda User Controlled Infrastructure For European Research is currently being established to address some of the key technical challenges that must be addressed to enable on-demand end-to-end network services across multiple domains. The LUCIFER network concept and testbed is directed at allowing applications to be aware of their complete Grid environment – all resources (computational and networking) and other capabilities. It also allows them to make dynamic, adaptive, and optimized use of heterogeneous network infrastructures connecting various high-end resources.

This infrastructure will enhance and demonstrate solutions that facilitate vertical and horizontal communication among applications middleware, existing network resource provisioning systems, and the proposed Grid-GMPLS control plane. The project's main goal is broken down into the following objectives. It will:

(1) Demonstrate on-demand service delivery across access-independent multidomain/multivendor research network testbed on a European and world-wide scale. The testbed will include (a) EU NRENs: SURFnet, CESNET, PIONIER as well national testbeds (VIOLA, OptiCAT, UKLight), (b) GN2, GLIF and cross-border dark fiber connectivity infrastructure, (c) GMPLS, UCLP, DRAC, and ARGON control and management plane, and (d) multivendor equipment environments.

(2) Develop integration between application middleware and transport networks, based on three planes.

   (i) A service plane will provide (a) a clear set of APIs aimed at facilitating the development of new applications requiring a combination of Grid and network services, (b) a set of fundamental service components (developed as extensions to UNICORE) that implement the necessary mechanisms allowing network and Grid resources to be exposed in an integrated fashion, and to make reservations of those resources, and (c) policy mechanisms for networks participating in a global hybrid network infrastructure, allowing both network resource owners and applications such as Grids to have a stake in the decision to allocate specific network resources.

   (ii) A network resource provisioning plane will provide (a) adaptation of existing Network Resource Provisioning Systems (NRPS) to support the framework of the project, (b) full integration and testing of representative NRPS to the testbed, and (c) implementation of interfaces between different NRPS to allow multidomain interoperability with the testbed's resource reservation system.

   (iii) A control plane will provide (a) enhancements to the GMPLS control plane (defined here as Grid-GMPLS, or G^2MPLS) to provide optical network resources as first-class Grid resource, (b) interworking of Grid-GMPLS-controlled network domains with NRPS-based domains, i.e., interoperability between Grid-GMPLS and UCLP, DRAC, and ARGON, and (c) development of network and application interfaces (e.g., Grid-OUNI, Grid-APIs) for on-demand and advanced lambda reservations, enhanced with Grid layer specific information/heuristics, such as CPU and storage scheduling.

(3) Disseminate the project experience and outcomes to the targeted actors: NRENs and research users.

This research project will rely on experimental activities on a distributed testbed interconnecting European and world-wide optical infrastructures. Specifically, the testbed involves European NRENs and national testbeds, as well as GÈANT2, cross-border dark fiber and GLIF infrastructures. A set of highly demanding applications will be adapted to prove the concept. Internationally, the project is a partner project to the US EnLIGHTened research project. Both projects will collaborate on developing a control plane for Grid computing research as well as extend functionality from each testbed to the other.

The LUCIFER project is a large collaborative interdisciplinary research and testbed effort involving 20 partners from nine countries, including (a) NRENs, including CESNET (Czech Republic), PIONEIR (Poland), SURFnet (Netherlands); (b) national testbeds, including VIOLA, OptiCAT, and UKLight; (c) commercial companies, including ADVA, Hitachi, and Nortel; (d) SMEs, including NextWorks – Consorzio Pisa Ricerche (CPR); and (d) research centers and universities, including Athens Information Technology Institute (AIT-Greece), Fraunhofer SCAI (Germany), Fraunhofer IMK (Germany), Fundaciò i2CAT (Spain), IBBT (Belgium), RACTI (Greece), Research Centre Jülich (Germany), University of Amsterdam (Netherlands), University of Bonn (Germany), University of Essex (UK), University of Wales-Swansea (UK), SARA (Netherlands), and non-EU participants, including MCNC (USA), and CCT@LSU (USA). The project is funded by the European Commission.

A.2.12 GLOBAL ENVIRONMENT FOR NETWORK INNOVATIONS (GENI)

At this time the US National Science Foundation and the network research community is conceptualizing the Global Environment for Network Innovations (GENI), a facility that would increase the quality and quantity of experimental research results in networking and distributed systems and would accelerate the transition of those results into products and services. A key goal of this initiative is to support research that would lead to a major transition from the current Internet to one that has significantly more capabilities and also provides for much greater reliability and security.

The research community is developing a proposed design for GENI, which would be a large-scale (national and international) persistent research testbed. Its design would allow for multiple research projects to be conducted simultaneously. The testbed design would be modularized so that various experiments could be conducted at various network layers without interference.

For physical infrastructure, GENI will use multiple resource components (termed the "GENI substrate"), including not only those generally associated with networks such as circuits, transport, transit equipment, and wireless devices, but also other types of resources, such as computational clusters and storage devices. A software management framework will overlay these substrate resources. This overlay will allow for the partitioning and use of the underlying resources in various combinations. This approach is similar to the one being developed by the Grid community. Furthermore, this infrastructure will contain attributes that are common to Grid environments – it will be programmable, virtualizable, accessible by multiple communities, and highly modular.

This initiative is being supported by the NSF Directorate for Computer and Information Science and Engineering (CISE) (www.geni.net).

A.2.13 DEPARTMENT OF ENERGY ULTRASCIENCE NET

The US Department of Energy's (DOE) UltraScience Net is an experimental, national-scale network research testbed. The UltraScience Net, which has been developed by the DOE Office of Science, was designed and implemented to prototype the architecture, services, and technologies that will be required by multiple next-generation science projects. These science project requirements have shaped the design and capabilities of the testbed. Although many of these science projects are data intensive and, therefore, require extremely high-performance, high-volume services, the design includes considerations of multiple other requirements, including signaling for capacity on demand. The UltraScience Net was implemented to prototype a future DOE advanced network and to assist that organization transition toward that network. It is expected to support highly distributed terascale and petascale applications.

The UltraScience Net recognizes the important of multilayer networking and integration among layers. Among the capabilities of the testbed are those that enable signaling for capacity on demand, for dedicated channels, including for end-to-end full lightpaths and subsegments. Experimenting with such nonrouted services is a special focus of the testbed research. Dynamic and scheduled provisioning as well as management functions are implemented through an out-of-band channel. UltraScience Net research activities include those focused on protocols, middleware, dynamic resource assembly, integration with instrumentation, and considerations of integration with edge resources, including mass storage devices and computational clusters. Experiments are conducted on the testbed using specific applications (www.csm.ornl.gov/ultranet).

A.3 PROTOTYPE IMPLEMENTATIONS

The advanced networking research community has formed partnerships with advanced data-intensive application developers, especially those involved with global science projects, to design and implement specialized facilities that provide services, including Grid network services, which are required by such applications. Several of these initiatives are described in the following sections.

A.3.1 STARLIGHT

Operational since 2001, the Chicago-based international StarLight facility (Science Technology and Research Light-Illuminated Gigabit High-performance Transport) was designed and developed *by* the research community *for* the research community. The facility is both a research testbed and an early production prototype of a next-generation communication services exchange. StarLight has been termed "the optical STAR TAP," because it evolved from the earlier Science Technology and Research Transit Access Point.

StarLight supports many of the world's largest and most resource-intensive science and engineering research and development initiatives. StarLight serves as a proving ground for innovative architectures and services, enabling researchers to advance global e-science application development using deterministic services, especially those based on addressable wavelength (IP-over-lightpath) provisioning and dynamic lightpath switching. This facility is part of a prototype large-scale computational Grid based on optical networks ("LambdaGrid"), incorporating novel protocols for high-performance, high-volume transport services nationally and internationally, and new types of distributed control and management planes. The StarLight facility supports multilayer services at 1 Gbps, 2.5 Gbps, 10 Gbps, and multiple 10 Gbps. These services can be directly integrated into Grid environments and they can be used as extended backplanes for those environments.

StarLight currently provides 1-Gbps, 2.5-Gbps, and 10-Gbps services to 25 national, international, state-wide and regional networks, which support diverse constituencies, including research and education, federal agency science, and experimental testbeds. StarLight's flexible architecture and technology supports multiple categories of advanced services, ranging from those based on novel research and experimental methods to those based on pre-production prototypes and full production implementations. It also provides co-location space for the placement of high-performance networking, database, visualization, and computer cluster equipment, much of which is shared by teams of researchers. The environment facilitates advancements in middleware (transport and signaling protocols) and network measurement and monitoring research for e-science applications. StarLight is a member of the GLIF consortium and has instantiated GLIF open lightpath exchange services.

StarLight is developed and managed by the Electronic Visualization Laboratory (EVL) at the University of Illinois at Chicago (UIC), the International Center for Advanced Internet Research (iCAIR) at Northwestern University, and the Mathematics and Computer Science Division at Argonne National Laboratory, in partnership with Canada's CANARIE and The Netherlands' SURFnet. StarLight is made possible by major funding from the National Science Foundation to the University of Illinois at Chicago (award ANI-9712283 for the period 1997–2000, award SCI-9980480 for the period 2000–2005, and award OCI-0229642 for the period 2002–2006), and Department of Energy funding to Argonne National Laboratory. (STAR TAP and StarLight are service marks of the Board of Trustees of the University of Illinois.) www.startap.net/starlight.

A.3.2 TRANSLIGHT

TransLight was the result of the NSF Euro-Link award, which funded high-performance connections between the USA and Europe from 1999 through June 2005. TransLight, conceived in 2001 and implemented in 2003, was a global partnership among institutions – organizations, consortia and National Research Networks (NRENs) – that wished to provide additional bandwidth on their networks to world-wide communities of researchers for scheduled, experimental use.

TransLight consisted of many provisioned 1–10 Gbit Ethernet (GE) lambdas among North America, Europe, and Asia through the StarLight facility in Chicago. This global-scale initiative provided a foundation for a distributed infrastructure through the collaborative development of optical networking tools and techniques and advanced middleware services. TransLight became the first persistent LambdaGrid.

TransLight created a model for application-empowered optical networks (AEONs). AEONs enable reconfigurable networking and user-controlled traffic engineering of lightpaths for optimum throughput and workflow among distributed facilities. AEON techniques could be used at layer 1 to provide for optical VPNs – for security and traffic segregation – as simple replacements for standard VPNs with easier means for management and configuration. Additional techniques explored methods to implement high-performance, low-cost transport by substituting direct path channels for routers at locations in which they were not required. AEONs could be routed at the network edge (or at individual servers), rather than in the network core, to guarantee performance.

TransLight evolved into a two-year experiment to develop a governance model of how US and international networking collaborators would work together to provide a common infrastructure in support of scientific research. In September 2004, as more and more countries began sharing bandwidth, TransLight members dissolved the TransLight governance body in favor of having the GLIF, an international body, assume this role.

TransLight was designed and implemented by the Electronic Visualization Laboratory of the University of Illinois at Chicago, with the StarLight research community, particularly CANARIE through CA*net4 and SURFnet through NetherLight. It was supported by funding from the USA National Science Foundation (award SCI-9730202 to UIC, which ended September 2005), the Netherlands' GigaPort Project (SURFnet) and Canada's CANARIE. www.startap.net/starlight.

A.3.3 NETHERLIGHT

NetherLight is the GLIF Open Lightpath Exchange (GOLE) in Amsterdam, The Netherlands, and it has been operational since January 2002. NetherLight is an advanced open optical infrastructure and proving ground for network services, optimized for high-performance applications – especially those that require Grid infrastructure. NetherLight is operated as a production facility, with active monitoring seven days a week, 24 hours a day.

The heart of NetherLight consists of a large optical cross-connect that interconnects multiple 10-Gbps lambdas in order to cross-connect lightpaths between multiple national and international network facilities, including StarLight, UKLight, CzechLight (Prague), NorthernLight (Stockholm), the CERN network, and related networks and facilities. For those applications requiring connectivity services at other layers, NetherLight also provides gigabit Ethernet switching capabilities.

With these facilities NetherLight enables multiple specialized international high-performance communication services, in partnership with other advanced research networks. NetherLight has supported multiple landmark data-intensive science experiments and demonstrations, and is pioneering new concepts for architecture that may find their way into other GOLEs worldwide and the traditional telecommunications world as they move to next-generation architectures.

One of the focal points for development of NetherLight is control plane technology, with the multidomain environment in mind as much as possible. These developments are being undertaken in close cooperation with leading projects world-wide, paving the way to services based on end-to-end user-provisioned lightpaths.

NetherLight is realized by SURFnet, The Netherlands' national network organization for higher education and research, in the context of the GigaPort project (www.netherlight.nl).

A.3.4 UKLIGHT

UKlight is an integral part of the United Kingdoms's e-science infrastructure, consisting of international 10-Gbps connections to StarLight in Chicago and NetherLight in Amsterdam and an internal infrastructure that interconnects the UK's major university research institutions. UKlight is based upon 10-Gbps SDH connections between primary points of presence and international peering points. These connections are multiplexed into Ethernet "lightpaths" of 1–10 Gbps that present as services to the research community. All core nodes are supported by high-performance optical switches.

The external connectivity to StarLight in Chicago and NetherLight in Amsterdam is terminated within the University of London Computing Centre (ULCC). The University College of London (UCL) and the Imperial College, as part of the University of London, are connected directly to ULCC.

UKLight has been used for multiple experiments demonstrating the enabling capabilities of high-performance lightpaths for Grid research. Grid applications that have been supported on the network include high-energy physics and radio-astronomy. UKLight is also being used for basic research on lightpath provisioning and on transportation protocols.

UKlight has been provisioned and managed by UKERNA, a not-for-profit company, which manages the UK NREN, SuperJANET (www.uklight.ac.uk/).

A.4 NATIONAL AND INTERNATIONAL NEXT GENERATION COMMUNICATIONS INFRASTRUCTURE

A number of countries are beginning to implement next-generation optical networking infrastructure that will be able to support Grid services nationally. Several of these national projects are described in the following sections.

A.4.1 CANARIE

CANARIE is Canada's advanced Internet development organization, a not-for-profit corporation supported by members, project partners, and the national government. In 1998, CANARIE deployed CA*net3, the world's first national optical Internet research and education network. When it was designed and implemented, CA*net3 was among the most advanced in the world, and its design has since been replicated by many commercial and research and education networks.

More recently, CANARIE has undertaken the design, deployment, and operation of CA*net4, a national advanced network with international peering points that has capabilities which exceed the earlier network. Ca*net4 is based a series of point-to-point optical wavelengths, almost all of which are provisioned at OC-192 (10 Gbps).

CA*net4 is pioneering, through research and production implementation, the concept of a "customer-empowered network." This architecture allows end-users to directly control the dynamic allocation of network resources. Providing such capabilities on a national scale enables multiple communities of end-users to innovate in the development of network-based applications. These applications, based upon the increasing use of computers and networks as the platform for research in many fields, are essential for the national and international collaboration, data access and analysis, distributed computing, Grid environments, and remote control of instrumentation required by researchers (www.canarie.ca).

A.4.2 SURFNET6

SURFnet, the advanced networking organization of The Netherlands, has designed and is implementing SURFnet6 a national network with international peering points that provides both standard Internet services and lightpath-based services, up to 10 Gbps. Using this national infrastructure, individual large-scale applications and projects can be provided with dedicated lightpaths as directly accessible and controllable persistent reliable resources.

SURFnet6 has been globally recognized as a leading-edge hybrid network providing both full Internet functionality and lightpaths. SURFnet6 was the world's first national research network to design and implement this type of hybrid network.

Currently, the lightpath capabilities of the network are being used to support multiple science projects, including Very Long Baseline Interferometry (VLBI), which allows astronomers to observe objects in space with extremely high resolution. Using the SURFnet6 lightpaths, researchers are combining datastreams from telescopes in many other countries to form a simulated telescope as large as a whole continent, a technique that can be extended to the entire globe.

SURFnet6 was developed within the GigaPort project, a successful public–private collaboration of the Dutch government, industry, and educational and academic institutions, directed at enhancing national knowledge infrastructure (www.surfnet.nl).

A.4.3 NATIONAL LAMBDA RAIL

The National Lambda Rail is a US national distributed facility that is based on leased optical fiber and was deliberately designed and developed not as a network but as an infrastructure that could support multiple different types of networks, including experimental research networks. A key characteristic of the NLR is its ability to support many distinct networks for the US research community with the same core infrastructure. Using the NLR, experimental and productions networks can be implemented within the same facilities but can be kept apart physically and operationally. For some communities, the NLR provides production-level services for advanced applications, with guaranteed levels of reliability, availability, and performance. For other communities, the NLR provides researchers with capabilities for implementing experimental networks, including national-scale testbeds, and deploying and testing of new networking technologies, without the usual restrictions of production networks.

The unique nature of the NLR infrastructure provides researchers with high-performance access to multiple facilities, networks, experimental data, science data repositories, computational clusters, specialized instruments, mass storage systems, etc. These resources can be combined and used in unique ways. These capabilities facilitate the development of new technology as well as its migration to production, because this transition can be accomplished within the same infrastructure.

These capabilities also allow research and development to be conducted at levels above individual networking technologies. The NLR design was created to enable research projects that have not been possible previously, for example exploring interrelationships among network resource components that have been separated within traditional network production environments. Using the NLR, it is possible to conduct research related to new types of resource integration including those related to applications, inter-network layer signaling, advanced switching methods, new protocols, end-to-end service provisioning, network middleware, and lightpath provisioning.

The NLR is a multi-institutional partnership and nonprofit corporation funded by the academic higher education community and partner organizations (www.nlr.net).

A.5 INTERNATIONAL FACILITIES

The advanced research community has established several initiatives that are creating international facilities to support data-intensive applications, global e-science and engineering, experimental network research, and related projects. Two such projects are described here along with an advanced international network initiative developed to support international physics and an emerging project that will provide advanced networking services to the European community.

A.5.1 GLOBAL LAMBDA INTEGRATED FACILITY (GLIF)

The Global Lambda Integrated Facility (GLIF) is an international virtual organization that promotes the paradigm of lightpath (lambda) networking. The GLIF is a collaborative initiative among world-wide NRENs, consortia, and institutions that manage facilities supporting addressable lightpaths. The GLIF is neither a network nor a testbed. However, GLIF provides lightpaths internationally as an integrated facility to support data-intensive scientific research, and supports middleware development for lambda networking. The resources that it provides are being used to support multiple testbed projects. This organization is also developing an international infrastructure by identifying equipment, connection requirements, and necessary engineering functions and services.

The term "facility" in the name of the organization is included to emphasize that participants are not merely creating a network. They are designing and implementing a fundamentally new distributed environment, consisting of networking infrastructure, network engineering, system integration, middleware, and applications, to support the world's more resource-intensive applications.

The GLIF initiative is pursuing a common vision of a new Grid computing paradigm, in which the central architectural element comprises optical networks, not computers. This paradigm is based on the use of wavelength parallelism. Just as a decade ago supercomputing migrated from single processors to parallel processing, advanced networking is migrating from single-stream transport to parallel transport using multiple wavelengths of light (lambdas) on optical fiber. This goal is motivated by the need to support this decade's most demanding, high-performance e-science applications.

GLIF is developing and showcasing "application-empowered" networks, in which the networks themselves are schedulable Grid resources. These application-empowered deterministic networks, or "LambdaGrids," complement the conventional networks, which provide a general infrastructure with a common set of services to the broader research and education community. A LambdaGrid requires the interconnectivity of optical links, each carrying one or more lambdas, or wavelengths, of data, to form on-demand, end-to-end lightpaths in order to meet the needs of very demanding e-science applications.

The secretariat services are provided by TERENA, and funding is provided by participant organizations (www.glif.is).

A.5.2 GLOBAL RING NETWORK FOR ADVANCED APPLICATION DEVELOPMENT (GLORIAD)

The Global Ring Network for Advanced Application Development (GLORIAD) is a facility that provides scientists around the world with advanced networking tools that improve communications and data exchange, enabling active, daily collaboration on common problems. GLORIAD provides large-scale applications support, communication services, large-scale data transport, access to unique scientific facilities, including Grid environments, and specialized network-based tools and technologies for diverse communities of scientists, engineers, and other researcher domains.

Currently, GLORIAD has undertaken an expansion plan directed both at building a 10-Gbps hybrid backbone to interconnect national science and engineering networks in the USA, Russia, China, Canada, Korea, and The Netherlands and at creating a switching infrastructure at layer 2 to serve special communities, unique applications with particularly heavy service requirements, and to provide an infrastructure for network research and experimentation. The layer 2 switching infrastructure will eventually evolve into a similar infrastructure for switching layer 1 lightpaths. The GLORIAD network includes special Points of Presence (PoPs) in Chicago, Amsterdam, Moscow, Novosibirsk, Khabarovsk, Beijing, Hong Kong, Hong Kong, and Daejon (South Korea). At layer 1, the transport core of is supported by Multiservice Provisioning Systems (MSPPs).

GLORIAD's supporters include the US National Science Foundation, a consortium of science organizations and ministries in Russia, the Chinese Academy of Sciences, the Ministry of Science and Technology of Korea, the Canadian CANARIE network, and The Netherlands' SURFnet organization along with some telecommunications services donated by the VSNL telecommunications company. GLORIAD's US home is at the Joint Institute for Computational Science of the University of Tennessee and Oak Ridge National Laboratory (www.gloriad.org).

A.5.3 ULTRALIGHT

The UltraLight research project consists of an initiative established by a partnership of network engineers and experimental physicists who are developing new methods for providing the advanced network services required to support the next generation of high-energy physics. Currently, the physics research community is designing and implementing new experimental facilities and instrumentation that will be the foundation for the next frontier of fundamental physics research. These facilities and instruments will generate multiple petabytes of particle physics data that will be analyzed by physicists world-wide. A key focus of the UltraLight project is the development of capabilities required for this petabyte-scale global analysis. The project is also developing novel monitoring tools for distributed high-performance networks based on the MonALISA project.

This community has recognized that current Grid-based infrastructure provides extensive computing and storage resources, but is restricted by networks, which are used as external, passive, and largely unmanaged resource as opposed to integrated resources. Therefore, the UltraLight initiative is designing, implementing, and operating an advanced transcontinental and intercontinental optical network testbed, which interconnects high-speed data repositories and high-performance computing clusters, within the USA and in Europe, Asia, and South America. The project is also developing and deploying prototype global services that expand existing Grid computing systems by integrating the network as an actively managed component. In addition, the project is integrating and testing UltraLight as a basis for Grid-based physics production and analysis systems currently under development as part of major physics research projects such as ATLAS and CMS.

The UltraLight project is directed by the California Institute of Technology, and managed by the University of Florida, with the University of Michigan (Network Engineering), California Institute of Technology (Applications Services), Florida International University (Education and Outreach), University of Florida (Physics Analysis User Community), and California Institute of Technology (WAN-In-Lab). This project is funded by the National Science Foundation (www.ultralight.org).

A.5.4 GEANT2

At this time, GEANT2 is establishing an initiative that will provide lightpaths to the science communities, including the network research communities in Europe. GEANT2 is the seventh generation of the pan-European research and education network, which interconnects all of the nations in Europe (34 countries through 30 national research and education networks), and provides for international connections. The GEANT2 initiative was officially established in 2004, and its topology was announced in 2005.

The core of the network will consist of multiple 10-Gbps paths, which can be subdivided into individual single Gbps channels. Currently, GEANT2 is developing methods by which it can provide dedicated lightpaths to individual research communities and projects for special initiatives, including those related to Grids. For example, the Enabling Grids for E-Science (EGEE) initiative is attempting to implement a large-scale Grid infrastructure along with related services that will be available to scientists 24 hours a day. EGEE is a major Grid project that will use GEANT2 and the (NRENs connected to it. These organizations will collaborate in defining Grid network services requirements, specifications, and interfaces.

The JRA4 project is an EGEE joint research activity that is integrating into the Grid middleware the appropriate components, such as interfaces, that will allow automated access to GEANT2 and NREN network services. The Grid middleware design includes modules that allow the specification of service levels and bandwidth reservations, including through the network control plane. These techniques will allow Grid applications to establish and manage connections and flows through the dynamic network configuration and reconfiguration. Other tools will measure performance parameters within the network to provide feedback information to application and middleware processes.

GEANT2 is co-funded by the European Commission and Europe's national research and education networks, and is managed by the DANTE organization. EGEE is funded by the European Commission (www.geant2.net, http://egee-jra4.web.cern.ch/EGEE-JRA4).

Index

802   202, 207
  see also Ethernet
802.1D   202
802.11a ("WiFi")   297, 298
802.11b ("WiFi")   297–8
802.11g ("WiFi")   298
802.16 ("WiMax")
  access technology   301
  standardization   76
802.17, see Resilient packet ring (RPR)
802.1ad ("Provider Bridges")
  features   203
  standardization   76
802.1ah ("Backbone Provider Bridges")
  features   203
  standardization   76
802.20   301
802.3ae (10 Gbps LAN), standard   76, 204

Access Grid   7
Ad hoc wireless networks   297
Additive increase multiplicative decrease (AIMD)   172
AEON, see Application-empowered optical network (AEON)
AIST, see National Institute of Advanced Industrial Science and Technology (AIST)
All-photonic   235–6
  network service   235

Application-empowered optical network (AEON)   325
Application Grid   7
Application-initiated connections   229
ARPANET   101
ASON, see Automatically switched optical networks (ASON)
Assured forwarding   107
Asynchronous transfer mode (ATM)   199
Authentication, authorization, and accounting (AAA)   59, 114, 136
  example of network service   114
  trust model   137
Automatically switched optical networks (ASON)   74, 313
Autonomous system (AS)   186, 189, 254

Bandwidth characteristics   256
  achievable bandwidth   256
  available bandwidth   256
  capacity   256
  utilization   256
Behavioral control   224–9
  L1   224
Berkeley sockets   179
BGP, see Border gateway protocol
BIC   153, 160
BIC TCP   160
Biomedical Informatics Research Network (BIRN)   316

Bit error rate (BER)   219
Border gateway protocol (BGP)   65, 189
  component of grid network infrastructure   135
  element of IP networks   198
BPEL4WS, see Business process execution language for web services (BPEL4WS)
Broadband over power lines (BPL)   301
Burst   240–50
Burst control packet (BCP)   242
Business Process Execution Language for Web Services (BPEL4WS)   70, 89, 139
  component of grid network infrastructure   139
  standardization   70

CA*net4   292, 326
California Institute of Telecommunications and Information Technology (Calit2)   34
CANARIE   326–7
CAVE   181
CAVERNsoft   181
CERN   25–6
CESNET   321
CHEETAH (circuit-switched high-speed end-to-end transport architecture)   316
CineGrid   33–6
Cipher block chaining (CBC)   194
Closeness characteristics   257–8
Cognitive radio   298–9
Collaboration   18–23
Command line interfaces (CLI)   127
Common information model
  bindings of a grid network infrastructure   113, 127
  standardization   73
Community scheduler framework (CSF)   119
Composable-UDT   178
Computational science   26–7
Congestion loss   268
  class-based QoS   271–2
  distributed deflection   270–1
  local deflection   270
  neighbor deflection   270
Control plane   225–6
  functions   226

DAME, see Distributed Aircraft Maintenance Environment (DAME)
Data Grid   7
Data mining   30–3
Decentralization   56, 280
Decreasing AIMD or DAIMD   176–7
Delay characteristics   256–7
Department of Energy (DOE)   323
Determinism   220
Differentiated services code point (DSCP)   187
Differentiated services field   86–7, 105–7, 187–8
DiffServ   87, 93, 105, 106–7, 188
Distributed Aircraft Maintenance Environment (DAME)   36–41
Distributed Management Task Force (DMTF), profile   73–4
Distributed optical testbed (DOT)   314
DOE, see Department of Energy (DOE)
DOT, see Distributed optical testbed (DOT)
DRAC service plane   65
DRAGON (Dynamic Resource Allocation via GMPLS Optical Networks)   316–17
DWDM-RAM   123–6, 134
  features   123–4
  overview   123–4
Dynamic Ethernet intelligent transit interface (DEITI)   94
Dynamic range   237

EarthScope   316
EGEE, see Enabling Grids for E-Science (EGEE)
Electronic Visualization Laboratory (EVL)   20, 23, 34, 324
Enabling Grids for E-Science (EGEE)   330
End-to-end principle   8, 101–2
Endpoint reference   129
EnLIGHTened   319
Enterprise Grid Alliance (EGA), profile   69–70
E-science applications   219
Ethernet
  Ethernet-over-SONET   206–7
  future wire rates   204
  metro and WAN challenges   202
  OAM   203
  profile   201–5
  provider-provisioned service   69

Index

Ethernet-over-SONET 109
EU datagrid 259
Euro-Link 324
EVL, *see* Electronic Visualization Laboratory (EVL)
eVLBI 284
Expedited forwarding (EF) 106
Explicit congestion control protocol (XCP) 166–7
Exterior BGP (EBGP) 189
Exterior gateway protocol (EGP) 189
Extra-Grid 7

FAST 153, 158, 159–60
Fault detection 263–4
  BGP 264
  GMPLS 264
  IGP 264
  LMP 264
  OSPF 264
  SNMP 264
Fault-tolerant Grid platform (FT-Grid) 263
Fiber switches (FXCs) 217
Fiber to the premises (FTTP) 300
Forward equivalent class (FEC) 268
Frame relay 198, 200
Free space optics (FSO) 301
FXCs, *see* Fiber switches (FXCs)

G.7042 109
G.709 109
GÈANT2 322
GEL, *see* Global E-science laboratory (GEL)
General-purpose architecture for reservation and allocation (GARA) 93, 107, 121
Generic framing procedure (GFP) 109
  features 207
  standard 74
Generic route encapsulation (GRE) 200
GENI, *see* Global Environment for Network Innovations (GENI)
GGF NM-WG 255
GigaPort 325, 327
GLIF open lightpath exchange (GOLE) 324, 325
GLIF, *see* Global Lambda Integrated Facility (GLIF)
Global Environment for Network Innovations (GENI) 322

Global E-science laboratory (GEL) 224
Global Grid Forum (GGF), profile, 10, 68
Global Lambda Grid 7
Global Lambda Integrated Facility (GLIF) 97, 291, 328–9
  multidomain considerations 137
Global Ring Network for Advanced Application Development (GLORIAD) 329
Globus toolkit 115–17
  container 117
  data management 119
  execution management 118–19
  extensible I/O 122
  monitoring and discovery 119–20
  security 120
GLORIAD, *see* Global Ring Network for Advanced Application Development (GLORIAD)
GMPLS (Generalized Multiprotocol Label Switching) 218
GOLE, *see* GLIF open lightpath exchange (GOLE)
Grid 2–3
Grid abstraction 2, 4, 5–6
Grid architectural principles 1
Grid customization 5, 14
Grid decentralization 4, 11–12
Grid determinism 4, 11
Grid dynamic integration 4, 12
Grid first class entities 1–2
Grid infrastructure software 114
Grid network infrastructure software 114–15
Grid network services 219, 235–40
Grid-OBS 246–50
Grid programmability 4, 10
Grid resource allocation and management (Globus' GRAM) 116
Grid resource sharing 4
Grid resources
  coordination 233–5
  discovery 231
  scheduler 233
Grid scalability 4, 12
Grid security 5, 13
Grid services 219
Grid types 7–14
Grid virtualization 4

Grid web services   83
GridFTP   25
   features   121–2
   standardization   68
GUNI (Grid UNI)   239

High Performance Networking Research Group (GHPN-RG)   64, 69, 114
High-speed TCP, standardization   161–2
Hoplist   255–6
Host channel adapter, see Infiniband
Hourglass design   186

iCAIR, see International Center for Advanced Internet Research (iCAIR)
Infiniband
   features   211–14
   MPI libraries   214
   remote direct memory access   214
   standard   75–6
   virtual lane   213
   web services representation   213
Infiniband Trade Association (IBTA), profile   75–6
Institute of Electrical and Electronics Engineers (IEEE)   76
Inter Grid   7
Interior BGP (IBGP)   189
Interior gateway protocol (IGP)   189
International Center for Advanced Internet Research (iCAIR)   324
International Organization for Standardization   51
International Telecommunication Union (ITU)   74–5
Internet   8, 11
Internet Engineering Task Force (IETF)   58, 71–3
Intra-Grid   7
IntServ   107
IP multicast   194
IP packet delay variation (IPDV)   257
IPsec   193
IPSphere Forum, profile   71
IPv6   57, 72, 192–3
ISO   51
ITU-T NGN   74, 278–80
I-WAY   8
I-WIRE   314–15

Japan Gigabit Network II (JGN II)   317
JGN II, see (Japan Gigabit Network II)
Job Submission Description Language (JSDL)   90
Joint wavelength assignment (SPP-JWA)   267
Just-enough-time   245
Just-in-time   239
JuxtaView   22

KISTI, see Korean Institute of Science and Technology Information (KISTI)
Korean Institute of Science and Technology Information (KISTI)   224

L1 (layer 1)
   benefits   219–21
   connection type   235
   high capacity   220
   network services   217, 219
   VPN   221–2
L1 QoS   109
Lambda User Controlled Infrastructure For European Research (LUCIFER)   320–2
LambdaGrid   19, 110, 224
LambdaStream   178–9
LambdaVision   20, 21
LAN-PHY   285, 287, 290
Large Hadron Collider (LHC)   25, 229, 234
Layer 1 Virtual Private Network (L1VPN), standard   73, 110, 221
Layer 1, see L1 (layer 1)
Layer 3   185–95
Layer 4   186, 192
Light-emitting diodes (LEDs)
   access technology   301
   edge technology   300
Lightpath   14, 217
The link capacity adjustment scheme (LCAS)   109
   features   207
   standard   74
Loss characteristics   257
LUCIFER, see Lambda User Controlled Infrastructure For European Research (LUCIFER)

Management plane   225
MaxNet   163–6
Measurement methodologies   254

Index

Media access control (MAC) 76, 109, 198
MEMs switching 248
Meta-schedulers 234
Metro Ethernet Forum 202, 204
Monitoring and discovery services (Globus' MDS) 116
MPI
 MPI forum profile 71
 with Infiniband 211–14
MPLS-TE, see Multiprotocol label switching with traffic engineering (MPLS-TE)
MSPP, see Multiservice provisioning systems (MSPP)
Multiprotocol label switching with traffic engineering (MPLS-TE) 227
Multi Router Traffic Grapher (MRTG) 258
Multiservice provisioning systems (MSPP) 316
Multiprotocol label switching (MPLS), 284
 profile 198–201
 in shared network infrastructure 200
 virtual private networks 200–1
 with grid network services 201

National Institute of Advanced Industrial Science and Technology (AIST) 316
National Institute of Information and Communications Technology (NICT) 317
National Lambda Rail (NLR) 327–8
National Science Foundation (NSF) 314, 316
NetBLT 173
NetFlow 129
NetherLight 286, 292, 325–6
Network abstraction 53
Network address port translation (NAPT) 57, 192
Network determinism 54–6
Network quality of service 100–1, 102
Network resource management system 64–5, 123
Network self-organization 279–80
Network user categories 50
Network virtualization 52–3
Network Weather Service 133
Network–network interfaces (NNI)
 component of grid network infrastructure 135
 standard 75

NewReno 151, 157–8
Next-generation network (NGN), see ITU-T NGN
NICT, see National Institute of Information and Communications Technology (NICT)
NLR, see National Lambda Rail (NLR)
NSF, see National Science Foundation (NSF)

O-NNI, see Optical network-network interface (O-NNI)
OBGP 96
OEO, see Optical-to-electrical-to-optical (OEO)
OFA, see Optical fiber amplifier (OFA)
Offset 242
OIF, see Optical Internetworking Forum (OIF)
OMNI, see Optical Metro Network Initiative (OMNI)
OMNInet 312–13
On-demand provisioning 220
OOO, see Optical-to-optical (OOO)
Open Communications Architecture Forum (OCAF), standardization 75
Open Grid optical exchanges 283–91
Open Grid services architecture (OGSA) 62
 standardization 68
Open Grid services exchange 281–2
Open services communications exchange 281–3
Open shortest path first (OSPF) 65, 189, 198
Open Systems Interconnect model 51–2
Open Systems Interconnection basic reference model 51, 279
OpEx (operational Expense) 220
Optical burst switching 240–50
Optical cross-connect (OXC) 248
Optical dynamic intelligent network (ODIN) 65, 95
Optical fiber amplifier (OFA) 313
Optical Internetworking Forum (OIF)
 demonstration 204
 NNI 221
 profile 75
 UNI 221

Optical Metro Network Initiative (OMNI) 312–13
Optical network–network interface (O-NNI) 313
Optical switches
　design　304–5
　futures　306–7
　high performance　302
　recent advances　303–4
　reliability　306
　use in core networks　305–6
Optical-to-electrical-to-optical (OEO)　217
Optical-to-optical (OOO)　217
OptiCAT　321–2
OptIPuter　19, 20–2, 34, 110, 315–16
Organization for the Advancement of Structured Information Standards (OASIS)　70, 89
　profile　70
OSI　51–2
OSPF, see Open shortest path first (OSPF)

Pacific Northwest GigaPoP (PNWGP)　317
Particle physics　24, 26
Passive optical networking (PON)　300
Path-based protection　266
　dedicated path protection (DPP)　266
　shared path protection (SPP)　266
Peer-to-peer (P2P)　50, 83
Per-hop behavior (PHB)　106, 187
Photonic integrated circuits (PIC)
　core technology　302
　general　294
Photonic switch node (PSN)　313
Physical layer, QoS　236
Physical layer impairments　238–9
　chromatic dispersion (CD)　239
　linear impairments　238
　nonlinear　239
　polarized mode dispersion (PMD)　238
PIONEIR　322
PNWGP, see Pacific Northwest GigaPoP (PNWGP)
Pseudo-wire emulation　205
PSN, see Photonic switch node (PSN)

QoS, see Quality of service (QoS)
Quality of service (QoS)　87, 217, 218, 220, 235
Quanta　180–2

Radio astronomy　24, 30
Radio frequency identification (RFID)　59, 299
RealityGrid　27
Reconfigurable optical add-drop multiplexers (ROADMs)　218
Recovery　228–9, 264–72
　ARQ　264
　dynamic restoration　265
　preplanned protection　265
　protection　229
　restoration　229
Reliable blast UDP (RBUDP)　173
Remote direct data placement (RDDP), standardization　72
Remote direct memory access in Infiniband　214
Resilient packet ring (RPR)
　features　207–8
　requirements posed by Ethernet services　203
Resource reservation protocol (RSVP)　105
Resource reservation protocol with traffic engineering (RSVP-TE)　227
Restoration　264–72
　loss　268
RFC 1958　103
RFC 2474　103
RFC 2598　106
RFC 2768　60–1
RFC 3439　103
RFC 768　172
RFID, see Radio frequency identification (RFID)
ROADMs, see Reconfigurable optical add-drop multiplexers (ROADMs)
Routing models　226
Routing information bases (RIBs)　190
Routing information protocol (RIP)　189
Routing metrics　191–2
Routing policies　190
Routing topologies　190
RSVP-TE, see Resource reservation protocol with traffic engineering (RSVP-TE)

SABUL (simple available bandwidth utilization library)　174–5
Scalable adaptive graphics environment (SAGE)　20
Scalable TCP　162

Index

Science Technology and Research Transit Access Point (STAR TAP) 323
SCTP 88
Semantic web
  distributed service awareness 139
  ontologies 131
  standardization 71
Semiconductor optical amplifiers (SOAs) 232
Sensors 299
Separate wavelength assignment (SPP-SWA) 267
Service Grid 6
Service level agreements (SLAs) 53
Service matrix 289
Service-oriented architecture (SOA) 61
Session initiation protocol 138
Shared path protection with joint wavelength assignment (SPP-JWA) 267
Shared path protection with separate wavelength assignment (SPP-SWA) 267
Signal to noise ratio (SNR) 237
Silicon optical amplifiers (SOA) 232
Simple network management protocol (SNMP) 128, 258
Simple path control protocol (SPC) 93, 313
SNMP, see Simple network management protocol (SNMP)
SOA, see Silicon optical amplifiers (SOA)
Software-defined radio (SDR) 298–9
SONET/SDH 109, 312, 316
  Ethernet over SONET 206
  features exposed via UNIs 208
  profile 206
  with regard to Ethernet services 202
SPC, see Simple path control protocol (SPC)
SPICE 29–30
SPP-SWA, see Joint wavelength assignment (SPP-JWA); Separate wavelength assignment (SPP-SWA)
STAR TAP 323, 324
StarLight 28, 292, 323–4
StarPlane 318–19
Study group 13 74, 75, 110
SuperJANET 24, 27
SURFnet 27, 28, 327
SURFnet6 138

Switching granularity 235
System to intermediate system (IS-IS) 189

Target channel adapter, see Infiniband
TCP, see Transmission Control Protocol (TCP)
TDM, see Time division multiplexing (TDM)
TE (traffic engineering) 199
Telecommunication management network (TMN) 133
Tell and wait 243
TeraGrid 7
TeraGyroid 27–9
TERENA 329
Testbeds 218, 312–23
Time division multiplexing (TDM) 217
T-LEX (Tokyo Lambda Exchange) 317
Token-based security 136
TOS octet 187
Transaction language 1 (TL1) 53, 128
TransLight 324–5
Transmission Control Protocol (TCP) 88, 147
  AIMD 172
  binary increase congestion (BIC)
    metrics 153
    profile 160
  congestion control
    delay-based 149
    explicit signal 149
    fairness 152
    feedback system 150–2
    loss-based 149
    loss recovery 153
    queuing delay 153
    responsiveness 152–3
    stability 152
    throughput 152
  fairness 149
  FAST
    metrics 153
    profile 159
  high-speed (HS)
    metrics 156
    profile 161–2
  NewReno
    metrics 153
    profile 157–8

Transmission Control
    Protocol (TCP) (*Continued*)
  profile   147
  proxies   180
  Reno   157
  scalable
    metrics   153
    profile   162
  Vegas   158–9
  Westwood   162–3
  window flow control   147–9
Tsunami   178

UCLP, *see* User-controlled lightpaths (UCLP)
UDP   88, 146, 167
UDP-based data transfer protocol (UDT)   174
UK e-science   24, 27
UKLight   24, 26, 326
UltraLight   330
UltraScience Net   323
Uniform resource identifier (URI)   90
Universal description, discovery, and integration (UDDI)   90
User-controlled lightpath architecture (UCLP)   95–6
User-controlled lightpaths (UCLP)   30, 32, 321
  example of grid network infrastructure   123
  views on resource scheduling   135
User datagram protocol (UDP)   105, 146, 171–8
  profile   146–7
User–Network Interfaces (UNI)   53, 89
  bindings of a grid network infrastructure   127
  definition   208–10
  fit with SOA   209
  relationship with other interfaces   208–10
  requirements posed by grid applications   208–10
  standard   75

Vertically integrated optical testbed for large scale applications (VIOLA)   65, 123, 318
Virtual bridged local area networks   94
Virtual concatenation (VCAT)   109, 207
  features   207
  standard   74

Virtual organization   50, 51, 53
Virtual private LAN services   205
Virtual private networks
  BGP/MPLS IP VPNs   201
  general considerations for layer 2   197–8
  layer 1   210
  layer 3   205
  MPLS based   283, 284
  OAM   204
  with UNI   208–9
Visualization   18–23, 55
vLANs   76, 93, 95
Vol-a-Tile   22

W3C   71
WS-Agreement
  component of grid network infrastructure   131
  standardization   68
WAN-PHY   76, 287
Wavebands   217
Wavelength division multiplexing (WDM)   218, 240
Web Based Enterprise Management (WBEM)   53, 65, 73
  bindings of a grid network infrastructure   127, 128
  standardization   73
Web services business process execution language   89
Web Services Definition Language (WSDL)   89–90
Web Services Resource Framework (WSRF)   10, 89
WS-Notification
  component of grid network infrastructure   130
  standardization   70
World Wide Web Consortium (W3C), profile   71
WS Resource Framework
  component of grid network infrastructure   129–30
  standardization   70
WSBPEL   89

XML   71, 90

Y.1312, *see* Layer 1 Virtual Private Network (LIVPN)